DATE DUE

PETROLEUM GEOLOGY

Elk Basin Field, Wyoming-Montana. The Elk Basin anticline can be recognized by the opposing dip directions of the scarp making sandstones. *Courtesy Pan-American Oil Co.*

PETROLEUM GEOLOGY

SECOND EDITION

KENNETH K. LANDES

Professor of Geology
University of Michigan

 ROBERT E. KRIEGER PUBLISHING COMPANY
MALABAR, FLORIDA

Original edition 1959
Reprint 1975,1976,1978,1981

Printed and Published by
ROBERT E. KRIEGER PUBLISHING CO., INC.
Krieger Drive
Malabar, Florida 32950

Printed in the United States of America

Library of Congress Cataloging in Publication Data

Landes, Kenneth Knight, 1899–
  Petroleum geology.

  Reprint of the ed. published by Wiley, New York.
  Bibliography: p.
  1. Petroleum--Geology. I. Title.
[TN870.L25 1975] 553'.282   74-26700
ISBN 0-88275-226-X

PREFACE

TO THE

REPRINT EDITION

The purpose of making changes in this edition from the Second Edition of Petroleum Geology has been (1) to correct actual errors (both typographical and self-inflicted), and (2) modernize some areas that either were glaringly out-of-date or were subjects of such interest to me that I could not resist updating.

Unfortunately I had neither the time nor the stamina to do an 80% plus or minus revision as I did in writing the Second Edition. During the 16 year span between that and this partial revision there has been a flood of vital papers in petroleum geology, which I have not been able to incorporate. However, the completely rewritten Appendix, "References," cites many places where the interested student can find papers of significance.

The principal subjects new or updated that I have added include discussion of the importance of courses in petroleum geology *per se* (Chapter 1), newer information on the origin and evolution of petroleum with special emphasis on the role played by earth heat, and the overriding essentiality of source rocks, all in Chapter 5.

In the second edition of Petroleum Geology I promised the reader that the lengthy chapters in the first edition on geographic and geologic distribution of oil and gas fields around the world would be expanded and published in a separate volume, "Petroleum Geology of World Oil Fields." It developed that my ambition far exceeded my physical and intellectual endurance, and I had to settle for "Petroleum Geology of the United States," which was published by John Wiley and Sons in 1970.

Kenneth K. Landes

Ann Arbor, Michigan
2 December, 1974

PREFACE

TO THE

SECOND EDITION

It was just eight years ago that I wrote the preface to the first edition of *Petroleum Geology*. During those eight years geology and other facets of oil finding have evolved and developed to such an extent that I found it necessary to rewrite seventy to seventy-five per cent of the first edition.

The new book can be divided into three parts. The first two chapters acquaint the reader with the background of professional petroleum geology and with oil and gas drilling and production methods. The latter methods also provide background material, with emphasis on the ways in which petroleum technology impinges upon the activities of the geologist.

Chapters 3 and 4 are concerned with the techniques of the petroleum geologist. Perhaps this discussion should come after the geologic occurrence, for it is logical that the oil should be formed and trapped before the search for it is begun. Practice, however, is presented before theory in this book so that the beginner will learn the language of the specialist and observe the sources of evidence and methods of study used in developing the theoretical concepts. Nevertheless, there is nothing to prevent the reader or teacher from reversing this suggested order if he desires. The technical aspects of oil finding are not presented in professional detail in Chapters 3 and 4. These techniques are best learned on the job. The objective in this book is merely to give a start to the embryo petroleum geologist.

The remainder of *Petroleum Geology* is concerned with the geologic occurrence of petroleum. This material is the most vital, in my opinion. The more we learn of the origin, migration, and accumulation of oil, the more intelligent will be the prospecting and the more widespread the results. This "academic" side of petroleum geology is actually intensely practical. The emphasis, wherever occurrence of oil is discussed in this book, is on the manner in which the hydrocarbons have been trapped rather than on details of stratigraphic nomenclature.

An appendix, "The Literature of Petroleum Geology," is included in order to give the student a concise list of modern references.

A considerable part of the first edition, which concerned the distribution of oil and gas fields throughout the world and future oil supplies, has been eliminated from the second edition. This material will be expanded and published in a separate volume, *Petroleum Geology of World Oil Fields*, which is now in preparation.

This book is written for the student of petroleum geology regardless of whether he is in the classroom or on the job. Petroleum geology evolves with such rapidity that few can keep up with all phases even while practicing the profession. There is no resemblance, except in the name of the course, between the petroleum geology I expounded to students thirty-three years ago and the current offering which has crystallized into this edition. The research papers referred to in this synthesis had not even been written in 1925.

May you enjoy this book, whether you browse through it or read it!

KENNETH K. LANDES

Ann Arbor, Michigan
January 1, 1959

ACKNOWLEDGMENTS

I acknowledge with pleasure and gratitude the help of the following friends who read and criticized constructively different sections of the manuscript: Merrill Haas, Carter Oil Company, Tulsa, Oklahoma, Chapter 1; D. T. Hoenshell and John S. Hegwer, Mobil Producing Company, Billings, Montana, Chapter 2; Forest B. Vick, Baroid Division, National Lead Company, Houston, Texas, Chapter 2 (part); W. E. Hassebroek, Halliburton Oil Well Cementing Company, Duncan, Oklahoma, Chapter 2 (part); George H. Norton, Atlantic Refining Company, Dallas, Texas, Chapter 3; Milton B. Dobrin, Triad Oil Company, Calgary, Alberta, Chapter 3 (part); George Dickinson, Shell Development Company, Houston, Texas, Chapter 4; Benjamin T. Brooks, Consulting Chemist, New York, N. Y., Chapter 5; Hollis Hedberg, Gulf Oil Corporation, Pittsburgh, Pennsylvania, Chapter 6; George Dickinson and E. K. Schluntz, Shell Development Company, Houston, Texas, Chapter 7; Donald L. Katz, Professor of Chemical Engineering, University of Michigan, Ann Arbor, Michigan, Chapter 7 (part); K. C. Heald, Consulting Geologist, Fort Worth, Texas, Chapter 8; Ralph Grim, Professor of Geology, University of Illinois, Urbana, Illinois, Chapter 8; Clarence L. Moody, Consulting Geologist, Houston, Texas, Chapter 9; John T. Rouse, Mobil Producing Company, Billings, Montana, Chapters 10 to 14 inclusive; M. King Hubbert, Shell Development Company, Houston, Texas, Chapter 13 (part); G. M. Knebel, Standard Oil Company (N. J.), New York, N. Y., Chapter 15; Samuel P. Ellison, Jr., Professor of Geology, University of Texas, Austin, Texas, all chapters. These are all busy men, but thanks to their willingness to spend time and thought on my chapters the material therein has been improved markedly.

I wish to express also my sincere thanks to the following companies and individuals for both hospitality and information while I was gathering material for this edition: Magnolia Petroleum Company and Charles A.

Reinke, Jr., Midland, Texas; Arabian-American Oil Company and O. A. Seager of New York, and Richard A. Bramkamp (deceased, September 1, 1958), R. W. Powers, and associates of Dhahran, Saudi Arabia; Bahrain Petroleum Company and Charles Davis, Richard de Mestre, and John W. James of Awali, Bahrain Island; Kuwait Oil Company and Hollis Hedberg of Pittsburgh, Pennsylvania, and George Becker and associates of Ahmadi, Kuwait; and the Iraq Petroleum Company and F. R. S. Henson of London, England, and Richard Browne and H. V. Dunnington, formerly of Kirkuk, Iraq, but now of Baghdad and London, England, respectively.

Individuals and companies that have contributed material (in addition to illustrations) which has been incorporated into this edition are: Theodore A. Link, Consulting Geologist, Calgary, Alberta; Paul Kents, Department of Mineral Resources, Province of Saskatchewan, Regina, Saskatchewan; Shell Oil Company, through George Dickinson of Houston, Texas; and Schlumberger Well Surveying Corporation, through W. P. Briggs of Houston, Texas. This valuable aid is deeply appreciated.

Although the illustrations are acknowledged individually where they appear I wish at this time to thank collectively the many companies, house organs, trade journals, magazines, and individuals who so kindly either supplied me with illustrations or extended permission to reproduce illustrations. Many illustrations have been adapted from publications of the American Association of Petroleum Geologists, and I am indebted to J. P. D. Hull, Managing Editor of that organization, for general permission to reproduce these drawings. The drafting for the first edition was by John Jesse Hayes; Derwin Bell and Norbert Archbold contributed sketches to the second edition.

Last but not least has been the splendid assistance of my wife, Susan Beach Landes, who has functioned with equal efficiency as editor, critic, and typist.

K. K. L.

CONTENTS

THE PROFESSION
OF
PETROLEUM GEOLOGY

DEFINITION AND IMPORTANCE

Petroleum geology is the utilization of geology in the search for, and in the exploitation of, deposits of petroleum and natural gas. The liquid and gaseous hydrocarbons are so intimately associated in nature that it is customary to shorten the expression "petroleum and natural gas" to "petroleum" or "oil" and to assume the inclusion of natural gas. This practice is followed in *Petroleum Geology.*

The petroleum industry has profited much by its liaison with geology. According to figures collected annually by the American Association of Petroleum Geologists and published in the June issue of the bulletin of that society, an average of five out of every six wildcat wells (wells drilled outside the boundary of known oil fields) are located upon geological advice. The chances of oil being discovered in these wells are two to four times greater than they are for the one well out of six located without benefit of geology. For this reason, every producing company of any magnitude has a geological staff, and more geologists are employed in the oil industry than in any other branch of the profession. In 1958 there were probably considerably more than 20,000 professional petroleum geologists throughout the world.

The practical value of geology[1] in oil finding is fairly well known, but the benefits that have accrued to geology through exploration for oil have

[1] Ira H. Cram. "Geology Is Useful," *Bull. Am. Assoc. Petrol. Geol.,* Vol. 32 (January, 1948), pp. 1–10.

not been adequately appreciated. The search for new deposits of oil and gas has led to geologic mapping in all parts of the globe, thereby greatly increasing our knowledge of world geology. By drilling thousands of wells, and by sharing with others the geological records of those wells, the petroleum industry has made possible a far more accurate picture of the third dimension in the earth's crust than ever could have been achieved by extrapolation of surface observations. Koester calls attention to the fact that every modern drilled well provides a "measured section" of the stratigraphy from the grass roots to total depth, which may exceed four miles. "Over two hundred million feet (38,000 miles!) of section in wells will be measured by the bit, the microscope, and the electrical logger this year. These rocks, generally, will receive much more detailed study in the field, the office, and the laboratory than is given the normal outcrop in the field." [2]

DeGolyer,[3] as long ago as 1921, pointed out some of the geological discoveries that were due to the exploration for oil, and many more could be added to the list today. Examples are the discovery of (1) buried crystalline rock mountains and uplifts, such as the Nemaha granite ridge of Kansas and Oklahoma; (2) the great thickening of the sedimentary rock column in basins of deposition and in embayment areas such as the Gulf Coast; (3) enormous vertical prisms of salt ("salt domes") in the same region; and (4) the existence and location of valuable deposits of helium, carbon dioxide, uranium, coal, limestone, natural brines, rock salt, potash, and sulfur, as well as oil and gas.[4] From the records of wells drilled for oil, geologists have been able to add many new chapters to the sedimentation and diastrophic histories of the earth's crust.[5] Old shorelines and buried stream beds have been mapped; ancient reefs have been discovered; hidden faults have been recognized; and evidence of volcanism in supposedly "quiet" periods has been found. From a purely scientific standpoint, deep wells drilled for oil have made possible the measurement of earth temperatures in the uppermost five miles of crust.[6]

Illing[7] has aptly stated, ". . . if geology has contributed greatly to the

[2] Edward A. Koester, "Look Back—Then Look Forward," *Bull. Am. Assoc. Petrol. Geol.*, Vol. 39 (July, 1955), p. 1193.

[3] E. DeGolyer, "Debt of Geology to the Petroleum Industry," *Bull. Am. Assoc. Petrol. Geol.*, Vol. 5 (May-June, 1921), pp. 394–398.

[4] Hugh D. Miser, "Some Notes on Geology and Geologists, 1907–1947," *Bull. Am. Assoc. Petrol. Geol.*, Vol. 32 (July, 1948), pp. 1340–1348.

[5] F. H. Lahee, "Contributions of Petroleum Geology to Pure Geology in the Southern Mid-Continent Area," *Bull. Am. Assoc. Petrol. Geol.*, Vol. 43 (1932), pp. 953–964.

[6] Hugh D. Miser, "Our Petroleum Supply," *Jour. Washington Acad. Sciences*, Vol. 29 (March, 1939), pp. 93–109.

[7] Vincent C. Illing, "Influence of Oil on Geology," *Oil Weekly* (September, 1946), p. 176.

growth of the oil industry the debt is not a one-sided one. Geology owes a great deal to the oil industry in the expansion of its knowledge and the increased efficiency of its methods . . . . It would be well to bury the terms 'Pure' and 'Applied' geology so long as they are used in a separatist sense. There is only one geology. Its purity depends on its truth, not on its application."

BEGINNINGS OF THE PETROLEUM INDUSTRY [8]

Oil and gas were used by man for many years before their production became an industry. The use of seep oil and asphalt in the Mediterranean area in the calking of water craft extends back into earliest history and no doubt into prehistory. Noah, of Old Testament fame, was a pitch peddler. He delivered asphalt from seeps to markets along the Euphrates River; [9] naturally he calked his ark with his own product. Likewise the amphibious cradle of the infant Moses was made water tight with seep "pitch." In the New World seep oil in northern Pennsylvania and western New York State was collected by both Indians and pioneer whites for use as lubricant, liniment, and laxative.

In the later pre-industrial history of oil and gas these substances were found by accident while wells were being drilled for water or salt brine. The art of drilling a well is an ancient one; Goodrich [10] reports that a well was drilled in China for salt to a depth of 450 feet *circa* 221 B.C. Brine wells in the Allegheny Mountains were reaching depths of 800 feet by the middle 1800's, and an artesian water well at St. Louis was bored to a depth of 2193 feet in 1854.

During the first half of the nineteenth century the most actively drilled area in the United States was a belt extending from New York State across western Pennsylvania and eastern Ohio into West Virginia and Kentucky. Here natural brines, from which salt was obtained by evaporation, could be found at relatively shallow depths. In a few places gas was struck. It was used as a fuel in salt processing, and in 1821 it was used to light the community of Fredonia, New York. Other wells struck oil, to the great disappointment of the operators. In a few instances this oil was bottled and sold in the same manner as seep oil, but most of it was allowed to go to waste.

[8] Paul Giddens, *Beginnings of the Petroleum Industry* (Pennsylvania Historical Commission, 1941).

[9] Merrill Haas, *informal communication.*

[10] Harold B. Goodrich, "Early Exploration Methods," *World Petroleum*, Vol. 10 (May, 1939), pp. 32–37.

The curtain could rise on the petroleum industry only after the stage was properly set. By the beginning of the nineteenth century man was rebelling against going to bed at dusk. He was learning to read, and that required a better illuminant than an open fire. The urbanization movement was under way, and street lighting became desirable. A satisfactory lubricant was necessary for the machinery of a growing mechanized civilization. All these demands were to increase many fold during ensuing decades, and to them new demands were added.

During previous centuries the relatively modest demand for oil had been met largely by the whaling industry. But the growing demands of the 1800's soon depleted the whale supply, so that by 1850 whale oil was selling for as much as $2.50 a gallon. The rising price had accelerated the search for a substitute, and it was found in kerosene or "coal oil" obtained from coal by distillation. During the decade 1850–1860, fifty-six plants were built in the United States alone to make kerosene from coal. Although coal oil could compete successfully with whale oil, it was still a relatively expensive product. The coal had to be mined, destructively distilled, the resulting oil refined, and the waste ash disposed of. The stage was set for an oil direct from the earth which only needed refining in order to qualify for a rapidly growing market. And the coal oil refineries could be, and many of them were, converted readily to the refining of crude oil. At least one of these refineries, which started with coal in 1854 and became the world's largest kerosene producer before switching to oil about 1860, is still operating at the original site in Brooklyn, N. Y. [11]

The commercial production of oil from hand dug wells preceded commercial production from bored wells. Pierre Belon,[12] a botanist, observed in a paper published in 1553 that some of the people in northwestern Asia Minor had "dug wells as far down as 60 feet" in order to increase petroleum production over that obtained by seep skimming. Burma is reported to have had 520 active dug wells in 1797, producing the equivalent of 600,000 barrels annually, much of which was exported.[13] In the 1850's Romania also produced oil commercially from hand-dug wells. Canada became a producer in 1857 through discoveries near Oil Springs, Ontario. History is vague as to whether or not all of the early wells in this field were hand dug. In any event, it wasn't long before both Canada and Romania were also producing from drilled wells.

[11] "Brooklyn," *Flying Red Horse* (Autumn, 1954), p. 18.

[12] R. J. Forbes, "Pierre Belon and Petroleum," *Mededelingen van de Koninklijke Vlaamse Academie voor Wetenschappen, Letteren en Schone Kunsten van Belgie*, Klasse der Wetenschappen, Jaargang XX, 1958, No. 1, pp. 3–18. (Reference through the courtesy of Professor W. R. Taylor, Department of Botany, University of Michigan, Ann Arbor, Michigan.)

[13] Sir Boverton Redwood, *Petroleum and Its Products*, Vol. I (1906), pp. 45–46.

Fig. 1-1. Then and now (1859–1959). *Courtesy American Petroleum Institute.*

The discovery well for the United States, drilled for the purpose of finding oil, is generally considered to be the Drake well,[14] which although located in western Pennsylvania was promoted by an East Coast group. Among the members of this group were George H. Bissell, youthful New York lawyer; James M. Townsend, New Haven banker; Benjamin Silliman, professor of chemistry, geology, and mineralogy at Yale University; Edwin L. Drake, colonel by courtesy and ex-railroad conductor by vocation; and William A. Smith, brine well driller. In modern terminology these men were respectively promoter, financier, technical adviser, tool pusher (drilling superintendent), and driller. It would be pleasant to record that the location for the well was made by geologist Silliman, but that is not the case. Professor Silliman's contribution was in the form of a chemical analysis and an optimistic report concerning the potential usefulness of the oil which enabled banker Townsend to raise the needed capital, mainly in New Haven. The well (Fig. 1–1) was located in June of 1859 near Titusville, Pennsylvania, close by an old oil spring, and on August 27, at a depth of $69\frac{1}{2}$ feet, oil began entering the well, and the first oil boom was under way. Literally thousands of wells were drilled in western Pennsylvania within a few months of the completion of the Drake well, and the search soon extended into neighboring states and far beyond. By the turn of the century oil was being produced commercially in the United States in Ohio, West Virginia, Kansas, Oklahoma, Texas, Colorado, Wyoming, and California. In the meantime oil had been discovered and was being produced in quantity in Russia, Netherlands East Indies, and Poland.

Offshore oil and gas production started in California in 1896, and it too spread around the world, with the greatest activity in Russia's Caspian Sea, the Gulf of Mexico off the United States and Mexico, Lake Maracaibo in Venezuela, the Caribbean off Trinidad, the North Sea of western Europe, the Persian Gulf, and the seas off southeastern Asia and Australia.

EVOLUTION OF OIL FINDING[15]

Early Oil Finding. The Drake well and its immediate successors were drilled in creek bottoms. Therefore, the first rule of thumb to be followed in prospecting for new supplies was to stick to creek bottoms.

---

[14]Parke A. Dickey, "The First Oil Well," *Jour. Petrol. Technol.*, Vol. 9 (January, 1959), pp. 14–26.

[15]J. V. Howell, "Historical Development of Structural Theory of Accumulation of Oil and Gas," *Problems of Petroleum Geology* (American Association of Petroleum Geology 1934), pp. 1–23; Wm. B. Heroy, "Petroleum Geology; Geology, 1888–1938," *Geol. Soc. Am. 50th Anniversary Vol.* (1941), pp. 512–548; Ralph Arnold, "Two Decades of Petroleum Geology, 1903-1922," *Bull. Am. Assoc. Petrol. Geol.*, Vol. 7 (November-December, 1923), pp. 603–624; E. DeGolyer, "Development of the Technique of Prospecting," *Science of*

"Creekology" reigned supreme until some rugged individualists drilled successful wells on the uplands. It was observed next that the Appalachian oil pools were much longer than they were wide and that their long axes trended about N 30 E (which is the strike of the rocks in this region). Consequently, it became the vogue to follow the trend ("ruler geology"), and it is still good practice when the traps tend to be elongated. The discovery of the prolific Spindletop oil field on the Texas Gulf Coast led prospectors to seek topographic mounds similar to the one overlying the Spindletop salt plug, and other large oil fields resulted.

All these methods have had a fair degree of success because they have scientific basis, although the prospector using such a method may know little and care less about any possible scientific connection. Even "creekology" can be explained, for in the Appalachian region trellis drainage is common, and many of the streams are subsequent, paralleling the strike of the rock formations.

However, the most successful early oil finding method was to drill in the vicinity of seeps. "Seepology," where it can be applied, is still the best of all discovery techniques, for it starts with definite evidence of the presence of oil or gas. As described in Chapter 3, many of the great oil fields of the world owe their discovery, in part at least, to the presence of seeps.

Birth and Infancy of Professional Petroleum Geology. Howell[16] has published a detailed account of the development of the structural theory of oil and gas accumulation. In 1842 William Logan, pioneer Canadian geologist, noted the presence of oil seeps on anticlinal crests in the Gaspé Peninsula. Nine months after the drilling of the Drake well, in 1860, Professor Henry D. Rogers called attention to the anticlinal location of the new wells in Pennsylvania. The Drake well itself was drilled on an anticline mapped years before by Rogers and published in 1858. The first clear statement of the anticlinal theory was made by another Canadian, T. Sterry Hunt, in 1861. In the immediately succeeding years many others wrote regarding the anticlinal theory. In the meantime, in 1860, Alexander Winchell noted that unfractured sandstones were sufficiently porous to contain oil.

Two men stand out as pioneers of professional petroleum geology in the United States. These men are John F. Carll and I. C. White. Carll (1828–1904), a civil engineer who became a geologist through experience in the western Pennsylvania oil fields, joined the Pennsylvania Geological

Petroleum (Oxford University Press, 1938), Vol. 1, p. 268 et seq.; Paul H. Price, "Evolution of Geologic Thought in Prospecting for Oil and Natural Gas," *Bull. Am. Assoc. Petrol. Geol.*, Vol. 31 (April, 1947), pp. 673–697; Parke A. Dickey, "The First Oil Well," *Jour. Petrol. Technol.*, Vol. 9 (January, 1959), pp. 14–26.

[16]*Op. cit.*

ırvey in 1874 as geologist in charge of petroleum and natural gas surveys. His seven reports were major contributions to petroleum geology. "He saw the role that gas played in the movement of the oil to the well bore and realized that gas should be conserved in order to produce the maximum amount of oil. He also explained that oil did not occur in an underground lake or pool in the producing formation, but in the pores of a sandstone. He showed that the normal porosity of sandstone was great enough to account for the most productive wells discovered up until that time."[17]

In the early 1880's Professor I. C. White (Fig. 1–2) rediscovered the structural theory of oil and gas accumulation and then proceeded to apply his knowledge of geology to the discovery of oil and gas fields. He was singularly successful in this regard. Between 1884 and 1889 he located the discovery wells for the Washington, Pennsylvania, oil and gas field, the Grapefield gas field, the Belle Vernon field, the Mannington field of West Virginia, and others.[18] Later geologists, working in this same area, have noted the extremely erratic nature of the reservoir sandstones and have questioned the applicability of White's theories to this province, but the fact remains that he did use geology in the successful search for new oil and gas fields.

During the remaining years of the nineteenth century, Dutch geologists were employed professionally in the search for oil in the East Indies, and British geologists were similarly engaged in Mexico. In 1897–1898 geological departments were started in two companies in California. During the first fifteen years of the twentieth century, the U. S. Geological Survey published an imposing number of papers and monographs on the geology of various oil fields. At the same time the oil companies became convinced of the utilitarian value of geology, and those companies actively sought geologists to establish geological staffs within their organizations. A considerable group was recruited from the ranks of federal and state geological surveys. So new is the profession of petroleum geology that many of these "pioneers" are still alive.

Subsequent Development. The period 1911 to 1921 was one of great discoveries in the Mid-Continent, Gulf Coast, California, and elsewhere. Most, but not all, of these discoveries were due to the application of geology. By 1915 geological staffs were the rule rather than the exception in company organizations. The American Association of Petroleum Geologists (the Southwestern Association of Petroleum Geologists, at first) was born in 1917. The geologist was engaged in mapping anticlines or, especially on the Gulf Coast, in searching for surface indications. For this work a general training in geology was adequate.

[17] William S. Lytle, "John F. Carll," *Geotimes*, Vol. 1 (March, 1957), pp. 8–14.
[18] I. C. White, "The Mannington Oil Field and the History of Its Development," *Bull. Geol. Soc. Am.*, Vol. 3 (April, 1892), p. 195.

Fig. 1-2. Professor I. C. White (1848-1927), who first used structure in the successful search for oil and gas. *Courtesy Geological Society of America.*

Beginning in the early 1920's the geologist engaged in the search for new supplies of oil and gas had to become a specialist in petroleum geology. In the first place, most of the anticlines discernible at the surface were mapped, and, secondly, it became increasingly obvious that anticlinal folding was only one of a number of factors important to the localization of oil and gas in commercial deposits. New techniques were developed, and knowledge was extended into hitherto unknown or neglected fields. Micropaleontology, geophysics, sedimentation, stratigraphy, and paleogeology have all become essential sciences to the modern petroleum geologist.

Many of the laboratories, both academic and industrial, of subsurface geology and applied micropaleontology started between 1915 and 1920.[19] The peak years for plane table surveying were 1920 and 1921. Structure

[19]Hubert G. Schenck, "Applied Paleontology," *Bull. Am. Assoc. Petrol. Geol.*, Vol. 24 (October, 1940), pp. 1752-1778.

drilling enables the geologist to map the bedrock structures when, for one reason or another (usually because of inadequate outcrops) such mapping is not possible at the surface. This innovation arrived in 1919 and reached a peak in the 1920's. The torsion balance was introduced in 1920 and the seismograph in 1923. It is interesting to note that the possible applicability of the seismograph to structure mapping was discussed by J. A. Udden, State Geologist of Texas, in a paper read in 1920. The geophysical instruments had their period of maximum return inland between the late 1920's and the early 1940's. The great expansion in offshore petroleum search and discovery, which started in the 1950's, led to geophysical surveying from shipboard. This made it possible to determine the area, thickness, and structure of the sedimentary rocks beneath the waters of the world's oceans and seas.

Since 1917, subsurface geology, which is the study of the lithology, structure, and geologic history of the sedimentary section by means of well records, including both logs and samples, has grown steadily and is now an important part of the organization of every geological department. Today many oil and gas field discoveries are credited to subsurface studies.

The scientific method has not eliminated entirely random drilling. Actual wildcat well sites within a lease block may be chosen because of commitments made upon leasing, or on account of accessibility, or even on hunch. Such tests are often successful, primarily because the block was in a geologically favorable area or it would not have been leased in the first place.

PRESENT TRENDS IN PETROLEUM GEOLOGY

Petroleum geologists are currently thinking and working in the direction of: (1) rising above the prejudices which have shackled oil-finding from the start; (2) developing new oil territory by extending exploration far beyond previous lateral and vertical boundaries; and (3) continued improvement of the tools and techniques of exploration and better integration of geological and geophysical data.

Overcoming the Mental Hazards to Oil Exploration. From the very outset of the oil industry one of the greatest barriers to exploration has been ultraconservative thinking. "There would be mirth-provoking irony in a map of the United States showing the boundaries, lateral and horizontal, beyond which dogmatists have at one time or another said oil could not be found—which mental barbed-wire fences have snapped under the irrepressible urge of the wildcatter."[20] Throughout the relatively brief history of oil exploration such man-created barriers as

[20]Samuel W. Tait, Jr., *The Wildcatters* (Princeton University Press, Princeton, N. J., 1946), p. xiii.

"commercial oil deposits cannot occur in limestones," "structural basins cannot contain oil fields," "there is no oil below the Mississippi lime (or the Dundee, or what have you)," "there is no porosity below 15,000 feet," "the structure is too complicated," "red-bed areas are barren," and "there is no oil west of the Nemaha granite ridge" have been demolished by men who had both vision and a capability for courageous thinking. Discoveries of oil in the so-called Weber "quartzite" and in sediments of continental origin have blasted two more firmly entrenched prejudices.[21]

Another handicap to discovery has been the tendency to permit a dry hole to condemn an entire township, or county, or even larger area. Actually, or course, a dry hole condemns little more than its own area, and many cases are on record in which producing wells have completely surrounded an earlier drilled dry hole.

The fact that the mental hazards listed above, and countless others, have been overcome does not mean that there are no prejudices today. Every district contains geologists who dogmatically confine the occurrence of oil to certain areas and stratigraphic depths. "Sometimes a geologist becomes convinced that oil cannot occur in an area and finally says he will drink all the oil found there. When this happens, he should ask for a transfer. He has been in the area too long. Negative thinking never found a barrel of oil."[22]

Fortunately, the present trend in petroleum geology appears to be in the direction of less prejudiced thinking. The company geologist must also, of course, weigh the costs of a new exploring venture against the probable returns; therefore economic factors as well as prejudices may lead to the rejection of a proposed exploration. Unfortunately a tendency to underestimate the possible returns has also been responsible for rejection of projects which later proved to be highly profitable. Ultraconservative economic estimating is a discovery deterrent as well as ultraconservative geological thinking.

The Development of New Provinces. For many years the spread of oil exploration was hampered by marketing considerations such as absence of pipe lines into the area of possible interest. Now oil has become a commodity of such value that no area, no matter how remote, is considered "out of bounds," and geologists are no longer so closely restricted in exploration. As a result, oil is now being sought in such distant places as the polar regions, at greater depths than ever before reached, and beneath "geological masks" heretofore considered too expensive to penetrate. Noble[23] lists, as examples of geological masks: water, thrust sheets, volcanic rocks, high-velocity limestones, and unconformities. Much activity

[21]Earl Noble, "Geological Masks and Prejudices," *Bull. Am. Assoc. Petrol. Geol.*, Vol. 31 (July, 1947), pp. 109–147.

[22]Merrill W. Haas. "The Oil Finder," *The Link*, Vol. 23 (May-June, 1958), p. 1.

[23]Earl Noble, *op. cit.*

has already been shown in exploring for new oil deposits beneath the waters overlapping the continental shelves. This costly exploration has met with success, and submerged areas will continue to be investigated as potential oil "lands."

Improving the Techniques of Oil Finding. In the preceding section on the evolution of oil-finding it was noted that the emphasis in exploration method has shifted from surface structure mapping through structure drilling and geophysical surveying to subsurface geology and integration of geology and geophysics. Each one of the first three of these oil-finding techniques reached a relative peak and then declined fairly rapidly. This decline has occurred for two reasons: (1) the fact, brought out by DeGolyer,[24] that a technique (a single operation in the search for oil) exhausts its possibilities by being used; and (2) the initial success and resulting popularity of the succeeding new method. However, in regard to the first point, it should be realized that complete exhaustion of possible application is more of a theoretical than an actual end point. Whenever a shift of scale, as from reconnaissance to detailed, is introduced, the entire area can be rerun with the old method, but with the occupation of many more stations. Likewise the opening up of a new territory for exploration, whether it be lateral or vertical, will give further impetus to old methods. Therefore, even though the applicability of an old technique may and does diminish, it may be a long time before it disappears.

The phenomenal successes of geophysical prospecting led to the crowding out of surface geological methods (including photogeology) in some areas long before it had exhausted its possibilities. Both in the Gulf Coast and Rocky Mountain provinces considerable chagrin has been caused by the realization that oil fields discovered by geophysical means could have been found just as easily by the far cheaper surface method. According to Russell,[25] fields in central and northern Louisiana could have been found by surface mapping, but only 2 actually were so discovered up to 1941. Surface mapping is credited with assisting in the detection of an additional 6 fields, leaving 16 in which the possibilities of discovery by surface geology were ignored. The first oil discovery in Mississippi was brought about through surface geology, although the state had previously been covered several times by geophysical surveys. Russell also points out that in 1941 detailed surface structural studies had just made a beginning in the northern Gulf Coast region, including parts of Texas, Louisiana, Arkansas, Mississippi, Alabama, and Florida.

[24] E. DeGolyer, "Future Position of Petroleum Geology in the Oil Industry," *Bull. Am. Assoc. Petrol. Geol.*, Vol. 24 (August, 1940), p. 1389.

[25] R. Dana Russell, "Future of Field Geology," *Bull. Am. Assoc. Petrol. Geol.*, Vol. 25 (February, 1941), pp. 324–326.

A highly significant trend in petroleum geology has been a growing awareness of the value of combining and coordinating oil-finding techniques. It is now realized that the various exploration methods are not mutually exclusive and should not be considered to be competitive. The exploration department of a modern oil company is so organized that it can support one oil-finding technique by the use of any others that may be applicable. In 1942 approximately 200 petroleum geologists contributed opinions to a symposium on petroleum discovery methods.[26] A summary of the replies shows that more geologists favored "improving and coordinating all methods" as a better approach to the problem of oil and gas discovery than any other method.[27] More recently Dennison and Warman have stated: "The main feature of oil exploration at present is the absence, and apparent unlikelihood, of the development in the near future of any revolutionary techniques. The search for new and more obscure reserves of oil is pursued largely by refinements of earlier methods."[28]

Weaver[29] has pointed out the possibilities of reinterpreting existing geological and geophysical data in order to find potential hydrocarbon traps that may have been missed during earlier analyses of the same information, especially if the earlier analyses were made at a time of less advanced knowledge.

The point has been reached in the older producing regions of the world at which the obvious places to explore for oil, determined by air photograph, plane table, or seismograph have been exhausted. The so-called stratigraphic traps for oil accumulations are rarely visible at the surface, no matter how good the bedrock exposures, and most of these traps elude the mechanical devices so far developed for determining the hidden and perhaps deeply buried geology. Therefore it becomes necessary for the geologist to learn how to locate the more difficult stratigraphic and combination type oil and gas traps. For this work he must be more than a recorder of formation elevations; he must be an interpreter, a speculative geologist. Interpretation and speculation of practical value can be built only upon accurate knowledge of geologic processes and geologic history, of sedimentation and stratigraphy, of diagenesis and diastrophism. Only by detailed paleogeologic studies can the presence and location of such potential traps as facies changes, porosity wedges, overlaps, and uncon-

[26] *A Symposium on Petroleum Discovery Methods* (Research Committee, American Association of Petroleum Geologists, 1942).

[27] C. Don Hughes, "Graphic Arrangement of a Symposium on Petroleum Methods," *Bull. Am. Assoc. Petrol. Geol.*, Vol. 26 (August, 1942), pp. 1410–1412.

[28] A. T. Dennison and H. R. Warman, "Recent Trends in Exploration Methods," *Inst. Petrol.*, Vol. 43 (July, 1957), pp. 187–197.

[29] Paul Weaver, "Avenues for Progress in Geology and Geophysics," *World Oil* (April, 1951), pp. 102 et seq.

formities be determined or the relative timing of petroleum generation and trap creation be ascertained. Such paleogeologic studies constitute research of the purest type.

The oil companies have recognized the place of fundamental research in oil exploration and as a result many company geologists are engaged in research today, and laboratories have been built to house this activity. The American Petroleum Institute has underwritten several research projects, including one (No. 51) on recent marine sediments.[30] Dutch geologists have also been studying modern sediments.[31] Paleoecology[32] has become of great importance as a means of locating potential traps of the varying permeability type. Micropaleontology[33] is used not only as a means of correlating sedimentary strata, but also in reconstructing the paleogeology.

Without doubt the most important oil-finding technique is unfettered thinking on the part of geologists. Cram[34] calls for greater imagination in locating sites for wildcat wells. According to DeGolyer ". . . as much oil may be found in the future with new viewpoints as with new techniques. The most important tool of which any geologist is possessed is his mind, the only tool which is sharpened and not dulled by use."[35]

Pratt warns us to never cease cerebrating: "Where oil is first found, in the final analysis, is in the minds of men. The undiscovered oil field exists only as an idea in the mind of some oil-finder. When no man any longer believes more oil is left to be found, no more oil fields will be discovered, but so long as a single oil-finder remains with a mental vision of a new oil field to cherish, along with freedom and incentive to explore, just so long new oil fields may continue to be discovered.[36]

Dickey sums it up neatly: "Several times in the past we have thought we were running out of oil when actually we were only running out of ideas."[37]

---

[30]Francis P. Shepard and Clarence L. Moody, "Guides to Future Oil Traps," *Oil and Gas Jour.* (Nov. 16, 1953), pp. 228 et. seq.

[31]D. J. Doeglas, "Progress of Regional Sedimentology," *Proc. Fourth World Petrol. Congress*, Sec. 1/D, Paper 1, pp. 417–426 (1955).

[32]Raymond C. Moore, "Modern Methods of Paleoecology," *Bull. Am. Assoc. Petrol. Geol.*, Vol. 41 (August, 1951).

[33]C. C. Church, "Foraminifera, an Evaluation," *Bull. Am. Assoc. Petrol. Geol.*, Vol. 37 (July, 1953), pp. 1553–1559; William S. Hoffmeister, "Microfossils Provide New Technique in Exploration," *World Oil* (April, 1955), pp. 156 et seq.

[34]Ira H. Cram, "Excuses to Drill," *World Oil* (April, 1951), pp. 73 et seq.

[35]E. DeGolyer, "Plea for Loose Thinking," *Bull. Am. Assoc. Petrol. Geol.*, Vol. 34 (July, 1950), p. 1609.

[36]Wallace E. Pratt, "Toward a Philosophy of Oil Finding," *Bull. Am. Assoc. Petrol. Geol.*, Vol. 36 (December, 1952), p. 2236.

[37]Parke A. Dickey, "Oil is Found With Ideas," *Oil and Gas Jour.* (Sept. 15, 1958), pp. 284 et seq.

But no matter how great the improvement in the science of oil-finding it is doubtful that the element of chance can ever be eliminated. "Success in exploration depends upon luck and skill. What the proper proportion of each may be, I do not know. Success can be due, as it has been due many times, to luck alone, but I doubt whether it can ever be due solely to skill. The most perfect of prospects, selected by the most refined and exact of techniques, may be a failure. On the other hand, a prospect drilled at random for mistaken reasons or no reason at all may result in the discovery of a new and important oil field."[38]

FUTURE OF GEOLOGY IN THE OIL INDUSTRY[39]

There is no reason to believe that the demand for oil and gas will not continue to increase each year as it has, with few exceptions, from the very beginning. There are two reasons for this ever increasing demand. One is the world wide population increase, averaging a little over ten per cent each decade. The other (and more important) reason is the great increase year after year in the per capita consumption of petroleum products. In the highly mechanized United States the average consumption for every man, woman, and child is 15 barrels per year. Outside of the United States the annual per capita consumption is about one barrel.[40] The rapidly expanding use of petroleum products for transportation alone is increasing the demand by leaps and bounds throughout the world. It has been estimated that the world consumption will be in the neighborhood of 50 million barrels of oil daily by 1975,[41] over *three times* the figure for 1957. Most of this oil will have to be discovered between now and 1975, for the greater part of the present known world reserve, estimated at 260 billion barrels, will be used up by then.

The development and growth of the use of atomic energy should not be a cause of concern to the oil and gas industry. In the first place atomic energy, at present and for some time to come, is far from being competitive on a cost basis with the hydrocarbons; secondly, by the time the two energy sources do become competitive, if ever, the energy requirements of the world will be at such astronomical levels that the help of radioactive material in meeting the demand will be welcome.

[38] E. DeGolyer, "Foreword," in Carl Coke Rister's, *Oil! Titan of the Southwest* (University of Oklahoma Press, Norman, Oklahoma, 1949), pp. viii–ix.

[39] G. M. Knebel, "Blueprint for the Future," *Bull. Am. Assoc. Petrol. Geol.*, Vol. 40 (July, 1956), p. 1450.

[40] Frank A. Morgan, "Opportunity for Petroleum Geology," *A.G.I. Newsletter* (April, 1952), p. 2.

[41] Merrill W. Haas, "Your Future as a Petroleum Geologist," *lecture*, University of Michigan (Oct. 10, 1957).

It is obvious that if the world demand for petroleum is going to increase threefold in eighteen years, the industry is going to need all the help from geology that it can get. Not only is the demand for oil going up rapidly, but also the geological man-hours spent in discovering each million barrels of oil is rising sharply. This increase is due of course to the fact that we have progressed from more or less obvious traps to hidden traps, and as time goes on, every thousand new traps discovered are better hidden than were the preceding thousand. By the close of 1957 it had become evident in the United States that: (1) the percentage of successful wildcats was declining; (2) the volume of oil discovered by each successful wildcat was declining; (3) exploratory costs were rising; and (4) the reserves found per exploration dollar were declining sharply.

Some have expressed the fear that the discovery of a direct method for locating oil might do away with petroleum geology as a profession. Such a discovery is always a possibility, but it is highly improbable that the device would be so perfect that no interpretation on the part of the geologist would be necessary. The invention of the electric log has not displaced the subsurface geologist; on the contrary, it has given him an additional and valuable tool.

Pratt[42] has pointed out the permeation of the oil industry by geology, and the trend toward geologic administrators in company organizations.

TRAINING FOR PETROLEUM GEOLOGY[43]

Much time has been spent in recent years in discussing the academic background considered essential to a career in petroleum geology (American Association Petroleum Geology Committee) or to geology in general (Conferences on Training in Geology). Before considering any specific recommendations emanating from these groups, it is desirable to empha-

[42]Wallace Pratt, "Geology in the Petroleum Industry," *Bull. Am. Assoc. Petrol. Geol.*, Vol. 24 (July, 1940), pp. 1209–1213.

[43]Reports of Committee on College Curricula, *Bull. Am. Assoc. Petrol. Geol.*, Vol. 25 (May, 1941), pp. 969–972; Vol. 26 (May, 1942), pp. 942–946; Vol. 27 (May, 1943), pp. 694–697; Vol. 28 (May, 1944), pp. 670–675; "Conferences on Training in Geology," *Geol. Soc. Am. Interim Proceedings* (March, 1946); (June, 1946); (March, 1947) (Fourth and Fifth Conferences); (August, 1947, Final Report and Recommendations); "Proceedings of Joint Conference of Committee on Geologic Education and Association of Geology Teachers," *Geol. Soc. Am. Interim Proc.* (March, 1948, Part 2); "Report of the Committee on Geologic Education of the Geological Society of America," *Geol. Soc. Am. Interim Proc.* (July, 1949, Part 2), pp. 17–21; Bradford Willard, "Post-War Geology," *Science*, Vol. 100 (Oct. 20, 1944), pp. 348–350; James T. Wilson, "Preparation for Petroleum Geology and Geophysics," *Jour. of Eng. Educ.*, Vol. 42 (January, 1952), pp. 295–296; John D. Haun, "Training the Petroleum Geologist," *Subsurface Geology in Petroleum Exploration* (Colorado School of Mines, 1958), pp. 3–8.

size the usually neglected fact that the academic training of any geologist, and especially a petroleum geologist, has maximum value at the time of starting professional employment, after which the value of his schooling decreases steadily, approaching the vanishing point twenty years later. Other factors, including native intelligence, industry, personality, and on-the-job learning and experience increase in importance until they become paramount.

However, the fact remains that the nature of the geologist's scholastic training is important in first securing and then in retaining a professional appointment. All agree that the embryo geologist should not neglect to obtain adequate background in the cognate fields of mathematics, physics, and chemistry. Biology is a "must" subject for those intending to specialize in paleontology. English composition and report writing are decidedly helpful, for to reach and remain in a position of responsibility almost always requires the ability to write intelligibly. A majority of the conferees on this subject also recommended acquiring skill in at least one foreign language. Work in speed reading is advisable, for the geologist today has to assimilate a vast amount of literature.

The petroleum geologist must have a thorough training in the fundamental courses in geology. These include physical, historical, structural (both elementary and advanced), and field geology; mineralogy and petrology; stratigraphy, paleontology, and sedimentation; and physiography and geomorphology. A field course of at least one summer's duration is of utmost importance. In addition to, and much more important than, learning techniques of geologic surveying, the student in field geology learns fundamental geology itself. How handicapped is the subsurface geologist who attempts to study from samples or electric logs a 50-foot bed of limestone but who has never worked over such a section in quarry wall of cliff face! Another course especially recommended for geologists pointing toward a career in petroleum is elementary geophysics.

Finally, there is the course in petroleum geology *per se*. In the early days of petroleum exploration, oil finding consisted of seep hunting, and searching the surface for anticlines. The latter involved a knowledge of stratigraphy, and the ability to carry out plane table or compass and clinometer traverses. It compares with the much older infancy of ore finding, which meant searching the surface for telltale oxidation products, and panning for heavy and resistant metallic minerals. Subsequently these fields have developed, both instrumentally and scientifically, into sophisticated methods of search for hidden deposits of valuable oil, gas, and mineral deposits. The mature approach to economic geology is to learn the *reasons* for the occurrence of these particular precious substances in localized parts of the earth's crust. This means determining the *origin* and later natural history of mineral deposits. In the case of petroleum many factors are involved, including probable source organisms, the natural environment in which these biota throve in the geological past, the type of sediment in which they were deposited on the seafloor, the wedding of organic hydrogen and carbon into petroleum and the essential conditions for this union, the continued evolution of petroleum, its migration through permeable rocks overlain by seal rocks, and its capture in a trap. The trap itself may be a structural feature, locatable by instrumental means, or it may be an elusive updip vanishing of permeability in a reservoir rock, to be searched out by subsurface geology with the help of the best that geophysics has to offer. This is modern petroleum geology. May the old canard, which still surfaces now and then, "petroleum geology as an academic course is not necessary; other geology courses cover it" rest in peace!

A perennial subject for discussion is the extent to which the educational institutions should go in attempting to teach professional procedures such as sample logging and electric log interpretation. Such training in techniques can be acquired only at the expense of other, more fundamental, subject matter. A majority of oil-company chief geologists have expressed a preference for men broadly trained in the fundamentals rather than narrowly trained in highly specialistic subjects.[44]

The feeling is practically universal that four years is an inadequate period for proper coverage of the general educational subjects and the essential fundamental courses in geology and cognate fields. A minimum of five years of academic training is recommended.

The academic career is of course followed by on-the-job training. This training varies between companies, ranging from a formal two-year apprenticeship program to informal learning under more experienced geologists. Recommendations for the training of geologists to be assigned to subsurface problems have been made by LeRoy[45] and others.

---

[44]"Report of the Committee on College Curricula," *Bull. Am. Assoc. Petrol. Geol.* (May, 1942), p. 942.

[45]L. W. LeRoy, "Training Subsurface Geologists," *World Oil* (Feb. 1, 1952), pp. 77–80; "Discussion," *World Oil* (May, 1952), p. 76.

2

DRILLING, COMPLETING, AND PRODUCING OIL AND GAS WELLS

During the drilling of virtually all wildcat wells, geologists are assigned to "sit on the well." [1] The geologist's job is to examine the samples as they are brought to the surface and thereby keep in touch with the stratigraphic progress downward. To implement his cuttings examinations, he may request that a core or cores be taken. He is the man who has the primary responsibility for discovering the oil. He recommends casing points and warns the driller where the pay zones may be expected. His stratigraphic and lithologic identifications are a basis for decisions regarding coring for permeability and saturation data, and testing for potential oil and gas yields.

When the drilling has been completed the geologist may be consulted by the engineering department regarding the completion practices to be followed. If a new field is discovered, production geologists are assigned to advise in the development of that field. Therefore, although the drilling and exploitation programs are carried out by drilling crews and petroleum engineers, geologists are also involved either as active participants or consultants. At the very least they are interested observers in these operations.

The logging activities that are carried on concurrently with drilling are described in Chapter 4. The variations in magnitude and the possible causes of reservoir pressures are described in Chapter 7.

[1] C. A. Caswell, "Sitting on a Well," *World Oil*, Vol. 130 (Feb. 1, 1950), pp. 100 et seq.

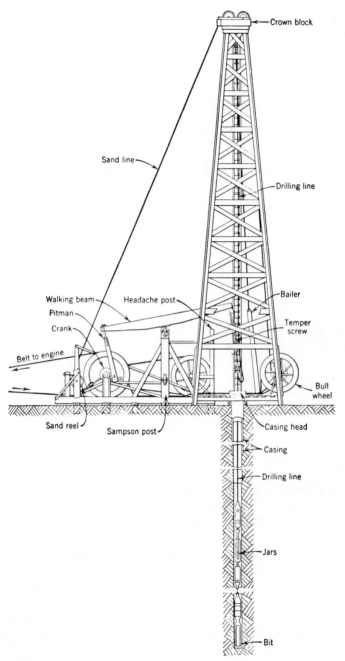

Fig. 2-1. Cable tool drilling rig. *Courtesy United States Steel News.*

Drilled Wells.[2] The geologist is interested in the techniques of well drilling for at least two reasons: (1) if he recommended the site for a test ("located the well"), he is much concerned in the adequacy of the well to prove or disprove the presence of oil or gas, and (2) regardless of the success or failure of a well as a producer, it yields geological information, the value of which depends both upon the driller and the drilling method. The geological record obtainable during a drilling of a well consists of logs and rock samples, both cuttings and cores. The quantity and character of the water, and of any oil or gas, struck during the drilling operations also contribute to the geological record. The proper interpretation of the record depends upon an intimate knowledge of the mechanics of cutting the formation and recovering the rock and fluid samples.

There are two distinct methods of drilling oil wells: the standard and the rotary. The standard method is the older, and rotary drilling has superseded it in most of the newer fields.

Standard drilling. The standard method is also referred to as cable-tool, churn, and percussion drilling. The hole is driven downward by the repeated dropping of heavy "tools." The blows struck by the bit at the lower end of the tool string chip off pieces of rock, the "cuttings," which are removed periodically so that deepening of the hole can continue.

The standard drilling rig is illustrated in Fig. 2–1. It consists of a derrick, usually 84 feet in height, with a sheave arrangement, the *crown block*, at the top. Other essential parts of the rig include the *bit* and other *drilling tools, drilling line, clamps, temper screw, walking beam, pitman, band wheel, bull wheel, sand reel,* and *power plant.* The power plant is some feet away from the base of the derrick. It operates by internal-combustion fuels, steam, or electricity. A belt from the power plant turns the band wheel, and an eccentric on the band wheel operates the pitman, which gives an up-and-down motion to the walking beam. The end of the walking beam opposite the pitman is directly over the hole being drilled. From this end are suspended the temper screw and clamps. The sand reel lies in back of the band wheel and is driven by the band wheel. Across the derrick floor is the bull wheel, which is also turned by the band wheel by means of a large rope belt.

The string of tools consists of the bit, with a V-shaped cutting edge that is sharpened periodically; the stem, into which the bit fits; the jars; and the socket. The jars consist of an inner and an outer section. When the

[2] J. E. Brantly, "Oil Well Drilling Machinery and Practices," *Elements of the Petroleum Industry* (American Institute of Mining Engineers, New York, 1940), Chapter 10, pp. 116–159. *A Primer of Oil Well Drilling,* 2nd edition, (1957), Petroleum Extension Service, University of Texas, and American Association of Oil Well Drilling Contractors, Dallas, Texas.

tools are lifted off the bottom, the inner section slides the length of the tool before it takes hold and lifts the underlying tools in the string. This arrangement makes a strong upward jerk possible in the event that the bit gets caught in the mud in the bottom of the hole. The socket at the top of the tool string is the means of attachment of the drilling line, which may be either wire cable or manila rope. The drilling line extends from the tool string through the clamps beneath the temper screw at the end of the walking beam and from there over the crown block and down to the bull wheel on which it is spooled.

The up-and-down movement of the walking beam raises and drops the bit at the bottom of the hole. Because of the elasticity of the drilling line, however, the vertical oscillation of the tool string is considerably greater than that of the walking beam. This difference produces a whiplash effect to the bit, increasing its cutting power on the rock. A few feet of water is kept in the bottom of the hole to soften and emulsify the cuttings. As the hole is deepened, the driller adds to the length of the drilling line below the walking beam by turning the temper screw or by slipping the line through the clamps. To "come out of the hole" for any purpose the drilling line is disconnected from the clamps, the rope belt which operates the bull wheel is put in place, and the line is reeled on the bull wheel until the tools emerge from the hole.

To bail out accumulated rock cuttings and excess water, the bailer, a long steel tube with a valve in the bottom, is dropped down the hole by means of a line spooled on the sand reel. It is brought to the surface and dumped as many times as may be necessary. If a core is desired, the bit may be replaced by a cylindrical core-barrel which consists of an outer drilling barrel and an inner core retaining tube.[3] The outer barrel has cutting teeth around the lower end. It rises and falls as the walking beam moves up and down, but the inner tube does not move upward until the tool is pulled from the hole. As the outer barrel cuts away the rock around the bottom of the inner tube, the latter tube moves down over an undisturbed core of rock. A trap ring at the bottom of the tube breaks off and retains the core when the tool is jerked upward to bring it to the surface.

The drilling well may encounter a water flow of such magnitude that it is impossible to "dry up" the hole by bailing. Sometimes the water in these aquifers is under so much artesian pressure that it will overflow at the surface. As it is not possible to drill efficiently in a hole full of water, each water vein has to be "cased off." The well must first be drilled below the water sand into a fairly impervious rock such as a shale. Then casing (pipe) is placed in the hole, from the bottom to the top. After this operation, drilling operations are resumed, but a smaller-diameter bit

[3] R. C. Glover, "Cable Tool Coring," World Oil (April, 1951), pp. 147 et seq.

must be used in order to pass through the casing. There is, of course, another reason for the use of casing besides shutting off ground-water flows. Without casing, the wall of a hole may cave. In extreme instances this may block the hole, trapping the tools beneath, and necessitating time-consuming and expensive "fishing" operations. In any event, cavings from above the point where the preceding sample was taken are out of place in the cuttings bailed to the surface, and they tend to confuse the record. For this reason the geologist, in examining cable-tool cuttings, must keep in mind the distance below the casing point at which each sample was obtained.

In addition to collecting and labeling the samples, the driller keeps a log in which he enters the depths to the top of each different rock type. He also includes the casing history of the well, the water zones, and any "shows" of oil or gas that may have appeared. Other types of logs also may be made for cable-tool wells. These are discussed in a subsequent section.

If the well is successful in finding oil, an innermost line of pipe, the "oil string," is run into the well, higher water is cemented off, and as many of the outer strings of casing are salvaged as possible. The oil flows or is pumped to the surface through a string of tubing.

Rotary drilling. The turning of the drill pipe in the rotary method of drilling rotates the bit at the bottom of the hole, chipping and abrading off pieces of rock as it turns. The cuttings are removed continuously by a stream of either mud, air, or natural gas, which is pumped from the surface down through the drill pipe, out an opening in the bit, and back up to the surface between the drill pipe and the walls of the hole.

The modern derrick for rotary drilling (Fig. 2–2) may be 176 feet or more in height. It can be distinguished from the standard derrick not only by its greater size but also by the presence of a catwalk, or "forble board," around the outside of the derrick, about two-thirds of the distance above the floor. The rotary derrick, like the standard, has a *crown block* at the top and a *power plant*. The largest machine that has to be powered is the *draw works*, by means of which the drill pipe is hoisted in and out of the hole. Operating by either a "take off" from the draw works or by independent power are the *mud pumps* and the *rotary table*. The latter is circular but has a square slot in the center which turns the kelly (described below). The table rotates in a horizontal plane. Above it, suspended by a cable from the crown block, is the *traveling block*, to which the huge *hook* is attached.

The traveling block and hook carry the weight of the tools and drill pipe in the hole. The cable that loops back and forth between the traveling block and the crown block ends up spooled around the reel on the draw

1. Steel Derrick	11. Drive Bushing	20. Triple Engine Drive
2. Crown Block	12. Tongs	21. Duplex Slush Pump
3. Traveling Block	13. Rotary Drilling Unit	22. Suction Strainer
4. Rotary Hose	14. Draw Works	23. Discharge Line
5. Standpipe	15. Brantly Hydraulic Feed	24. Mud Tank
6. Hook	Control	25. Mud Mixing Pump
7. Swivel	16. Derrick Substructure	26. Mud Mixing Jets
8. Elevators	17. Drilling Line	27. Drill Pipe
9. Kelly Cock	18. Dog House	28. Sand (Coring) Line
10. Kelly	19. Automatic Weight Indicator	

Fig. 2–2. Rotary drilling. *Courtesy Oil Well Supply Company.*

works. At the top of the string that extends downward to the bit at the bottom of the hole is the giant *swivel;* the non-rotating section of the swivel is suspended from the hook by steel *elevators.* Next below is the *kelly,* which is square or hexagonal in cross section so that it just fits the slot in the *drive bushing* in the rotary table. As the table rotates, at speeds of 50 to 300 revolutions per minute, it turns the kelly which turns the drill pipe which rotates the bit at the bottom of the hole. The *drill pipe,* into the top section of which the kelly is screwed, is circular instead of square or hexagonal in cross section. It serves the dual purpose of turning the bit and acting as a conduit for the drilling mud that is pumped down from the surface. At the bottom of the string of drill pipe, separating the pipe from the bit, are one or more *drill collars.* A drill collar is an extra heavy walled rigid tube which is used to provide the weight applied to the bit and withstand the compressive shocks from the drilling bit. The bit itself is screwed into the base of the lowest drill collar. Rotary bits[4] are of several types; bits commonly used in cutting rock are shown in Figs. 2–3 and 2–4. As the bit is rotated on the bottom of the hole, the teeth chip away at the rock. At the same time, a stream of mud issues from the bottom of the bit, keeping the bit and rock formation cool so as to prevent vitrification and simultaneously carrying the cuttings away from the bit and up the hole toward the surface. Upon reaching the surface the drilling mud is led onto a tilted vibrating screen (the "shale shaker") which separates the coarser rock cuttings from the drilling fluid. The cuttings are discharged at the lower end of the shaker screen and samples of the cuttings are caught in a bucket or on a board; the drilling fluid itself passes through the screen and flows to a settling pit prior to recirculation (Fig. 2–5).

In modern practice the drilling fluid is mixed to definite specifications prepared by a mud engineer.[5] The liquid and solid ingredients are mixed in a mud-mixing machine ("hopper") located on the edge of the slush pit where the fluid is stored. The drilling fluid is picked up by powerful pumps which circulate it through pipe and flexible hose into the top of the swivel, and from there it travels ("circulates") down the drill pipe to the bit at the bottom and then back up the annular space between drill pipe and rock wall of the hole to the surface. The average speed of the drilling fluid up the annulus is about three feet per second, but the actual

[4] L. L. Payne, "Conventional Rock Bits," Subsurface Geology in Petroleum Exploration (Colorado School of Mines, 1958), pp. 637–652; W. M. Booth and R. M. Borden, "Jet Bits," *ibid.,* pp. 653–663.

[5] "Syllabus on Drilling Muds," Revised Chapter 14 of J. E. Brantley's *Rotary Drilling Handbook,* 5th edition (Baroid Division National Land Company, February, 1957); T. N. Dunn, "Drilling Fluids," Subsurface Geology in Petroleum Exploration (Colorado School of Mines, 1958), pp. 715–733.

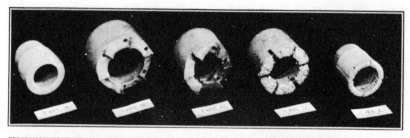

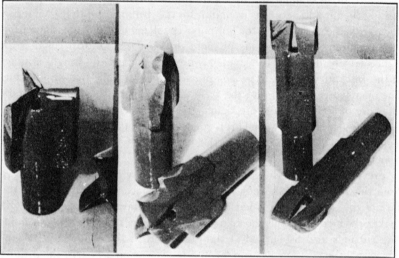

Fig. 2-3. Types of drilling bits. *Top:* A, blank bit; B, insert type (soft forma-tions); C, saw-tooth type (soft formations); D, crackerjack type (medium hard formations); E, diamond cast set (hard formations). *Bottom:* left, three-way bits; center, pilot type; right, fish tail bits. *Courtesy George E. Failing Supply Company.*

uphole velocity varies widely. After passing through the shale shaker the drilling fluid may be passed through a large centrifuge, which for the purpose of recovering expensive weighting ingredients and removing harmful ingredients such as drilled solids, operates like cream separators.

The drilling fluid has several important functions in addition to lubri-cating the bit and transporting the cuttings to the surface. In soft rock the fluid itself, jetted out through the bit, may act as a drilling agent by eroding the material ahead of the bit. If the circulation is interrupted, the fluid, because of the presence of colloidal materials, gels so that the cuttings in transit upward do not settle down around the bit and stick it. The drilling mud, where it passes through permeable layers, tends to

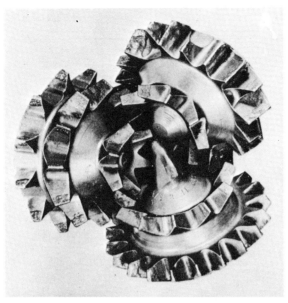

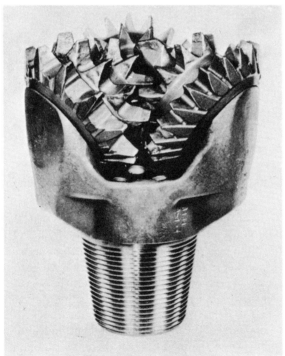

Fig. 2–4. Underside (top) and side (bottom) views of a tri-cone rock bit. *Courtesy Hughes Tool Company.*

27

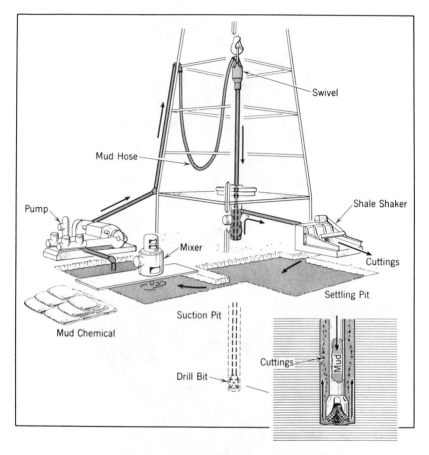

Fig. 2–5. The mud circulation system in a rotary drilling unit.

plaster the walls of the hole with "filter cake", which seals the hole so that
further drilling fluids are not lost into these formations.   Unfortunately,
the circulating fluid may seal off oil and gas formations as well, and no one
can guess the number of reservoir rocks that may have been missed in the
past in this way.   It is now possible to contract the service of a trained
crew and a specially built trailer to aid in detecting pay zones.   Various
devices are used to detect the presence of hydrocarbons, even in exceedingly minute amounts, in the drilling fluid and cuttings returning from the
well.   The influence of the mud cake on the records obtained by electric
logging, especially micrologging, is considerable.   This influence is
described in Chapter 4.   The function of the *weight* of the mud column
in sealing off reservoir rocks is a major technological advance that enables

drillers to penetrate high pressure oil and gas reservoirs which formerly caused disastrous blowouts and the consequent loss of life and property. Drilling muds are of two main types, water-base and oil-base. The former are used much more extensively, as oil-base drilling fluids are confined largely to special purpose drilling. Water base drilling fluids are a mixture of various clays and chemicals in water. Chemicals are added to assist in the control of flow properties of these suspensions of clay particles in water. Other chemicals and special clays may be added for special purposes.

The drilling fluid to be used is tailor-made to fit the conditions anticipated. To make the fluid gel if circulation is interrupted, bentonite or other clays may be added. Drilling fluids that contain a low percentage of clays, solids, or colloids may have high filtration losses, which result in the deposition of a thick filter cake. This situation can be controlled by adding bentonite, starch, or cellulose to the drilling fluid. The drilling of cavernous limestone formations or undersaturated sands may result in the drilling fluid being lost to these porous sections. This "lost circulation" is a serious problem in rotary drilling. Usually these losses can be overcome by adding such fibrous or bulky materials as beet pulp, sawdust, tattered sacks, special cements, cellophane, fragmented plastic foil, hay by the ton, and even stranger material.

Drilling through thick salt sections with fresh-water drilling fluids may create caverns through solution of the salt by the drilling fluid. Therefore, where rock salt is anticipated the drilling fluid should be saturated with salt.[6] This operation not only tends to prevent future casing failure, but also permits securing a complete sample record.

When the reservoirs contain high pressure gas, costly "blowouts" can be prevented by preparing a drilling fluid of such density that the weight of the column of fluid in the hole exceeds the pressures encountered when the drill penetrates the high pressure formation. Water alone weighs about 8.3 pounds per gallon. Low specific gravity clays can be added until the weight reaches about 11.0 pounds per gallon, but after that the drilling fluid becomes too viscous to be pumpable. For drilling high-pressure formations, a weighting material, usually the mineral barite, is added to the drilling fluid. In this way the weight can be increased to as much as 20 pounds per gallon. This amount has proven to be adequate for most deep drilling situations.

In order to obtain and maintain the drilling fluid specifications called for, dependable instruments have been developed for use on the job to determine various fluid characteristics such as weight, viscosity, gel strength, filtration, and salinity. Drilling fluid costs per well vary widely,

[6] George R. Gray, "Chemicals in Drilling Mud," *Oil and Gas Jour.* (Dec. 15, 1958), p. 96.

depending on well depth, maximum drilling fluid weight, lost circulation etc. As an example, a 12,000 foot well in Oklahoma might budget $30,000 for drilling fluid, and a 12,000 foot well in the Gulf Coast might budget $75,000. Oil-well drilling in the United States alone consumes about a million tons of barite, a half million tons of bentonite, and a quarter of a million tons of other clays annually.

Oil-base drilling fluid uses oil instead of water as the continuous liquid phase, or clay carrier. These drilling fluids are used principally for drilling into low-pressure reservoirs, drilling into producing formations which contain clays that become water-wet and thus restrict production, and for recovering native-state cores. Oil-base drilling fluids are generally used in low-pressure areas where weighted fluids are not required. These drilling fluids are also useful in drilling through formations such as salt, anhydrite, and heaving shales that react with water. The clay in oil-base drilling fluid is less likely to clog casing perforations than in the case of water-base drilling fluids. Drilling and research have resulted in the development of a water-in-oil emulsion drilling fluid which is designed for the same uses as straight oil-base drilling fluid. Its advantages are said to be easier handling and non-inflammability.[7]

The use of drilling fluid in rotary drilling has been superseded in some areas by air or gas.[8] One or the other of these media is circulated under pressure down inside the drilling string in the same way as mud. The cuttings are flushed out of the hole by the velocity of the stream of air or gas and are discharged through a pipe leading away from the drilling rig, building up a cone of powder-size cuttings at the outlet. The advantages of this method of drilling are: (1) elimination of the problem of lost circulation; (2) greatly ("phenomenally") increased drilling rates; (3) savings in drilling costs ranging from 20 to 60 per cent; and (4) easier identification of oil and gas bearing zones. Although most of the saving in drilling expenses is due to the much greater footage per rig per hour, other economies have been noted, especially in bit costs. Disadvantages are the increased fire and explosion hazard, (one rig burned down after setting a new rate-of-penetration record) the powder size of the cuttings, and the limitation of water-bearing formations. Air or gas drilling are practicable

[7] J. L. Lummus, "Multipurpose Water-in-Oil Emulsion Mud," Oil and Gas Jour. (Dec. 13, 1954), p. 106 et seq; G. G. Priest and T. O. Allen, "Non-plugging Emulsions Useful as Completion and Well Servicing Fluids," Jour. Petrol. Technol., Vol. 10 (March, 1958), pp. 11–14.

[8] George E. Cannon and Ralph A. Watson, "Review of Air and Gas Drilling," Jour. Petrol. Technol., Vol. 8 (October, 1956), pp. 15–19; C. L. Moore and V. A. LaFave, "Air and Gas Drilling," Jour. Petrol. Technol., Vol. 8 (February, 1956), pp. 15–16; M. M. Brantly, "Drilling with Gas and Air," Subsurface Geology in Petroleum Exploration (Colorado School of Mines, 1958), pp. 735–741.

only where (1) the water-bearing strata are limited both in number and volume; (2) none of the formations penetrated have a tendency to slough; and (3) reservoir pressures are low enough to make safe operation possible. Compressed rig floor exhaust gases, which also have been used successfully, lessen the fire and explosion hazard. Considerable experimentation was being carried on in 1958 in the use of "foaming" agents and chemical precipitates to control water influx in air and gas drilling. The foaming agents convert the formation water to a mist and the chemicals precipitate solids in the pores of the aquifer adjacent to the borehole.

When the hole is deep, a round trip to change bits consumes many hours. In coming out of the hole, the string is first raised above the bottom of the kelly and steel wedges or "slips" placed in position to prevent the string from dropping back down. The kelly with overlying swivel is disconnected from the string and dropped into a slanting hole, the "rathole" prepared for this purpose, where it is out of the way. Subsequently, the string is raised three or four lengths of drill pipe at a time above the derrick floor, the pipe below is suspended by the slips, and the three- or four-length sections are unscrewed and racked in a corner of the derrick. One member of the crew, the derrick man, mounted on the forble board high up in the rig, assists in the racking. Eventually the bit is brought to the surface and replaced, and the entire procedure is repeated in reverse.

During the drilling of a rotary well, the driller watches a gauge and by means of controls regulates the pressure on the bit at the bottom of the hole in accordance with the character and the structure of the rock. The bit in a rotary-drilled hole can wander off from the vertical for a considerable distance and still be rotated by the drill pipe. Crooked holes in the Seminole District, Oklahoma, reached such magnitude that the depth figures were as much as 800 feet (in 4000 feet) greater than the actual depth. The lateral distance across which the bit wandered was as much as the distance between wells.[9] Obviously, geological records, such as structure contour maps, made from the logs of crooked holes are of questionable value. Since the hectic days at Seminole, however, the technique of drill-hole surveying[10] has developed to such an extent that it is now possible to determine the exact position, both horizontally and vertically, of any point in a rotary-drilled well. Modern practice is to prevent or control deviation in the first place by using correct combinations of weight on the bit, revolutions per minute, and type of bit. However, if the deviation becomes too serious to cure in this manner, a vertical

[9] F. H. Lahee, "Crooked Holes — Next Important Problem," *Oil and Gas Jour.* (March 28, 1929), pp. 38 et seq.

[10] J. B. Murdock, Jr., "Oil Well Surveying," *Subsurface Geologic Methods* (Colorado School of Mines, 1950), pp. 548–591.

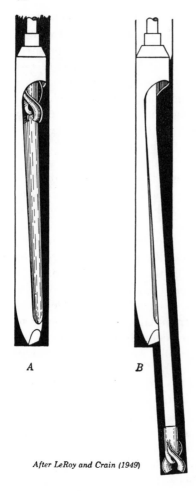

Fig. 2–6. Whipstock. *Left:* bit coming into whipstock at top. *Right:* bit and drill stem, diverted by whipstock, digging new hole. *Courtesy Colorado School of Mines.*

A   B

After LeRoy and Crain (1949)

course can be resumed by use of a *whipstock* (Fig. 2–6), which is placed in the hole to divert the bit back into the desired direction.[11]

Not only have the petroleum engineers learned how to prevent "crooked" holes, but also they have developed a technique for diverting holes[12] when desired. A very practical application of this technique has been in killing "wild" wells by drilling a new well so as to intercept the blowing

[11]John R. Suman, "Drilling, Testing and Completion," *Elements of the Petroleum Industry* (American Institute of Mining and Metallurgical Engineers, New York, 1940), pp. 161–180.

[12]Harry C. Kent, "Directional Drilling," *Subsurface Geology in Petroleum Exploration* (Colorado School of Mines, 1958), pp. 677–694.

or burning well at depth and then pumping water or mud into the wild well until it is finally subdued.[13] By means of directional drilling one drilling site, such as an expensive offshore crib, a location in rugged terrain, or a secluded spot in an area of high priced country estates, may be used for the drilling of many holes. Onshore wells are used to tap offshore oil pools in California, the Gulf Coast, and elsewhere. One producer was drilled beneath the Pacific for a total bore length of 11,440 feet of which the horizontal deflection was 10,244 feet and the vertical depth 4240 feet; the maximum angle from the vertical was 75°.[14] Other slant holes have been drilled beneath Los Angeles harbor, the Mississippi River, and the Corpus Christi ship channel. Directional drilling has also been used to cut across the steeply dipping oil-yielding fractures in the Spraberry formation of west Texas, and to slide under the inclined fault plane in the Dominguez fault trap field in California.

The rock samples obtained in rotary drilling are of two types, cuttings and cores.[15] Rotary drill cores can be obtained in three ways: by conventional core barrel, by wireline core barrel, and by a side-wall corer. The conventional core barrel is placed at the end of the drill-pipe string in place of the usual bit. As the barrel rotates, teeth at the bottom cut out a core which rises inside the barrel as the hole is deepened. When the barrel is filled, the driller must come out of the hole and remove the core. This process is time-consuming, but it does result in a core of maximum possible thickness.

The wire-line method of coring removes the necessity of making a round trip in order to remove the core. With this method, an outer barrel, or drill collar, with fish-tail (or diamond[16]) bit on the end is kept at the bottom of the drill pipe. "Ordinary drilling may be done in the usual manner by dropping a bit plug inside the drill pipe to shut off the main core-barrel passage through the bit and divert circulation through ports to the bit blades. When a core is to be taken, the bit plug is removed by means of an overshot run inside the drill pipe on a wire line. The inner core barrel with a core-catcher assembly in the bottom and a vent valve in the top is then dropped inside the drill pipe and automatically latches into place in the drill collar. After the core has been cut, the inner core

---

[13] Gordon B. Nicholson, "Directional Hole Tames Canadian Wild Well," *World Oil,* Vol. 128 (October, 1948), pp. 213 et seq.

[14] Gordon Jackson, "Directional Drilling Today," *Jour. Petrol. Technol.,* Vol. 5 (September, 1953), p. 28.

[15] H. L. Landua, "Applications of Coring," *Subsurface Geology in Petroleum Exploration* (Colorado School of Mines, 1958), pp. 711–714; William M. Koch, "Core Drilling," *ibid.,* pp. 695–710.

[16] Hoskins Hartwell, "New Wire Line Diamond Core, Drill Saves Round Trip Costs," *Oil and Gas Jour.* (Dec. 27, 1951), pp. 76 et seq.

barrel is removed by means of the same overshot and wire line used to recover the bit plug."[17]

Side-wall coring has proved to be highly practicable, and as a result, several different methods of coring and types of side-wall corers have been developed.[18] The percussion or bullet type of side-wall corer is the most widely used. It shoots a core barrel into the wall, whence it is retrieved by a wire line initially attached to one end of the "bullet." Any type of side-wall core has the disadvantage of being more thoroughly flushed by mud filtrate than one taken while first penetrating the formation. Harder formations are not generally cored with side wall equipment.

In actual practice, side-wall coring is usually done only after the hole has been drilled and an electric log made. Then side-wall cores are obtained at points which may be indicated by a study of the electric log, or at points where bottom-hole coring was inadequate because of missed sections or incomplete recovery. The advantages of this method are several.[19] Complete bottom-hole coring is expensive and often unnecessary. By first completing the hole and then examining critical parts of the section (as determined from the electric log) by side-wall cores, adequate data regarding porosity, saturation, contacts, facies, and so forth can be obtained at minimum expense. Furthermore, portable core-testing equipment and personnel can be brought to the well for use during this relatively brief period of coring, instead of periodically shipping cores in to the main laboratory.

In addition to noting the data for a time log, the rotary driller also writes up a log based on his identification of the cutting and core samples produced. But much more important is the electric log. The running and use of electric and other logs are discussed in Chapter 4.

The usual practice in rotary drilling is to case off the surface water by a relatively short string of casing and then go the rest of the way without casing. However, if the well is to be drilled to depths in the neighborhood of 15,000 feet or more, an intermediate string of casing may be set. The mud fluid kept in the hole at all times prevents caving. If oil or gas is struck, casing is run to the producing zone and cemented into place, and the mud is pumped out.

The advantages of the rotary method over standard drilling are such that this method has displaced the standard in most areas, especially

[17] John R. Suman, "Drilling, Testing and Completion," *Elements of the Petroleum Industry* (American Institute of Mining and Metallurgical Engineers, New York, 1940), pp. 181–183.

[18] J. E. Kastrop, "Sidewall Sampling Tools and Techniques," *World Oil*, Vol. 128, Part 1 (November, 1948), pp. 100 et seq.; Part 2 (December, 1948), pp. 76 et seq.

[19] H. Lee Flood, "Sidewall Coring in the Gulf Coast Area," *Petrol. Engineer*, Vol. 12 (March, 1941), pp. 156–157.

where deep drilling is practiced. The necessity of repeatedly inserting a string of casing and then proceeding with a smaller hole places a severe limitation on the depths that can be reached with cable tools. Very few standard wells have been drilled over 5000 feet, and the deepest well on record of this type was bottomed at 10,096 feet.[20] On the other hand, there is no known absolute limit to the depth to which rotary wells can be drilled. Deeper wells than those now being drilled will require larger, heavier, and costlier equipment. The limitation on drilling deeper is purely economic; when drilling costs exceed possible returns there is no point in continuing further.

Other advantages of the rotary method include the shutting off of water sands without resorting to casing, much less caving, lesser blow-out hazard, greater completion diameter, less cost per foot for deep wells, greater speed (as much as 2500 feet per day) in soft rocks, and the ready availability of cores and electric logs.

Research is constant in attempting to improve the efficiency of drilling. For years the laboratories have been working on a bit that would combine the advantages of percussion drilling (mainly faster penetration rate in hard formations) with those of rotary drilling. The newest development is the hammer drill in which the bit, in addition to its conventional rotation, strikes the bottom of the hole from 300 to 1000 times a minute. The percussion is powered by the drilling fluid; the blows are delivered by a valve-piston-hammer arrangement.

The deeper the hole the greater the loss of energy involved in rotating a bit at the bottom of a string of drill pipe by turning the drill pipe at the top. Much more efficient in this regard is the turbodrill, in which the bit is rotated solely by the power of the stream of drilling fluid pumped down through non-rotating pipe. The hydraulic power of the mud stream is converted to rotating power by a turbine housed in a cylinder below the drill pipe. Although pioneering work was carried on in California over 30 years ago[21] it was the Russians who perfected the turbodrill to the point of becoming the standard drilling tool in their country. An American observer of comparative drilling rates in a Russian oil field reported the turbodrill from four to twelve times faster in making hole than the rotary drill.[22] Subsequently the Russian turbodrill was introduced into the United States, along with French adaptions thereof, and it is now

[20] F. M. Van Tuyl and W. S. Levings, "Review of Petroleum Geology in 1945," *Quarterly Colo. School of Mines*, Vol. 41 (July, 1947), p. 10.

[21] W. R. Postlewaite, "Whats Happening to Turbodrill Development?" *World Oil* (October, 1955), pp. 156–160.

[22] John B. O'Connor, "Turbodrills for American Drilling," *Monitor*, Vol. 1 (November, 1956), p. 4.

being used in test drilling in this country. Tentative conclusions are that penetration rates are faster, but bit footages are lower.[23] There are those who believe that the turbodrill will displace the rotary drill just as the rotary has displaced standard drilling.

Drilling depths. The date and depth of the first drilled well are unknown. Goodrich[24] mentions a brine well in China drilled to 450 feet about 221 B. C. By A. D. 1854, wells were being drilled below 2000 feet; an artesian water well in St. Louis reached a depth of 2193 feet that year. By 1928 the deepest well was an oil producer in west Texas that had penetrated 8523 feet into the earth's crust. The deepest hole in the world in September, 1974 was drilled in a deep basin in Washita County, western Oklahoma. It reached a total depth of 31,441 ft in Cambro-Ordovician carbonate rocks where molten sulfur was encountered. The deepest producer at that time was a gas well in Beckham County in southwestern Oklahoma. The reservoir depth was 24,065 to 24,585 ft. An oil depth record was established temporarily by a well drilled in 1956 in Plaquemines Parish, Louisiana delta, which reached a total depth of 21,465 ft, but the well never did produce for the market, and was plugged and abandoned in 1962. Apparently the current record holder for oil production is a well in Terrebonne Parish, also on the Louisiana Gulf Coast; the reservoir here is at a depth of 20,218 to 20,230 ft.

Apparently nothing (besides the necessity for heavier and costlier equipment) will prevent the continuing establishment of new drilling-depth records except the complete penetration of the sedimentary rock veneer. Inasmuch as the maximum depth to the basement rocks in the Gulf Coast and California provinces probably lies somewhere between 30,000 and 50,000 feet or deeper, we can expect the record-breaking testing to continue as long as the price of oil justifies such expensive exploration.

"Drilling in" and Completing a Well. After a well has been drilled into the potential reservoir rock, as much as two-thirds of the original drilling cost may be spent in obtaining reservoir data and testing the formation for production possibilities.[25] Specific items of testing include the porosity, the permeability, the thickness of productive parts of the reservoir, analyses of the hydrocarbons present, and the volume of connate water. These basic reservoir data are obtained largely by the

---

[23] J. A. Mitchell, "How Are Turbodrills Performing?", *Oil and Gas Jour.* (July 8, 1957), pp. 106 et seq.; E. Lawson Lomax, "Comparison: Turbodrill and Conventional Bit," *Oil and Gas Jour.* (March 3, 1958), pp. 133 et seq.; J. B. O'Connor, "Turbine Drilling," *Subsurface Geology in Petroleum Exploration,* (Colorado School of Mines, 1958), pp. 665–675.

[24] H. B. Goodrich, "Early Exploration Methods," *World Petrol.,* Vol. 10 (May, 1939), p. 32.

[25] "Testing Costs Money," *The Link,* Vol. 12 (April, 1947), pp. 12–15; J. E. Eckel, "Well Completion Methods," *Subsurface Geology in Petroleum Exploration* (Colorado School of Mines, 1958), pp. 763–775.

laboratory examination of cores cut from the reservoir rock. It is an increasingly common procedure to core the reservoir from top to bottom. The information thus obtained is invaluable in determining the capacity of the well to produce and in estimating the reserve situation.

The actual testing of a well for initial production is a routine operation. If the well flows, its production is gauged over a given period of time, and the yield for 24 hours is determined or estimated. The record for initial production appears to be held by a well in the Yates pool of west Texas drilled in 1926, which had a reported potential of over 200,000 barrels per day. A cable-tool well which does not flow can be tested by running the bailer and noting the quantity that can be withdrawn per hour without lowering the static fluid level in the hole. The testing procedure in a rotary well is of necessity quite different. At the time of penetration of the reservoir rock, the well is usually uncased below the surface water zone and the hole is full of mud. In the earlier days of rotary drilling it was necessary to run casing in the well and remove all the mud before the formation could be tested for its capacity to produce. Because of the bother and expense, such testing was often omitted, and without doubt many reservoirs capable of yielding oil or gas in commericial quantities were bypassed in the wildcatting of that period. During the 1920's the method of *drill-stem testing* which is in use today was developed.

"A drill-stem test is a temporary completion of the well. Drill-stem tests are usually made for one or both of the following reasons: (1) to determine the producible fluid content of a formation, and (2) to determine the ability of a formation to produce."[26] They also permit the determination of the reservoir fluid pressure. Drill-stem testing[27] uses the drill pipe as a conduit and obviates the necessity of casing the hole and pumping out the mud. An empty string of drill pipe with a valve at the bottom and a packer set so that it lands just above the formation to be tested is run into the hole. The packer seals off the mud from above so that, when the valve at the bottom is opened, the fluids in the reservoir flow into the empty drill stem. This flow is brought about by the pressure differential between the reservoir fluid which approximates 0.46 pound per foot of depth,[28] and the inside of the drill pipe, where the pressure is atmospheric.

Other things being equal, the amount of fluid entering a well is directly proportional to the permeability of the reservoir. Three methods of

[26] W. Marshall Black, "A Review of Drill-Stem Testing Techniques and Analysis," *Jour. Petrol. Technol.*, Vol. 8 (June, 1956), p. 21.

[27] W. A. Wallace, "Formation Testing," *Subsurface Geologic Methods*, 1950, pp. 731–746.

[28] John R. Suman, "Drilling, Testing, and Completion," *Elements of the Petroleum Industry* (American Institute of Mining and Metallurgical Engineers, New York, 1940), p. 201.

increasing the permeability of the rock near the well are shooting,[29] acidizing,[30] and hydraulic fracturing.[31] Not only do these three processes produce additional permeability in the nearby rock, but they also may break into a fracture system which was missed by the borehole. Joining the well with the natural interconnecting fractures in the reservoir rock increases the flow of oil into the well. In the event that the reservoir is not the rock itself but fractures penetrating the rock, shooting, acidizing, or formation fracturing can convert a dry hole into a commercial well.

Shooting is the detonating of a charge of nitroglycerin which has been lowered into the well. The explosion shatters the rock for a considerable distance, producing many rock faces from which the fluid can drain into the well. Acidizing is the pumping of large quantities of hydrochloric acid into carbonate rock reservoirs. This acid contains an inhibitor which deters corrosion of steel, such as in casing and tools. The carbonate rock, however, is leached outward some distance from the well by the action of the acid. The development of this technique in the early 1930's led to the revival of many declining limestone pools, with a second stage of flush production comparable in magnitude to the initial stage.

Hydraulic fracturing of reservoir formations is by far the newest of the three methods of extending the bleeding surface, but it made up for lost time with phenomenal rapidity. The first extensive use of fracturing was in 1949, but before its fifth birthday there had been 30,000 field applications of this process. Reservoir fracturing consists of pumping into the

[29]E. DeGolyer and Harold Vance, "Bibliography on Petroleum Industry," Bull. Agr. and Mech. Coll. Texas, 83 (1944), pp. 460–461 (25 references).

[30]Ibid., pp. 528–530 (69 references); Paul E. Fitzgerald, "A Review of the Chemical Treatment of Wells," Jour. Petrol. Technol., Vol. 5 (September, 1953), pp. 11–13; L. W. LeRoy, "Well Acidization," Subsurface Geologic Methods (1950), pp. 750–760.

[31]M. King Hubbert and David G. Willis, "Mechanics of Hydraulic Fracturing," Jour. Petrol. Technol., Vol. 9 (June, 1957), pp. 153–166; Roscoe C. Clark, "What Size Fractures," World Oil (May, 1956), pp. 159 et seq; R. E. Hurst, J. M. Moore, and D. E. Ramsey, "Development and Application of 'Frac' Treatments in the Permian Basin," Jour. Petrol. Technol., Vol. 7 (April, 1955), pp. 58–65; L. E. Wilsey and W. G. Bearden, "Reservoir Fracturing—a Method of Oil Recovery from Extremely Low Permeability Formations," Jour. Petrol. Technol., Vol. 6 (August, 1954), pp. 21–27; Warren L. Sallee and Fred E. Rugg, "Artificial Formation Fracturing in Southern Oklahoma and North-Central Texas," Bull. Am. Assoc. Petrol. Geol., Vol. 37 (November, 1953), pp. 2539–2550; W. E. Hassebroek, "Hydrafrac Treatment," Subsurface Geologic Methods (1950), pp. 723–731; C. J. Hauck, "Formation Fracture Operations in Ohio," Gas Supply, Transmission and Storage Conference (May 9–10, 1955), Pittsburgh, Pennsylvania, Operating Section, American Gas Association, pp. 1–12; Wendell S. Moore, "Fracturing in Eastern United States," Am. Petrol. Inst., Spring Meeting of Eastern District (May 11–13, 1955), Reprint Paper 826–26–B; "Elements of Hydraulic Fracturing," Halliburton Oil Well Cementing Company, 2nd edition, Duncan, Oklahoma, (January, 1958), pp. 1–23.

reservoir under high pressure a fluid and a "propping agent," usually sand. This mixture either produces new fractures or is forced into pre-existing cracks and fractures under such pressure that these fissures are enlarged and extended. The sand remains in the cracks, propping them open so that fluids can flow through them freely.

Initially hydraulic fracturing was done by a viscous gel, which later returned from the formation as a thin fluid having been broken by contact with a "gel-breaker" solution. Although still used to some extent, gel has been largely displaced by other fluids. These include refined oil, lease crude, water, water-oil emulsion, and various forms of hydrochloric acid. Although acid is sometimes used in sandstones where the cement is calcium carbonate, its predominant application is in carbonate-rock reservoirs. In the latter instance the treatment combines acidizing with fracturing. The usual procedure in a fracturing job is to flow the oil or water (and perhaps acid) from tanks or trucks, and convey the sand from a special sand carrying and handling truck, to a mixer. From the mixing machine the fluid-sand mixture passes through powerful pumps which force it down the well and into the reservoir. Both the mixer and the pumps are mounted on trucks. Surface pressures for injection vary widely from a few hundred to as much as 10,000 pounds per square inch. Bottom hole pressures vary in accordance with density of fluid, depth of well, and friction involved in pumping through tubular goods.

Reservoir fracturing has achieved notable success in both renewing old fields and in increasing the initial production of new fields. The shoe-string and shallow sand pools of eastern Kansas and Oklahoma have had a rebirth. Stripper wells down to a daily production of 4 barrels or less have jumped their daily outputs from 50 to 100 barrels. The once abandoned discovery well of the old Cushing field in north-eastern Oklahoma was reopened and given a fracture treatment whereupon it resumed production at an initial rate of 648 barrels per day! [32] As a general rule the flush production which immediately follows a fracture treatment declines rapidly, just as does the initial daily production of a discovery well. For this reason such rates cannot be extrapolated into the future.

Some economically marginal and even submarginal pools have been made to pay by fracturing. The Spraberry formation, a siltstone underlying thousands of square miles in west Texas, contains oil within steeply dipping fractures; the advent of fracturing increased enormously the economic potential of this reservoir.

It has been estimated that over 13,000 fracture jobs in the Permian Basin as of mid-1955 had increased the known recoverable reserve of that

[32] Anthony Gibbon, "Old Faithful Flows Again," *World Oil* (October, 1956), p. 108.

area by 120 million barrels. The downward trend of production in Illinois reversed itself and rose over 13 per cent with the application of formation fracturing in that state.

We know that fracture systems can and do occur in brittle rocks. We have learned that such brittle rocks as granite and chert, as well as tight sandstones and limestones, may contain oil in fractures where the conditions necessary for supply and accumulation are present. Commercial oil wells producing from Precambrian basement rocks in Kansas and from pre-Cretaceous basement rocks in Venezuela and California testify to this fact. Fracture treatments in the future may increase the relative percentage of production from this type of accumulation.

Completing a well includes all the steps taken to transform an exploration test into a producing oil or gas well. Once it has been demonstrated that the well can produce enough hydrocarbon to justify the completion costs, the necessary steps are begun. The simplest procedure can be followed if the reservoir is an indurated sandstone or competent limestone. Under these circumstances casing is run from the surface to the top of the reservoir or consolidated overlying rock, where it is cemented[33] into place. Then smaller-diameter tubing, through which the oil is to travel to the surface, is run inside the casing to near the bottom of the hole, which is open below the casing seat.

If the reservoir rock is incoherent it is not possible to produce from an open hole. One method, long employed in water wells, is the "gravel pack," in which gravel is poured into the hole below the casing seat in order to discourage collapse of the wall of the borehole. Another method is to cement loose sand grains together by a thermosetting resin (plastic) "in such a manner that sufficient porosity and permeability remain . . ."[34] One procedure is to attach screens or perforated pipe sections to the casing string at the proper point or points and cement the string into place before running the tubing. Another method with wide application is to run a solid string of casing from the surface through the producing interval, cement it into place, and then perforate the casing at the desired levels by means of a perforator,[35] which shoots bullets or directs shaped charges ("jets") through the casing into the wall rock.

Offshore Drilling. When the drilling rigs in a developing oil field first marched down to the water's edge and it was obvious that the boundary of the field was somewhere offshore naturally the drilling rigs moved out into

[33] William D. Owsley, "Twenty Years of Oil Well Cementing," *Jour. Petrol. Technol.*, Vol. 5 (September, 1953), pp. 17–18.

[34] G. G. Wrightsman and H. H. Spain, "Consolidation of Sands in Oil and Gas Wells," Standard Oil Company, N. J., *Tech. Publs.* (1947), pp. 141–148.

[35] Reginald L. Robinson and Blake M. Caldwell, "Two Decades of Gun Perforating," *Jour. Petrol. Technol.* Vol. 5 (September, 1953), pp. 14–16.

the water. Several methods were and are practiced to obtain the valuable hydrocarbons beneath lake, sea, or ocean floor; these are listed below.

1. Piers are built off the shore on which the rigs operate.

2. The land is extended outward by pushing soil and rock into an artificial peninsula.

3. Islands are built by pumping seafloor sediment into a doughnut built of large slabs of rock from onshore.

4. Slant holes are drilled from the shore area and from artificial islands and platforms so as to tap the reservoir out beneath the water.

5. In shallow water drilling rigs are moved onto barges, and the barges then are sunk so as to rest on the seafloor.

6. Separate platforms, some of them on legs several hundreds of feet long, are erected for drilling, production, and other purposes. These are serviced by boats and helicopters. A variation is the jack-up platform that is towed into place with legs straight up; these are then lowered to the seafloor, with a considerable length of legs still available up which the platform is jacked into place to above storm wave level.

7. "Semisubmersibles" which on reaching location float partly under water, and are secured by anchors.

8. Drill ships, with a rig mounted midships so as to drill through a well in the hull. These are likewise anchored, and can drill in water of any depth.

At first the offshore producing area was merely the extension of an onshore field. Subsequently many discrete offshore fields have been found, most of them out of sight of land. Exploration is taking place today hundreds of miles offshore, due to the availability of modern offshore drilling and production hardware. Both oil and gas are brought ashore by submarine pipelines.

Oil Production. After a successful well has been completed it is prepared for production. That a well produces at all is due to the pressure under which the fluids are held in the reservoir. The possible causes for reservoir pressure are discussed in Chapter 7; the following paragraphs describe the types and practical results of this pressure.

Dissolved gas drive. Petroleum is propelled out of the reservoir and into the well by one of three processes or by a combination of them: dissolved-gas drive, gas cap drive, and water drive. Early recognition of the type of drive involved is essential to the efficient development of an oil field. In dissolved-gas drive, the propulsive force is the gas which is in solution in the oil and which tends to come out of solution because of the pressure release at the point of penetration of a well. The movement of oil produced by dissolved-gas drive is analogous to the effervescence that results from the uncapping of an agitated bottle of soda water.

Dissolved-gas drive is the least efficient type of natural drive. Nothing can be done to control the gas-oil ratio; the bottom-hole pressure drops rapidly, and the total eventual recovery may be less than 20 per cent.

Gas-cap drive. If gas overlies the oil beneath the top of the trap, it is compressed and can be utilized to drive the oil into wells which are footed toward the bottom of the oil-bearing zone. By producing oil only from below the gas cap, it is possible to maintain a high gas-oil ratio in the reservoir until almost the very end of the life of the pool. If, however, the oil deposit is not systematically developed, so that bypassing of the gas occurs, an undue proportion of oil will be left behind. The usual recovery in a gas-cap field is 40 to 50 per cent.

Normally the gas in a gas cap contains in addition to methane other hydrocarbons in vapor form that can be separated out by compressing the gas. A well known example is natural gasoline which was formerly referred to as "casing-head gasoline" or "natural gas gasoline." However, at high pressures such as exist in the deeper fields of the Gulf Coast and other parts of the world, the density of the gas increases and of the oil decreases until they enter a single phase and are indistinguishable in the reservoir. These are the so-called retrograde condensate pools; retrograde because a decrease instead of an increase in pressure brings about condensation of the liquid hydrocarbons. When this reservoir fluid is brought to the surface and the "condensate" is removed, a large volume of residual gas remains. The modern practice is to "cycle" this gas by compressing and reinjecting it into the reservoir. By thus maintaining adequate pressure within the gas cap, condensation in the reservoir is prevented. Such condensation would prevent recovery of the oil, for the low percentage of liquid saturation in the reservoir precludes effective permeability.

Water drive. The most efficient propulsive force in driving oil into a well is natural water drive, in which the pressure of the edgewater forces the lighter oil ahead and upward until all the recoverable oil has been flushed out of the reservoir into the producing wells. In anticlinal accumulations, the structurally lowest wells around the flanks of the dome are the first to come into water, after which the oil-water contact plane moves upward until only the wells at the top of the anticline are still producing oil, and eventually these too have to be abandoned as the water displaces the oil. In a water-drive field it is essential that the removal rate be so adjusted that the water moves up evenly and as fast as space is made available for it by the removal of the hydrocarbons. An appreciable decline in bottom-hole pressure is necessary to provide the pressure gradient required to cause water influx. The pressure differential needed depends upon the reservoir permeability; the greater the permeability the less the difference in pressure necessary. The recovery in properly operated water-drive pools may run as high as 80 per cent. The force behind the water drive may be either hydrostatic pressure, the expansion of the reservoir water, or a combination of both.

Other drives. Gravity drive is an important factor where oil columns of several thousands of feet exist, as they do in some Rocky Mountain fields. Furthermore, the last bit of recoverable oil is produced in many pools by gravity drainage of the reservoir. Another source of energy during the early stages of withdrawal from a reservoir containing undersaturated oil is due to the expansion of that oil as the pressure reduction brings the oil to the bubble point (the pressure and temperature at which the gas starts to come out of solution).

Methods of producing wells.[36] Oil may be produced by natural flow, induced flow, or mechanical lift. In most reservoirs the energy is sufficient during the early stages of exploitation to force the petroleum into the wells and up the tubing to the surface, whence it flows to a tank battery. However, some fields initially, and almost all fields eventually, have to be produced by means of help from the surface. The simplest method is the gas lift, in which natural gas is pumped to the bottom of the well through the annular space between the tubing and the casing. The oil that collects in the boring is pushed up the tubing to the surface by the force of the compressed gas. Gas lift merely supplies the natural reservoir with energy that it never had or no longer possesses. The gas used is most often a by-product of the oil from which it has been separated at the surface. During the compression of the gas, any gasoline present ("wet" gas) is removed before the gas is reinjected into the reservoir. If the local gas supply is inadequate, natural gas from other fields may be used. Air can be used in the same manner but has the decided disadvantages of corroding all metal with which it comes in contact and of producing a highly explosive mixture.

By far the most common method of producing oil from non-flowing wells is by means of a pump which provides a mechanical lift to the fluids in the reservoir. A pump barrel is lowered into the well on a string of solid steel rods known as "sucker rods." Up-and-down movement of the sucker rods forces the oil up the tubing to the surface. This vertical movement may be supplied by a walking beam powered by a nearby engine, or it may be brought about through the use of a pump jack, which is connected with a central power source by means of "pull rods."

Some oil, especially in parts of California, is too heavy and viscous to

[36] C. V. Millikan, "Production Practice," *Elements of the Petroleum Industry*, (American Institute of Mining and Metallurgical Engineers, 1940), pp. 260–270; Morris Muskat, *Physical Principles of Oil Production* (McGraw-Hill, New York, 1949); S. J. Pirson, *Elements of Oil Reservoir Engineering* (McGraw-Hill, 1950); J. C. Calhoun, *Fundamentals of Reservoir Engineering* (University of Oklahoma Press, 1953); *Primer of Oil and Gas Production* (American Petroleum Institute, 1954).

flow in adequate volume into the wells. It has been found practicable to heat this oil in its underground reservoir so as to increase its fluidity and production rate.[37] The method most used is to lower a heating unit into the producing wells; if the yield is increased by as much as five barrels a day the project will pay out. The unit may be heated by electrical resistance or by circulating hot water. Another method is *in situ* combustion or "hot foot." The oil at the bottom of one well is ignited and the heat of combustion drives the oil through the formation into the circle of surrounding wells. Ignition is brought about by chemical means; the fire is maintained by forcing compressed air down the injection well. Underground nuclear explosions might also be used to heat sluggish oil.[38]

Substances which may be produced by a well. A considerable variety of materials may be produced by oil wells in addition to liquid and gaseous hydrocarbons. The natural gas itself may contain as impurities one or more non-hydrocarbon substances.[39] The most abundant of these impurities is hydrogen sulfide, which imparts a noticeable odor to the gas. A small amount of this compound is considered to be advantageous, for it gives warning of leaks when and where they occur. A larger amount, however, makes the gas obnoxious and difficult to market. Such gas is referred to as "sour gas," and much of it is used in the manufacture of carbon black. During World War II the sulfur shortage became so acute in the United States and Canada that sulfur recovery plants were installed in sour gas (and sour crude) areas in both countries.

A few natural gases contain helium. It is interesting to note that this element occurs in commercial quantities in certain gas fields in Texas, Oklahoma, Colorado, and Utah, but it has not yet been found in volume anywhere else in the world.[40] Nitrogen has also been found in some natural gases. The last few years have seen not only the discovery but also the commercial development of carbon dioxide gas occurring in natural reservoirs in the southwest.[41] Natural carbon dioxide is used in

[37] H. Walter, "Application of Heat for Recovery of Oil: Field Test Results and Possibility of Profitable Operation," *Jour. Petrol. Technol.* Vol. 9 (February, 1957), pp. 16–22; Bruce F. Grant and Stefan E. Szasz, "Development of Underground Heat Wave for Oil Recovery," *Jour. Petrol. Technol.* Vol. 6 (May, 1954), pp. 23–33.

[38] "Promise in Buried Atom Blasts," *Business Week*, (March 22, 1958), p. 28.

[39] C. E. Dobbin, "Geology of Natural Gases Rich in Helium, Nitrogen, Carbon Dioxide, and Hydrogen Sulfide," *Geology of Natural Gas* (American Association of Petroleum Geologists, 1935), pp. 1053–1072; C. C. Anderson and H. H. Hinson, "Helium-Bearing Natural Gases of the United States," *U. S. Bur. Mines Bull.* 486 (1951), pp. 1–141.

[40] Andrew Stewart, "About Helium," *U. S. Bur. Mines Information Circular* 6745 (September, 1933).

[41] Frank E. E. Germann, "The Occurrence of Carbon Dioxide," *Science*, Vol. 87 (June 10, 1938), pp. 513–521; J. Charles Miller, "Discussion of Origin, Occurrence, and the Use of Natural Carbon Dioxide in the United States," *Oil and Gas Jour.*, Vol. 25 (Nov. 9, 1933), pp. 19–20; Sterling B. Talmage and A. Andreas, "Carbon Dioxide in New Mexico," *N. Mex. Bur. Mines Bull.* 18 (1945?), pp. 301–307.

the manufacture of dry ice. Because of the great cooling effect of expanding carbon dioxide, the surface pipes and fittings of wells yielding any appreciable volume of this gas become coated with frost, which has led to the appellation "ice cream" wells. Two carbon dioxide fields have been developed in New Mexico and at least one field in each of the states of Colorado, Utah, and California. The origin of the non-hydrocarbon gases is discussed in Chapter 5 under "Origin of Natural Gas."

The liquid hydrocarbons produced by a well may include solid hydrocarbons in solution which separate upon reaching the surface and cause clogging of the pipes and fittings. Both the paraffin hydrocarbons and bitumen[42] are offenders in this regard.

By far the most abundant extraneous material is water. Many wells, especially during their declining years, produce vast quantities of salt water, and its disposal is both a serious and an expensive problem. Furthermore, the brine may be corrosive, which necessitates frequent replacement of casing, pipe, and valves, or it may be saturated so that the salts tend to precipitate upon reaching the surface. In either case the water produced with the oil is a source of continuing trouble.

If the reservoir rock is an incoherent sand or poorly cemented sandstone, large quantities of sand are produced along with the oil and gas. On its way to the surface sand has been known to scour its way completely through pipes and fittings. After it reaches the surface the sand presents a disposal problem. Some of the European oil wells have great mounds of sand immediately adjacent.

Reservoir pressure maintenance. Modern practice is to maintain pressures in the reservoir at somewhere near the original levels by pumping either gas or water into the reservoir as the hydrocarbons are withdrawn. This practice has the advantage not only of retarding the decline in production of individual wells and increasing considerably the ultimate yield, but also it may bring about the conservation of gas that otherwise would be wasted, and the disposal of brines that otherwise might pollute surface and near surface potable waters.

Since pressure was not maintained in the older fields, it is necessary to go back into those areas with *secondary recovery* projects, which are discussed in the chapter entitled "Future Oil Supplies" in the companion volume to this one, "Petroleum Geology of World Oil Fields." However, the leading secondary recovery methods also involve the pumping of gas (*"gas repressuring"*) or water (*"water flooding"*) into the reservoir. In fact, secondary recovery has been defined as "delayed pressure maintenance."[43]

---

[42] G. W. Preckshot, N. G. DeLisle, C. E. Cottrell, and D. L. Katz, "Asphaltic Substances in Crude Oils," *Am. Inst. Mining Met. Engrs.* T. P. 1514 (1942).

[43] Morris Muskat, "Secondary Recovery," *Physical Principles of Oil Production* (McGraw-Hill, New York, 1949), p. 645.

Pressure maintenance by *gas injection* involves converting some wells in the field to input wells and pumping gas down them. The gas is usually field gas, but outside gas may also be used. A most efficient arrangement is followed on Bahrain Island in the Persian Gulf. There gas (for which there is no market) is brought from a reservoir below the oil to the surface and immediately injected into the oil reservoir. As the natural pressure of the gas is greater than that in the oil reservoir, no additional compression at the surface is necessary.

The time interval between field development and first pressure maintenance activity is shortening. Thus gas injection has been carried on in the relatively new Abqaiq field of Saudi Arabia since 1954, and injection began in 1958 in the still newer Ghawar field about 20 miles to the southeast. The largest reinjection compressor plant so far built is seven miles off shore in Lake Maracaibo, Venezuela. It operates on a concrete platform nearly as large as a football field; the water depth is 60 feet. About 134 million cubic feet of gas daily are pumped down five injection wells into an oil reservoir 4500 feet below the lake surface. Oil is produced through 59 wells. "Not only will this project result in a much greater recovery of oil from the reservoir, but also a huge quantity of gas is being conserved for the future."[44] In the gas-cap drive-type of reservoir, it is expected that pressure maintenance by gas injection will increase eventual recovery as much as twofold.[45]

Considerable experimentation has been carried on in the use of different types of input gas. Examples are wet-casing head gas, enriched gas, LPG (liquid petroleum gases) such as butane and propane,[46] high pressure gas, and even nitrogen. High-pressure gas not only pushes oil through the reservoir, but also may produce a hydrocarbon exchange so that the concentration of LPG in the oil is increased.[47] Atmospheric nitrogen has been used successfully in the Elk Basin field along the Montana-Wyoming

[44]T. O. Edison, "Gas Injection Performance Review of the LL370 Reservoir in the Bolivar Coastal Field, Venezuela," *Jour. Petrol. Technol.*, Vol. 9 (June, 1957), p. 19.

[45]Sheldon F. Craddock, "An Evaluation of Gas Injection in the Emery West Pool, West Coyote Field, California," *Jour. Petrol. Technol.*, Vol. 7 (April, 1956), pp. 25–30; B. T. Milliken, Jr., "Twelve Years of Gas Injection into Frio Sand," *Jour. Petrol. Technol.*, Vol. 6 (June, 1954), pp. 11–15; J. B. Justus, R. V. Cassingham, C. R. Blomberg, W. H. Ashby, "Pressure Maintenance by Gas Injection in the Brookhaven Field, Mississippi," *Jour. Petrol. Technol.*, Vol. 6 (April, 1954), pp. 43–53; C. A. Davis, "The North Coles Levee Field Pressure Maintenance Project," *Jour. Petrol. Technol.*, Vol. 4 (August, 1952), pp. 11–21.

[46]Loren H. Jenks, John B. Campbell and George G. Binder, Jr., "A Field Test of the Gas-driven Liquid Propane Method of Oil Recovery," *Jour. Petrol. Technol.*, Vol. 9 (February, 1957), pp. 34–39.

[47]H. A. Koch, Jr., "High Pressure Gas Injection Is a Success," *World Oil* (October, 1956), pp. 260–264.

border.[48] It was used because the volume of field gas was inadequate, and the nearest available natural gas was 100 miles away. The nitrogen is obtained from the atmosphere by burning local field gas; one volume of the latter yields nine volumes of nitrogen.

Water injection is still predominantly a secondary recovery process ("water flood"). Probably the principal reason for this fact is that reservoir-formation water is ordinarily not available in volume during the early years of an oil field, and pressure-maintenance water from outside the field may be too expensive. When a young field does produce considerable water, it may be injected back into the reservoir primarily for the purpose of nuisance abatement, but reservoir-pressure maintenance is a valuable by-product. The same is true when water is pumped into a reservoir, as at Wilmington, California, to control surface subsidence.

Water injection for the purpose of pressure maintenance may be carried on concurrently with gas injection. Examples are Abqaiq in Saudi Arabia, Poza Rica, Mexico,[49] and Hackberry Field, Louisiana.[50]

[48]J. E. Lang, "Performance of the Elk Basin Tensleep Reservoir with Nitrogen Injection," *Mines Mag.*, Vol. 43 (October, 1953), pp. 70–73, 80.

[49]John C. Reidel, "Simultaneous Gas and Water Injection in Large Poza Rica Reservoir," *Oil and Gas Jour.* (November, 1955), pp. 120–123.

[50]R. L. Evans and A. E. Barry, "Unitization, and Water and Gas Injection, Hackberry Field, Louisiana," *Oil and Gas Jour.* (April, 1954), pp. 112–116.

3

EXPLORATION

Exploration is the term used in the petroleum industry for oil hunting. It is the pre-discovery phase; post-discovery activities are classified under exploitation, or development and production. This chapter discusses the role of the petroleum geologist in exploration, whereas the next chapter (4) takes up production and subsurface geology. However, no sharp line can be drawn between exploration and exploitation so far as subsurface geology is concerned, for this branch of petroleum geology is exceedingly useful both in the development of oil production and in the discovery of new deposits.

SURFACE OBSERVATIONS

Surface observations are still employed whenever possible; they are used extensively in the early prospecting of new territory. Surface methods include: (1) noting or determining direct indications of the presence of underlying hydrocarbons; (2) geological surveying; and (3) the utilization of maps and air photographs.

Direct Indications.[1] The outstanding direct indication of the presence of oil or gas in the rocks is a seep at the surface. Other possible indications are those peculiar to the Gulf Coast salt domes, and "micro-seeps."

Seeps. Benjamin Silliman once defined oil seeps as "natural outcrops of rock oil." Other surface occurrences of natural hydrocarbons include gas seeps, oil-impregnated rocks, and solid and semisolid asphalts and waxes.

---

[1] E. DeGolyer, "Direct Indications of the Occurrence of Oil and Gas," *Elements of the Petroleum Industry* (American Institute of Mining and Metallurgical Engineers, New York, 1940), pp. 21–25; E. DeGolyer and Harold Vance, "Bibliography of the Petroleum Industry," *Bull. Agr. Mech. Coll. Texas*, 83 (1944), pp. 360–361 (21 references).

An active oil seep is in reality an oil spring out of which issues anywhere from a few drops to several barrels of oil daily. Almost invariably such seeps are at topographically low spots where water has also accumulated. The lighter oil rises to the top of the water and covers it with an iridescent ("rainbow") film. In rare cases, as at Kirkuk in Iraq, the seeps are large enough to make separation and marketing of the oil a profitable enterprise.

Gas seeps may be detected by: (1) odor, due to the presence of sulfur compounds; (2) yellow colors in the rock surrounding the orifice, caused by sublimation; (3) a whistling noise when the gas emerges under pressure through a restricted opening; (4) "eternal fire" when the gas has become ignited by natural or artificial means; (5) exploding bubbles when the seep is covered by water or mud; and (6) mud volcanos.

We still know too little about mud volcanos. During four years of field work in the Sinu basin of northwest Colombia I saw hundreds of mud volcanos. There are two types, the regular mud volcano with its cone built up gradually (and partially destroyed by the next rain), and the explosive type. The regular mud volcanos are ant hill shaped and about the height of a man. While I was in the area a violent explosion created a new mud pie about 80 meters in diameter and one and one-half meters high. All of this mud boiled out in half a day. . . . The largest such mud-cake I have seen is east of Puerto Escondido on the Caribbean coast of northwestern Colombia. This one is about two kilometers across and 100 meters high. Embedded in the mud are big blocks and boulders up to five tons in weight. Just where the mud in these volcanos comes from is a mystery; the surface rock is a moderately hard sandstone, but a thick shale section lies at greater depth. In all mud volcanos I have seen the mud is similar in consistency and color to ordinary drilling mud.[2]

An explosive type of mud volcano appeared one and one half miles off the south coast of Trinidad in November, 1911, and in two days time created an island of more than eight acres. The mud carried with it fragments of sandstone and shale bedrock. At the height of the activity the escaping natural gas became ignited, intensifying the spectacular character of the eruption. The island was destroyed in a few months by subaerial and marine erosion.[3]

Oil-impregnated rock may be filled with seep oil that has worked its way through a considerable body of pervious rock instead of following a single fissure to the surface, or it may be a true reservoir rock that has been uncovered by erosion. In any event, the oil fills the interstices between the grains of sand or the cavities dispersed through the carbonate

[2] Paul Kents, extracted from a letter dated June 7, 1952, and published with his permission.
[3] Ralph Arnold and George A. Macready, "Island-Forming Mud Volcano in Trinidad, British West Indies," *Bull. Am. Assoc. Petrol. Geol.*, Vol. 40 (November, 1956), pp. 2748–2758.

rock. The Athabasca "tar" sands of northeastern Alberta illustrate oil-impregnated sandstone. The asphalt content of a limestone in Uvalde County, Texas may run as high as 50 per cent by volume.[4] Both asphalt-impregnated sandstones and limestones are quarried in various parts of the world, crushed, and used for road surfacing.

There are no lines of demarcation between very heavy oils, semisolid asphalts, and asphalts. These relatively heavy hydrocarbons may occur as impregnations, as small surface deposits, or in large "lakes." The impregnations are exactly the same as the oil impregnations, except that the hydrocarbons have lost their mobility. Small surface deposits of tarry hydrocarbons are common around the world in the areas of outcrop of marine Tertiary sediments. An example is the famous Rancho La Brea deposit in Los Angeles, in which so many animals, some of them now extinct, were trapped and entombed. Brea and tar are two of several synonyms applied to semisolid asphalts. *Pitch lakes* are large deposits of asphalt. Outstanding examples are the famous lake on the island of Trinidad in the West Indies and nearby Bermudez Lake on the mainland in eastern Venezuela. The latter pitch lake covers over 1100 acres and is estimated to contain 6 million tons of asphalt. It was actively worked for commercial asphalt between 1891 and 1931.

Natural asphalt is similar to the refinery product which supplies most of the commercial demand. Although very heavy oils may be "young" oils which have not yet evolved into the more common types of petroleum, it is probable that the seep asphalts are residual products from crude oil formed by the escape into the atmosphere of the lighter, more volatile, hydrocarbons. It is also possible that chemical reactions with near-surface ground waters hasten the asphaltization process.

Active or "live" seeps are those which continue to be supplied by oil rising from below. Even the asphalt lake on Trinidad is an active seep, as attested by the fact that more asphalt has been shipped from the lake than could possibly have been there in the first place. However, the asphalt impregnating a rock at the surface is usually the relic of a seep no longer active.

The geologist should also be cognizant of *false* indications of the presence of oil or gas. Films of iron oxide on water produce an iridescence like that of oil. Such films are easily identified, however, by the fact that they break into flakes when stirred, whereas true oil films cannot be severed, but instead the color bands trail out behind whatever is used as a stirrer.

[4] Donald Desmond Utterback, "A Study of Outcropping Bituminous Limestones and Sandstones with Reference to Porosity and to the Origin and Migration of Petroleum," *An Abstract of a Thesis Submitted in Partial Fulfillment of the Requirements for the Degree of Doctor of Philosophy in Geology in the Graduate School of the University of Illinois* (1936).

Oil seeps also may be created artificially, either by accident or design. Many short-lived oil "booms" have been started by the escape of gasoline or fuel oil from buried tanks and pipe lines. The oil "seeps" in many farm wells have been due to the generous use of lubricating oil on the pump. A book could be written on maliciously "salted" seeps in which vegetable cooking oil, refinery oil of various grades, and, very rarely, imported crude oil were used.

Marsh gas is difficult, if not impossible, to distinguish from natural gas. Both are dominantly methane, CH_4. Marsh gas is caused by the decay of plant material which has accumulated in marsh, bog, swamp, or lake bottom. All seeping gas which has passed through material of this type should be viewed with suspicion until and unless it can be proved that the gas is coming from underlying bed rock. Mud springs are not mud volcanoes, and the two should not be confused. Mud springs are merely springs in which the water emerges from a mud-covered opening.

Contrary to popular opinion, the presence of oil shale is *not* an indication of the probable presence of oil in commercial accumulations. It is so named because oil can be generated from the organic material within the shale by the application of heat; the same thing can be done with coal.

Seeps have been and can be of utmost importance in leading to the discovery of deposits of oil and gas. The value of seeps in oil finding has been emphasized by DeGolyer[6] and Link.[7] Areas where seeps have played an important role in discovery include Pennsylvania, southern Oklahoma,[8] the Gulf Coast, Wyoming, California, Mexico, Venezuela, Romania, Russia, the Middle East, the Netherlands East Indies, and elsewhere. Not every seep-containing province has been found to be productive of oil as yet, but the list of non-productive seep areas grows shorter with continued exploration. Madagascar and New Zealand are examples of areas containing seeps, but without commercial oil production at the time this book is being written. It has been pointed out[9] that seeps are qualitative but not quantitative indicators of the presence of oil.

Whether seeps are of immediate value in oil finding or not depends upon whether the seepage is parallel (Fig. 3-1) or transverse to the bedding, and, if parallel, upon whether the rocks are gently or steeply folded. Parallel seepage is movement through the reservoir rock by oil which has

[6] E. DeGolyer, "Direct Indications of the Occurrence of Oil and Gas," *Elements of the Petroleum Industry* (American Institute of Mining and Metallurgical Engineers, New York, 1940), pp. 21–25.

[7] Walter K. Link, "Significance of Oil and Gas Seeps in World Oil Exploration," *Bull. Am. Assoc. Petrol. Geol.*, Vol. 36 (August, 1952), pp. 1505–1540.

[8] Frank Gouin, "Surface Criteria of Southern Oklahoma Oil Fields, *Petroleum Geology of Southern Oklahoma* (American Association of Petroleum Geologists, 1956), pp. 14–35.

[9] E. DeGolyer, *op. cit.*

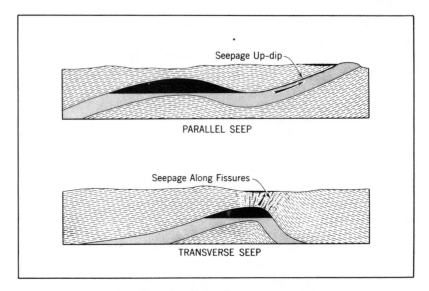

Fig. 3-1. Parallel and transverse seeps.

escaped from a down-dip trap, perhaps through recent tilting.   If the dips are very gentle, as in the Mid-Continent, the "mother lode" may be many miles away.   The seep merely affords evidence of the presence of oil in a certain formation, although it does not immediately lead to the discovery of an oil pool.   However, if the folding has been intense, as in most of the Tertiary oil provinces, the leaking trap may be tapped by a well drilled but a few yards from the seep.

Transverse or across-the-bedding seepage is brought about by recent fault or other fissure penetrating to the oil deposit itself.   In this situation the oil may be exploited by drilling in the immediate vicinity of the seep, the location depending upon the dip of the fissure and the depth of the reservoir.   Transverse movement of hydrocarbons through the rocks on one flank of the great Kirkuk anticline in Iraq has produced hundreds of seeps, many of which have been ignited so that they burn day and night. Direct connection between these seeps and the underlying oil reservoir has been proved by the fact that more and more of these flank seeps change from oil to gas as the gas cap expands downward with the removal of oil by producing wells.

As a general, but not invariable, rule the discovery cycle in any one province starts with exploration motivated and guided by the presence of seeps.   The next phase is the drilling of visible anticlines, and the final stage is the search for hidden traps.

Seeps may also occur offshore, and encourage exploration there as well, One notable example is the Santa Barbara, California, area where the three seep vents on the channel floor eight miles west of the city were known and recorded by explorers at least as early as 1776. A recent estimate of the average daily output of oil here is 50 to 70 barrels. Submarine seeps have been reported from many other localities, including Lake Maracaibo in Venezuela, the Gulf of Paria off Trinidad, the Gulf of Suez, Gulf of Mexico off Yucatan, the Pacific off northern Peru, and in Russia's Caspian Sea off the Apsheron Peninsula.[9a]

Surface indications in the Gulf Salt Dome Province. In addition to the usual criteria of the presence of oil, some of the salt domes in the Gulf Coast area are marked at the surface by signs peculiar to this type of structure. The salt domes, which are described subsequently, are immense prisms of salt usually topped by anhydrite-calcite cap rock which, in a few localities, contains native sulfur. The overlying rock is domed, and the adjacent rock has been upturned and truncated by the upward-moving prism. Oil may occur in the arched overlying strata, in the porous zones in the cap, or in the truncated flanking beds. Direct indications of the presence of salt domes beneath the surface include, besides oil and gas seeps, paraffin dirt, sulfur-bearing clay, salt springs, and outcropping cap rock.

Indirect evidence of the presence of salt domes has also proved of value. The upthrusting of the salt has been so recent in many places that the surface is arched in the form of a low hill rising above the flat coastal plain. Where the hill is too low to be recognizable the radial drainage pattern made by water running off the mound has supplied the clue. In other places leaching of the near-surface salt, followed by collapse into the salt caves, has produced a depression instead of a hill. When the uplift was not in the immediate past, erosion has truncated the dome producing inliers and a circular pattern to the geologic map, which is the best clue of all to the presence of a structural dome. Even when the different formations have not been recognized and mapped, the preference of vegetation for certain types of rock may lead to the development of circular vegetation patterns. Until the early 1920's, all the salt dome oil fields were discovered by one or another, or a combination, of these direct and indirect surface observations.

Soil analysis.[10] Soil analysis is an oil-finding technique that is also known as *geochemical prospecting*. It is an attempt to use what might be

[9a]Kenneth K. Landes, "Mother Nature as an Oil Polluter," *Amer. Assoc. Petroleum Geologists*, V. 57 (April), 1973, p. 637–641.

[10] Harold Bloom, "Geochemical Prospecting," *Subsurface Geology in Petroleum Exploration* (Colorado School of Mines, 1958), pp. 621–633; Leo Horvitz, "Near-surface Hydrocarbons and Petroleum Accumulation at Depth," *Mining Eng.*, Vol. 6 (December, 1954), pp. 1205–1209.

called *microseeps* as a guide to oil deposits in the same way that visible seeps have been used. Soil analysis is based on the theory that no rock is completely impervious, and so the lightest hydrocarbons, especially the gases, work through the cap rock and overburden to the surface. Geochemical prospecting is a natural evolution from the utilization of gas and oil seeps.

According to the theory underlying the technique of soil analysis, not only do gases escape to the surface but also liquid and solid hydrocarbons are formed from these gases, and their presence can be determined by means of microchemical techniques. Many different techniques have been applied in analyzing the soil for evidence of the existence of microseeps. The usual procedure is to obtain samples from varying depths by soil auger and then to test the samples chemically for various substances.

Because several million hydrocarbon-oxidizing bacteria per gram have been found in soil soaked in crude oil, it has been suggested that such bacteria might afford a clue to the presence of oil escaping from buried natural deposits.[11] This possible technique has been termed "microbiological" prospecting for oil.[12] It is still in the investigative phase in the United States, but Soli reports that microbiological prospecting has been pursued for some years in Russia "where it has met with outstanding success."[13] A wildcat well located on the basis of a favorable bacterial pattern was drilled in California in 1954, but it failed to find commercial oil.

For the first 20 years after the discovery of Spindletop on the Texas Gulf Coast in 1901 the presence of so-called paraffin dirt at the surface played an important role in the search for salt-dome oil deposits. This waxy material was assumed to contain paraffinic hydrocarbons that had seeped upward from an underlying oil accumulation. Now it is known that paraffin dirt is actually low in hydrocarbons, but is high in organic carbon and nitrogen. Furthermore, it is high in fungi, yeasts, and bacteria (including hydrocarbon consuming types), which probably give it its waxy appearance. Spectrometric examination of soil containing paraffin dirt has shown that petroleum gas is leaking through; "synthetic" paraffin dirt, apparently analogous to the natural material, has been produced by passing a stream of natural gas and air through a soil sample.[14] It can be concluded that paraffin dirt indicates the presence of a "micro" gas

[11] Claude E. ZoBell, "Influence of Bacterial Activity on Source Sediments," *Oil and Gas Jour.*, Vol. 109 (April 26, 1943), p. 24.

[12] R. J. Strawinski, "A Microbiological Method of Prospecting for Oil," *World Oil* (November, 1955), pp. 104–115; Giorgio G. Soli, "Geomicrobiological Prospecting," *Bull. Am. Assoc. Petrol. Geol.*, Vol. 38 (December, 1954), pp. 2555–2565.

[13] Giorgio G. Soli, "Micro-organism and Geochemical Methods of Oil Prospecting," *Bull. Am. Assoc. Petrol. Geol.*, Vol. 41 (January, 1957), p. 136.

[14] John B. Davis, "Studies on Soil Samples from 'Paraffine Dirt' Bed," *Bull. Am. Assoc. Petrol. Geol.*, Vol. 36 (November, 1952), pp. 2186–2188.

seep, but that the waxy material itself is not petroleum hydrocarbon, but was produced by micro-organisms that fed on petroleum hydrocarbons.

Besides running chemical and biological analyses of soil samples, investigators are also making gamma-ray surveys of the surface for the purpose of detecting gaseous hydrocarbon molecules seeping through the soil.[15] In addition to ground surveys, air-borne scintillation counters have been employed in the search for oil.[16] However, a recent study concluded that there is no direct correlation of surface radioactivity with oil at depth.[17] Although some discoveries have been claimed for geochemical prospecting, so far the application of these techniques has resulted in a discouraging number of failures. The theory on which geochemical prospecting is based is fundamentally sound. This technique has a major advantage, in theory at least, over all other exploration methods so far developed in that it purports to locate the actual presence of a buried oil accumulation, whereas all other oil-finding techniques are for the purpose of locating traps in which oil or gas may be stored. The inadequacies of soil analysis as a prospecting method may be overcome in the future by greater refinement of method.[18] If and when soil analysis does evolve into a practical method, it will probably have its greatest success in discovering oil accumulations in the less indurated rocks, especially those of Tertiary age. The tighter, older rocks are much less likely to permit the escape of hydrocarbons to the surface.

Geological Surveying.[19] Geological surveying, once the only technique of the petroleum geologist, is now but one of several skills that must be acquired.[20] It is doubtful that field geology will ever disappear as an

[15] John W. Merritt, "How to Avoid Costly Errors in Gamma Ray Surveying," *World Oil* (August, 1955), pp. 84–90; David S. Lobdell, E. F. Buckley, John W. Merritt, "Gamma Ray Exploration Comes of Age," *World Oil* (August, 1954), pp. 107–112; John W. Merritt, "Radioactive Oil Survey Technique," *World Oil* (July, 1952), pp. 78 et seq.

[16] R. W. Pringle, K. I. Roulston, G. W. Brownell, and H. T. F. Lundberg, "The Scintillation Counter in the Search for Oil," *Mining Eng.*, Vol. 5 (December, 1953), pp. 1255–1261; Hans Lundborg, "Air-borne Radioactivity Surveys," *Oil and Gas Jour.* (April, 1952), pp. 165 et seq.

[17] Alan F. Gregory, "Analysis of Radioactive Sources in Aeroradiometric Surveys Over Oil Fields," *Bull. Am. Assoc. Petrol. Geol.*, Vol. 40 (October, 1956), pp. 2457–2474.

[18] John W. Merritt, "Geochemistry as Aid to Successful Exploration," *Oil Weekly*, Vol. 115 (Oct. 9, 1944), pp. 35–38.

[19] Julian W. Low, *Plane Table Mapping*, (Harper & Brothers, New York, 1952); J. V. Harrison, "Method of Reconnaissance Geological Survey," *Bull. Am. Assoc. Petrol. Geol.*, Vol. 36 (October, 1952), pp. 2040–2043; F. H. Lahee, *Field Geology* (McGraw-Hill, New York, 1952), 5th edition; Kenneth K. Landes, *Plane Table Notebook* (George Wahr and Son, Ann Arbor, Michigan, 1947), 2nd edition.

[20] E. DeGolyer, "Historical Notes on the Development of the Technique of Prospecting for Petroleum," *Science of Petroleum* (Oxford University Press, 1938), Vol. 1, pp. 269–275.

aid to oil finding.   It is used intensively in exploring new areas, in the re-examination of older regions, and in checking geophysical prospects.

The methods and procedures followed in geologic surveying depend upon whether the survey is reconnaissance or detailed and upon the topography and structure of the country surveyed.

Reconnaissance surveying[21] is for the purpose of determining the oil prospects of a relatively large area in a limited space of time.   Fundamental to the survey is a base on which the observations can be entered. This base may be a map or it may be air photographs.   The ideal arrangement is to have both, using overlapping air photos ("stereo pairs") in the field, and transferring the field data onto a base map at camp.   When a base map is not available, an air-photo mosaic may be used instead.   If neither air photos nor adequate maps can be obtained, the geologist must make his own map by triangulation and plane tabling, by pace and compass, or by another method.

The observations made and entered on map or notebook include the following:

I. Environmental and economic.
   a. Accessibility of various parts of an area.
   b. Present and potential transportation media.
   c. Local supply and labor situation.
   d. Local market possibilities.
   e. Location of nearest oil or gas production.
   f. Location and logs of all wells drilled in area and environs.

II. Surface geologic.
   a. Direct or indirect evidences of oil or gas.
   b. Type, depth, and distribution of "masks"[22] obscuring the bed-rock geology.
   c. Distribution of sedimentary, igneous, and metamorphic rocks. Age of igneous activity and of metamorphism.

III. Stratigraphic.[23]
   a. Thickness of sedimentary rock veneer.
   b. Presence of and depth to possible source rocks in section.   Age.

---

[21] F. G. Clapp, "Fundamental Criteria for Oil Occurrence," *Bull. Am. Assoc. Petrol. Geol.*, Vol. 11 (July, 1927), pp. 683–703; J. J. Zunino, "Evaluation of Oil Exploration Methods," *World Oil*, Vol. 126 (July 7, 1947), pp. 13–20; K. C. Heald, "Geologic Engineering in the Petroleum Industry," *Engineering Geology* (Geological Society of America, 1950), pp. 251–271.

[22] Earl Noble, "Geological Masks and Prejudices," *Bull. Am. Assoc. Petrol. Geol.*, Vol. 31 (July, 1947), pp. 1109–1117.

[23] Raymond C. Moore, "Stratigraphical Considerations," *Science of Petroleum* (Oxford University Press, 1938), Vol. 1, pp. 304–305.

   c. Presence of and depth to possible reservoir rocks in section. Age.
   d. Presence of possible reservoir seals.
   e. Position and character of unconformities.
   f. Lateral changes in facies and especially in permeability.
   g. Convergence.
   h. Presence of soluble rocks in section.
   i. Carbon ratio of any coals present.

IV. Structural.
   a. Size and position of basins and other regional structural features.
   b. Presence of anticlines.
   c. Depths to possible reservoir rocks in different parts of area.
   d. Presence and nature of faults.
   e. Degree and character of jointing and fracturing in the brittle rocks within drillable depths.

The resultant illustrated report contains all the geologic and economic data that were obtained.[24] It also contains a classification of the oil possibilities[25] of different parts of the region and recommendations as to which areas, if any, merit detailed studies, or in some cases immediate testing without further geological exploration.

The purpose of *detailing* a relatively small area is to determine the areal extent of a surface structural anomaly, its vertical magnitude (closure), and the probable depths to the potential reservoir rocks. The geologist constructs an areal geologic map and a structure map. Positions and elevations are ascertained by plane table, alidade, and stadia rod (Fig. 3–2) or by compass and altimeter, with distance determined by pacing or automobile odometer. However, before or during the mapping, the geological party must prepare and learn a detailed stratigraphic section covering all the rocks involved in the surface work. By means of this section, it is possible to recognize several layers that can be used for elevation determinations and to adjust these elevations to a single datum on which the structure contours are drawn.

An alert field geologist learns many aids to geologic mapping. A soil cover does not necessarily preclude all knowledge as to the underlying bed rock. The vegetation differs in young soils, depending upon the lithology of the source formations.[26] Ant hills and burrowing animals supply

[24] George M. Johnson "Writing Scientific Reports" *Subsurface Geology in Petroleum Exploration* (Colorado School of Mines, 1958), pp. 779–791.

[25] Frank Reeves, "Outline for Regional Classification of Oil Possibilities," *Bull. Am. Assoc. Petrol. Geol.*, Vol. 30 (January, 1946), pp. 111–115.

[26] A. N. Murray, "Identification of Geological Formations by Growing Vegetation," *Tulsa Geological Society Digest*, Vol. 21 (1953), pp. 48–51; Robert H. Cuyler, "Vegetation as an Indicator of Geologic Formation," *Bull. Am. Assoc. Petrol. Geol.*, Vol. 15 (January, 1931), pp. 67–68.

Fig. 3-2. Plane table surveying in an area of "barren topography and naked geology." *Courtesy Standard Oil Company of California.*

samples of the material underlying the surface.

The growing importance of paleogeology and paleogeography in the search for stratigraphic traps means that the field geologist today must identify the environment of deposition of not only possible reservoir rocks, but of possible source and seal rocks as well. Furthermore he must attempt to determine the source of clastic sediment, the direction of transport, and the depositional limits. Determining and plotting the direction of dip of cross bedding[27] assists in this regard.

Many sedimentary basins are veneered, in part at least, by layered volcanic rocks including lava flows and pyroclastic deposits such as ash beds. The structure of the underlying sedimentary rocks will be reflected in the volcanics if the major folding has taken place since the volcanism. More and more field work can be anticipated in areas masked by volcanics. No longer can the petroleum geologist confine himself exclusively to sedimentary rocks.

Where the oil fields go out to sea the outcrops on the ocean floor con-

[27]William F. Tanner, "New Method for Mapping Old Shorelines," *World Oil* (April, 1956), pp. 123 et seq; Paul E. Potter and Raymond Siever, "Regional Crossbedding and Petrology as Source Area Indicators," *Science*, Vol. 122 (Nov. 25, 1955), pp. 1021–1022.

tain invaluable structural and stratigraphic information. Formerly professional divers were taught the art of taking contact strike and dip readings, but now skin diving geologists are used.[28] These geologists not only obtain structural information, but also they can sample the formations, measure sections, and carry on other aspects of geological surveying. Submarine geology has also been assisted by the fathometer, a shipboard device capable of obtaining accurate topographic profiles. The most recent development in this regard is the Sonoprobe,[29] a low frequency echo sounder with a high degree of accuracy, which also penetrates into the sea-floor sediment.

The report of the detail party includes the section and other stratigraphic data obtained, as well as the geologic and structural maps. At the same time other branches of the exploration department may have been at work so that concurrent reports are filed covering such possible items as a geophysical survey of the same area and a report from the subsurface department concerning previous tests drilled in the vicinity and their findings. Administrative officers collate these reports and decide (1) whether or not to proceed further; (2) what land to lease; and (3) where to drill the first test. Meanwhile, the geological party has moved to a new assignment.

Geologic maps. Maps showing the areal geology yield many structural clues. Not only by the presence of inliers but also by the distribution pattern of the outcrops of rocks of varying ages can the structure of the bedrock be determined (Fig. 3–3). In combination with good topographic maps it is possible to determine, by the intersection of contours with formation contacts, the elevation of a stratigraphic datum at various points, and from this information a structural contour map can be drawn.

Structure maps. Structure maps show, by contours (Fig. 3–4) or other means, anticlines, synclines, noses, and faults. Many noses that appear on structure maps with large contour interval become closed folds when mapped with smaller interval. Others remain structural noses in the near-surface formations but develop closure in deeper, possibly oil-bearing formations. For these reasons structural noses are favorable features for testing.

Many oil fields have been discovered through the testing of anticlines and other structural traps mapped and described by a governmental

[28] H. W. Menard, R. F. Dill, E. L. Hamilton, D. G. Moore, George Shumway, M. Silverman, and H. B. Stewart, "Underwater Mapping by Diving Geologists," *Bull. Am. Assoc. Petrol. Geol.*, Vol. 38 (January, 1954), pp. 129–147; George Shumway, "Compass-Inclinometer for Underwater Outcrop Mapping," *Bull. Am. Assoc. Petrol. Geol.*, Vol. 39 (July, 1955), pp. 1403–1404.

[29] C. D. McClure, H. F. Nelson, and W. B. Huckabay, "Marine Sonoprobe System, New Tool for Geologic Mapping," *Bull. Am. Assoc. Petrol. Geol.*, Vol. 42 (April, 1958), pp. 701–716.

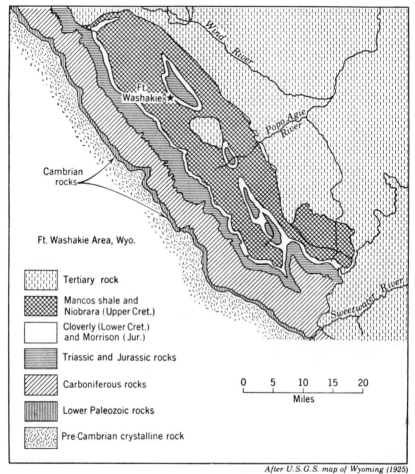

Fig. 3-3. Geologic map yielding structural information. Syncline, anticline, unconformable Tertiary overlap. *Courtesy U. S. Geological Survey.*

agency such as the Geological Survey, U. S. Department of the Interior. Even today, after years of exploration in the United States, criteria which lead to testing and discovery are being found by continued study of previously published maps of various types.

Topographic Maps. In addition to their value as base maps for geologic study, topographic maps may also yield structural information of value. The trained eye can recognize dip slopes (Fig. 3-5), and anticlines can be, and have been, recognized from observations of topographic maps alone. Faults may be traced by their scarps. The recognition of recent domes by circular drainage patterns has already been mentioned.

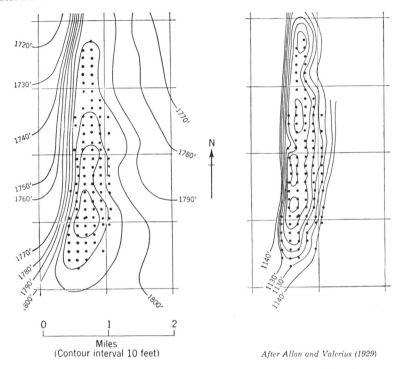

Miles
(Contour interval 10 feet)   *After Allan and Valerius (1929)*

Fig. 3–4. Structure contour maps. *Left:* contoured on surface limestone. *Right:* contoured on subsurface formation. Fairport field, Kansas. *Courtesy American Association of Petroleum Geologists.*

Geological interpretation of topographic maps can be aided by the use of "shaded relief" as shown by recently published maps of the U. S. Geological Survey. The combination of field work and a good topographic map may make possible a geologic map of much greater area than that actually covered in the field.

Photogeology.[30] Photogeology is the geological interpretation of air photographs. Although photographs of the ground taken from airplanes

[30] William A. Fischer, "Color Aerial Photography in Photogeologic Interpretation," *Photogrammetric Eng.*, Vol. 24 (September, 1958), pp. 545–549; Benjamin A. Tator, "The Aerial Photograph and Applied Geomorphology," *ibid.*, pp. 549–561; William R. Hemphill, "Small Scale Photographs in Photogeologic Interpretation," *ibid.*, pp. 562–567; Laurence H. Lattman, "Technique of Mapping Geologic Fracture Traces and Lineaments on Aerial Photographs," *ibid.*, pp. 568–576; Richard G. Ray, "Scale and Instrument Relationship in Photogeologic Study," *ibid.*, pp. 577–584; Richard G. Ray, "Photogeologic Procedures in Geologic Interpretation and Mapping," *U. S. Geological Survey, Bull.* 1043–A, (1956); D. J. Christensen, "Eagles of Geology," *Photogrammetric Eng.*, Vol. 22 (December, 1956), pp. 857–865; A Panel of Papers on Photo Interpretation, in two parts, *Photogrammetric Engineering*, Vol. 20 (American Society of Photogrammetry, Washington, D. C., June, 1954),

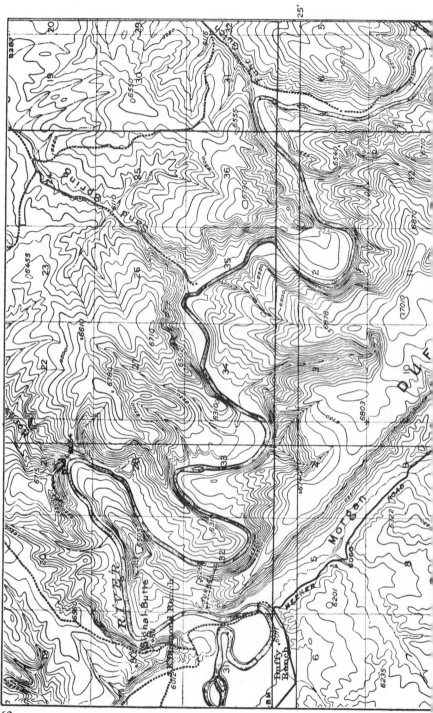

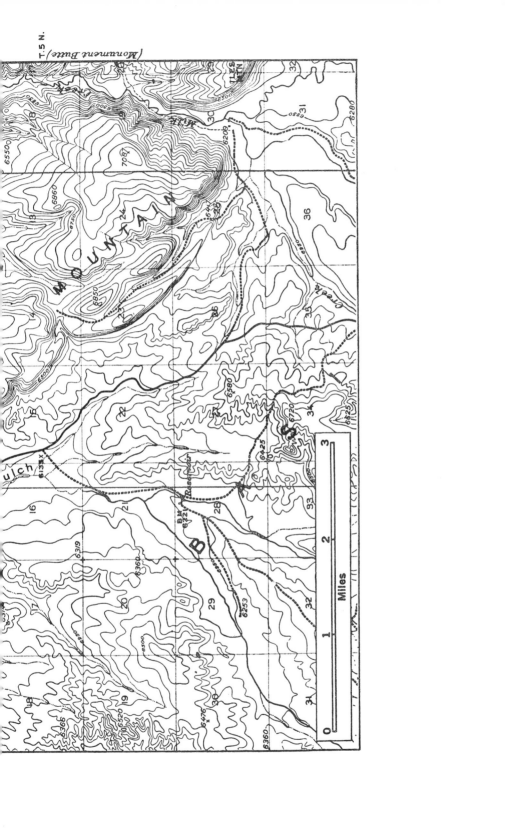

were scanned for possible use in petroleum exploration in Oklahoma as early as 1928, it was not until the years immediately preceding World War II that this technique became common in oil finding. There are two phases in the use of air photos: (1) photogrammetry, which is the transposition into a coherent and accurate whole of the data appearing on many single overlapping photos, and (2) the interpretation of the features appearing in the photographs in terms of geology, engineering, military operations, or other facets of human interest. Among the many geological applications of air photos are their employment in areal and structural mapping, both reconnaissance and detailed, in the study of shore features and processes, in stratigraphic work, and in mining and glacial geology. The explorer finds air photos invaluable, and the engineer uses them in the construction of topographic maps, for locating pipe lines and highways, and for other purposes. Air photos are used by the agronomist in soil mapping and by the oil company land department for correcting errors in ownership maps and for the inspection of proper-ties.[31]   The extent of world air-photo coverage is shown in Fig. 3–6.

Although photogeology is possible in regions of widely differing char-acter, including even heavily forested areas,[32] it is most valuable in such

pp. 395–466; "Selected Papers on Photogeology and Photo Interpretation," (Committee on Geophysics and Geography, Research and Development Board, Washington 25, D. C., April, 1953), pp. 225; Juan B. Puig, "La Fotogeologia aplicada a la exploracion petrolera (Application of Photogeology to the Exploration for Petroleum)" *Bol. asoc. mex. geol. petrol.*, Vol. 5 (November-December, 1953), pp. 369–423; H. T. U. Smith, editor, Sym-posium of 8 papers, *Photogrammetric Engineering* (December, 1951 and subsequent issues); Ronald R. Hartman and Kalman H. Isaacs, "System in Photogeology," *Bull. Am. Assoc. Petrol. Geol.*, Vol. 42 (May, 1958), pp. 1083–1093; William R. Hemphill, "Determination of Quantitative Geologic Data with Stereometer Type Instruments," *U. S. Geol. Survey, Bull.* 1043–C (1958), pp. 35–56; Louis Desjardins, "Techniques in Photogeology," *Bull. Am. Assoc. Petrol. Geol.,* Vol. 34 (December, 1950), pp. 2284–2317); Frank A. Melton, et al., *Photogrammetric Engineering*, Vol. 16 (December, 1950); Robert Helbling, "Studies in Photogeology, (in Connection with Geological Mapping in Switzerland)," *Art Institut Orell Fussli A-G.* (Zurich, 1949); Armand J. Eardley, *Aerial Photographs: Their Use and Interpretation* (Harper and Brothers, New York, 1942); H. T. U. Smith, *Aerial Photographs and Their Applications* (D. Appleton-Century, New York, 1943); G. C. Cobb, "Bibliography on the Interpretation of Aerial Photographs and Recent Bibliographies on Aerial Photog-raphy and Related Subjects," *Bull. Geol. Soc. Am.*, Vol. 54 (August, 1943), pp. 1195–1210; H. T. U. Smith et al., "Symposium of Information Relative to uses of Aerial Photographs by Geologists," *Photogrammetric Engineering*, Vol. 13 (December, 1947), pp. 531–628.

[31] Denis S. Sneiger, "Applications for Aerial Photography," *Oil Weekly*, Vol. 80 (Feb. 3, 1936), pp. 19 et seq.

[32] Laurence Brundall, "La Fotogeologia," *Petroleo Interamericano* (edición de enero, 1948).

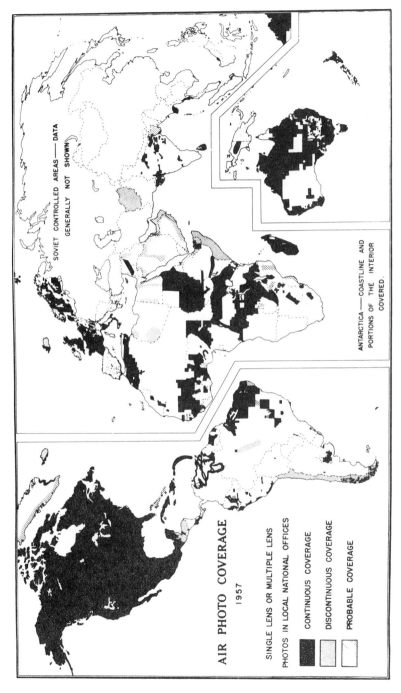

AIR PHOTO COVERAGE
1957

SINGLE LENS OR MULTIPLE LENS
PHOTOS IN LOCAL NATIONAL OFFICES

CONTINUOUS COVERAGE

DISCONTINUOUS COVERAGE

PROBABLE COVERAGE

SOVIET CONTROLLED AREAS — DATA
GENERALLY NOT SHOWN

ANTARCTICA — COASTLINE AND
PORTIONS OF THE INTERIOR
COVERED.

Fig. 3-6. Extent of world air-photo coverage, 1957. *Courtesy of Professor Kirk H. Stone, Department of Geography, University of Wisconsin.*

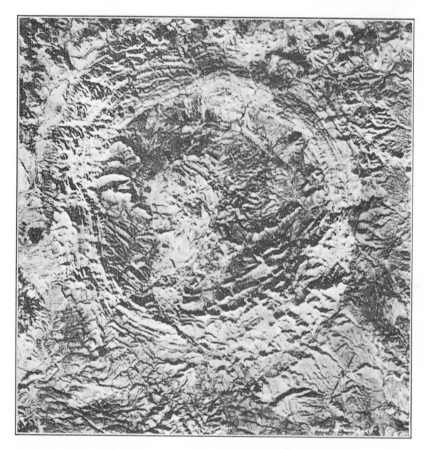

Fig. 3–7. Aerial photograph yielding structural information. Dome, Marfa Basin, Presidio and Brewster Counties, West Texas. *Courtesy Edgar Tobin Aerial Surveys.*

provinces as the basin and range where the terrain is one of "barren topography and naked geology."[33]   Under these conditions it is possible actually to see not only the parallel bands of sedimentary rock (Fig. 3–7) but also the direction and degree of dip of the rock strata (Fig. 3–8).   A structure map using strike and dip symbols can be plotted directly on the air-photo base.   By using photogrammetric techniques it is possible to obtain sufficient elevation and stratigraphic-thickness data to draw both structure-contour[34] and isopach maps.   Even when the rocks are not so

[33]"Photogeology Finds Structures," *The Link*, Vol. 12 (Carter Oil Company, Tulsa, Oklahoma, March, 1947), p. 8.

[34] L. E. Nugent, "Aerial Photographs in Structural Mapping of Sedimentary Formations," *Bull. Am. Assoc. Petrol. Geol.*, Vol. 31 (March, 1947), pp. 478–494.

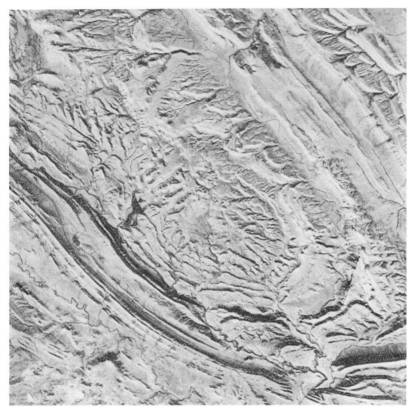

Fig. 3-8. Maverick Springs Anticline, Fremont County, Wyoming. Southeastern end of asymmetrical anticline. Part of oil field can be seen just south of northwest corner in structure apex area. Eroded anticlinal core exposes Triassic and Jurassic formations; flanking hogbacks are Cretaceous sandstones. Northeast flank of anticline dips about 18°, whereas the same formations on southwest flank are vertical and even overturned.° *U. S. Geological Survey Photograph.*

°David A. Andrews, "Geologic and Structure Contour Map of the Maverick Springs Area, Fremont County, Wyoming," *U. S. Geol. Survey,* Oil and Gas Investigations, Preliminary Map 13, 1944.

well exposed, structural information may be obtainable from air photos. Clues may be had from outcrop, stream, soil color, and vegetation (even submarine) patterns.[35] Some of these patterns, especially soil color and vegetation, may not be visible on the ground, but they are quite prominent in the air photographs. The major controls on patterns and tones that appear in photographs are differences in topography, soil moisture,

[35] Bernard M. Bench, "Discovery of Oil Structures by Aerial Photography," *Oil and Gas Jour.* (Aug. 26, 1948), pp. 98 et seq.; Charles de Blieux, "Photogeology in Gulf Coast Exploration," *Bull. Am. Assoc. Petrol. Geol.,* Vol. 33 (July, 1949), pp. 1251–1259.

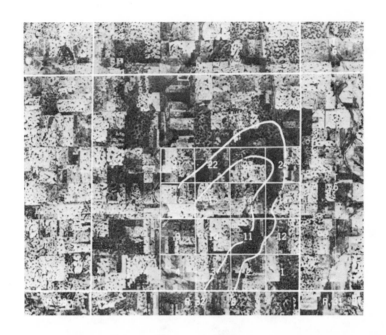

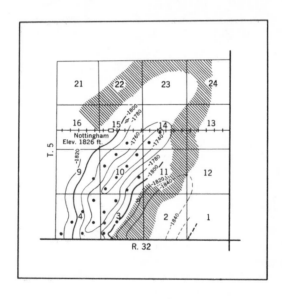

and bedrock lithology. In many places, such as the Middle East, the Gulf Coast salt-dome province, California, and Saskatchewan, the topography may reflect directly the underlying structure. Of course many of these hills-over-anticlines are of adequate relief to be visible with the stereoscope. Where the relief is small, radial drainage[36] off the higher area, stream adjustments that swing around the higher land, or tonal halos[37] indicate the presence of a low hill (Fig. 3–9). Tonal halos are dark bands surrounding a lighter area; they are caused by a higher soil moisture content around the foot of a hill. Faults may be indicated by lines of springs (or zones of higher soil moisture), or by lineaments.[38] Lineaments are linear features appearing on air photographs. They are due to topographic relief, stream adjustments, and vegetation, but these surface differences may be the result of the presence of faults cutting through the bedrock.

In recent years starting developments have taken place in the use of photo and other imagery for geological exploration and mapping. The seafloor can be mapped both by underwater photography and by advanced sonar (scanning by means of sound waves, or acoustic energy impulses). However, the greatest contributions to the geology of the surface of planet earth have been made by U.S. Army photoplanes and especially by orbiting space-craft of the National Aeronautics and Space Administration (Nasa). In addition to high altitude air photographs sophisticated "remote sensors" have been developed which produce their own imagery from which a surprising amount of geology can be interpreted. Commercial planes equipped for "side-look" radar imagery also supply data which make possible more detailed geologic mapping.

[36] B. A. Tator, "Drainage Anomalies in Coastal Plains Regions," *Photogrammetric Engineering* (June, 1954), pp. 412–417.

[37] W. O. Kupsch, "Submask Geology in Saskatchewan," *First Williston Basin Symposium* (Bismark, North Dakota, 1956), pp. 66–75, 1957.

[38] W. O. Kupsch and J. Wild, "Lineaments in Avonlea Area, Saskatchewan," *Bull. Am. Assoc. Petrol. Geol.*, Vol. 42 (January, 1958), pp. 127–134.

◄ **Fig. 3–9.** *Upper:* Air-photo mosaic of Nottingham oil field in southeastern Saskatchewan. Distinctive tonal semihalo has been marked on photograph. Each numbered section is one square mile. *Saskatchewan Government photo. Lower:* Subsurface contour map of 16-section area outlined on air photo, showing producing wells and relationship between tonal halo and structure. Contours drawn on top of Mississippian erosion surface by E. Hayes, Saskatchewan Department of Mineral Resources. *Courtesy Walter Kupsch and the Department of Geology, University of Saskatchewan.*

Increasing attention is being paid to fracture patterns[39] which in many areas produce a distinctive "grain" in the air photographs. Furthermore, these patterns also show up in soil covered areas, even when heavily forested. Local deviations in the pattern plan have been found to indicate structural or stratigraphic anomalies.

A striking application of air photos to oil discovery took place in eastern Venezuela. The Santa Rosa dome was originally discovered by seismic survey, but the first well, presumably drilled on the top of the dome, was a failure. "Inspection of aerial photographs showed a distinctive, oval patch of light-toned vegetation about two miles away from the dry hole. Future investigation proved that the light spot was the top of the Santa Rosa dome."[40] Some of the Gulf Coast salt domes are likewise visible in air photos because of the circular vegetation pattern.

Since World War II, air photos have been employed as a major tool in petroleum exploration in all parts of the world, including especially the Rocky Mountain province of the western United States, the Alberta foothills, northern Alaska,[41] and Venezuela.[42]

Color aerial photography[43] is the newest innovation in photogeology. Some stratigraphic contacts, marked by a color change too slight to be significant on a black and white picture, are clearly visible in color prints. On the other hand, most features of geologic value that can be picked up by color photography can also be seen in black and white. The greater cost of color pictures makes it unlikely that color will supersede black and white photography as a reconnaissance tool, but it may become useful in detailing the geology of limited areas where minor color distinctions are important in contact finding.

Prerequisite for acquiring skill in geologic interpretation of air photos are good eyes, training in geomorphology, and field experience. Good eyes are needed to use the various stereoscopic devices by means of which overlapping pairs of air photographs are viewed. A training in geomor-

---

[39] P. H. Blanchet, "Development of Fracture Analysis as Exploration Method," *Bull. Am. Assoc. Petrol. Geol.*, Vol. 41 (August 1957), pp. 1748-1759; Laurence H. Lattman and Wilds W. Olive, "Solution-widened Joints in Trans-Pecos Texas," *Bull. Am. Assoc. Petrol. Geol.*, Vol. 39 (October, 1955).

[40] Laurence Brundall, "La Fotogeologia," *Petroleo Interamericano* (edición de enero, 1948).

[41] Norman C. Smith and Sherman A. Wengerd, "Photogeology Aids Naval Petroleum Exploration," *Bull. Am. Assoc. Petrol. Geol.*, Vol. 31 (May, 1947), pp. 824-828.

[42] Sherman A. Wengerd, *op. cit.*

[43] J. I. Gimlett and Kenneth E. Hunter, "Color Aids Photogeological Interpretation," *World Oil* (April, 1958), pp. 123-125; "Here's a New Oil-finding Tool . . . It's Aerial Color Photography," *Oil and Gas Jour.*, Vol. 56 (March, 1958), pp. 122-124; B. H. Kent, "Experiments in the Use of Color Aerial Photographs for Geologic Study," *Photogrammetric Engineering*, Vol. 23 (December, 1957), pp. 865-868.

phology is essential to the recognition of dip slopes and other geologic features that may have topographic expression. Field experience is vital for the same reason.

SUBSURFACE OBSERVATIONS

Surface methods of oil exploration can be used only where the bedrock is adequately exposed. If the rock is hidden by a thin capping of soil, it may be possible by means of dug pits or trenches to expose the underlying bedrock for study. Some oil companies have succeeded in penetrating not only the soil but also recognizable layers of bedrock with hand augers, or with a jeep-mounted auger drill. But where natural outcrops are absent and hand-dug or drilled exposures not practicable, or where the sedimentary strata that are exposed cannot be correlated from outcrop to outcrop, structure mapping by surface observations is impossible. By the early 1920's much of the prospective oil territory in the Mid-Continent and the Gulf Coast that could be explored by surface methods has been mapped by some of the older, established oil companies. The oil companies then looked with hungry eyes at the much greater area in which the geology was a secret because of a veneer of soil, sand, loess, alluvium, or glacial drift, or because recognition of individual layers in a thick series of red beds or chalky shales was difficult if not impossible. Ever-increasing pressure to find more oil resulted in the development of exploratory tools which in effect give the geologist a "hammer with a longer handle" with which to penetrate the masks that hide the geological picture. The techniques developed to meet this situation include core drilling, geophysical prospecting, and the maximum utilization of the records obtained during the drilling of wells. Core drilling and geophysical prospecting are described in the following sections; the utilization of drilled well records is discussed under "Subsurface Geology" in the next chapter (4).

Structure Drilling.[44] One of the earliest uses of the structure core drill as a device to map possible oil-bearing anticlines was in Kay and Noble counties, Oklahoma, in the early part of 1922.[45] This survey

---

[44] E. J. Longyear Company, "Prospecting for Oil" (45 titles), *Bibliography of Diamond Drilling* (1948), pp. 19–22, and Supplement 1 (March 15, 1949); E. DeGolyer and Harold Vance, "Core Drilling for Structure" (17 titles), "Bibliography on the Petroleum Industry," *Bull. Agr. and Mech. Coll. Texas*, 83 (1944), pp. 437–438.

[45] H. G. Officer, Glenn C. Clark, and F. L. Aurin, "Core Drilling for Structure in the North Mid-Continent Area," *Bull. Am. Assoc. Petrol. Geol.*, Vol. 10 (May, 1926), pp. 513–530

resulted in the discovery of the Tonkawa field, and within a short space of time many core drills were at work in the Mid-Continent and elsewhere. A large number of discoveries, including some major oil fields, can be credited to the geological information obtained by structure drilling.

The structure core drill, as developed for use by the oil industry, is a lightweight rotary drill mounted on a truck. When changing locations, the mast folds forward over the cab of the truck (Fig. 3–10). Holes can be drilled to depths of over a thousand feet, but most structure drilling is carried on at lesser depths. Soft materials, such as shales, are drilled with a fish-tail bit; diamond or hard alloy bits are used to cut a core through rock which may contain the "key beds" which are to be used in determining the elevation of the stratigraphic datum.

The objective of the geologist assigned to a structure-drill survey is to obtain sufficient stratigraphic elevation data for the recognition and delineation of any structural anomalies that may be present. The usual procedure is to core-drill a series of holes along a straight line in the direction of the regional dip. If any reversal of this dip is noted, additional holes are drilled both along the line of the survey and on both sides in order to map the upfold. It is the duty of the geologist to run elevation lines to the structure-drill locations and to watch the cores and drill cuttings as they are brought to the surface for the particular stratum, or more often the sequence of strata, which he is using to determine the datum for a structure-contour map. Electric and gamma-ray logs,[46] described in Chapter 4, are also used to supplement and even replace rock samples in core tests, as well as in regular wells. Once the key horizon has been reached, its elevation is determined and entered on the map. At the same time drilling is discontinued, and the core drill is moved to the next location.

Stratigraphic ("slim hole") tests. The realization forcibly brought home by the development of the East Texas field in 1931, that oil could accumulate in stratigraphic as well as in structural traps, led to a greater interest in subsurface stratigraphy and especially in the tracing of lateral changes in the porosity and thickness of potential reservoir rocks. Inasmuch as most of this information lies beyond the practicable range of the portable core drill, the "slim-hole" rig was developed to obtain geological information at greater depths. With this machine, which is merely a lightweight rotary, it is possible to obtain a complete stratigraphic record through thousands of feet of section at lesser cost than with the conventional type of rotary rig.

Because of the cost advantage, slim holes (up to 6 or 7 inches in

---

[46] R. W. Biggart, "Modernized Core Drilling for Structure," *World Oil* (April, 1957), pp. 134–138.

diameter) are now drilled for production wells in some areas,[47] and slim holes drilled for geological information have become known as "strat" (stratigraphic) holes or tests.

Exploration Geophysics.[48] Geophysics is the physics of the earth; exploration geophysics is the application of physics to the search for new mineral deposits. Exploration geophysics has had a long and honorable history in mining geology. Many of the world's great iron-ore deposits were discovered by magnetic surveys, a technique that is now several hundred years old. The torsion balance, a gravity instrument, was first used in Czechoslovakia in 1915.[49]

Although it was not until the early 1920's that exploration geophysics became an important tool in oil finding, once it started the growth and development of this technique were phenomenal. The rapid expansion of the use of geophysical instruments was due to the many successes which followed. Before 1924, thirty-nine salt domes had been discovered in the Gulf Coast of Texas and Louisiana by surface methods. In the succeeding five years, sixty-four salt domes were discovered, and many more have been found since. The post-1924 discoveries have been due almost entirely to the use of geophysical instruments. Both gravity instruments and seismic surveys have been successful in this work, and one common procedure has been to use the cheaper gravity survey as a reconnaissance tool, checking the anomalies with the more expensive seismograph survey. The present trend in geophysical prospecting is to combine it with surface observations, using the geophysics to confirm and detail structures whose presence has been suspected by surface observations. Close coordination of geological and geophysical data is extremely important throughout an exploration program.[50]

There has been a rather distinct evolution in geophysical instruments since their first application in the search for oil. The dip-needle compass

[47] Symposia on Slim Hole Drilling, *World Oil* (May, 1954), pp. 117–174; *Oil and Gas Jour.* (Sept. 26, 1955), pp. 143–154.

[48] M. B. Dobrin and H. F. Dunlap, "Geophysical Research and Progress in Exploration," *Geophysics*, Vol. 22 (April, 1957), pp. 412–433; Herbert Hoover, Jr. "Twenty Years of Advancements in Geophysical Techniques," *Jour. Petrol. Technol.*, Vol. 5 (September, 1953), pp. 77–80; J. C. Karcher, "Exploration by Geophysical Methods," *Elements of the Petroleum Industry* (American Institute of Mining and Metallurgical Engineers, 1940), pp. 63–89; Donald C. Barton, "Exploration Geophysics," *Geol. Soc. of Am., 50th Anniversary Vol.* (New York, 1941), pp. 549–569; Harrison E. Stommel, "Subsurface Methods as Applied to Geophysics," *Subsurface Geologic Methods* (Colorado School of Mines, 1950), pp. 1035–1119.

[49] Sigmund Hammer, "Geophysical Exploration Comes of Age," *Bull. Am. Assoc. Petrol. Geol.*, Vol. 36 (July, 1952), p. 131.

[50] F. J. Agnich, "Exploration for Reefs by Geophysical Methods," *Proc. Fourth World Petrol. Congress* (1955), Section 1, pp. 619–634; D. C. Skeels, "Correlation of Geological and Geophysical Data," *ibid*, pp. 665–674.

(a)

(b)

Fig. 3-10. Truck-mounted core drills. (a) Shot-hole drilling rig. *Courtesy Seismograph Service Corporation.* (b) Rig mast folded over cab. *Courtesy Engineering Laboratories, Inc.* (c) Heavy-duty portable rotary rig. *Courtesy George E. Failing Supply Company.*

(c)

of the iron-ore surveyor was followed by the magnetometer. The ground
magnetometer in turn has been superseded by special types used in air-
borne surveys. The pendulum and its first offspring, the torsion balance,
have been followed by the readily portable gravity meter or gravimeter,
which can be carried in automobiles, canoes, and even strapped to a pack
board. The refraction method of seismograph surveying has been followed, but not entirely superseded, by the reflection method. In the
recording end, the conventional seismic paper record is being displaced
by the corrected record section made from magnetic tape.

The first important event in the application of exploration geophysics
to the petroleum industry was the discovery in March, 1924, by means of
the torsion balance, of the Nash salt dome in Fort Bend and Brazoria
counties, Texas. Within less than a year four additional salt domes were
discovered by companies using geophysical methods. Nearly one-half
billion dollars worth of oil and gas have been produced from fields discovered by geophysical methods in the Gulf Coast district alone. The
number of parties that are out today at any one time engaged in seismic
and other types of geophysical surveys can be counted in the hundreds.

Geophysical prospecting is the only feasible method of exploring large
areas that are water covered, such as the continental shelves.[51] Magnetic, gravimetric, and seismic methods have been adapted to offshore
work. As a general rule prospecting is cheaper on water than on land.

Magnetic methods.[55] Where the fairly common iron-ore mineral
magnetite is present in the underlying rocks, the magnetometer or other
magnetic instrument records abnormally high readings in that vicinity.
"Since sedimentary rocks are very weakly magnetic, if not completely
non-magnetic, a magnetic survey pictures conditions in the underlying
igneous, so-called basement rocks. This makes it possible to determine
the depths to the magnetic basement and thus to map sedimentary
basins."[56] The magnetometer also serves to determine the presence of
basic igneous rocks, especially dikes and sills, intruded into the sedimentary rock section.

The air-borne magnetometer was developed during World War II as a
means for detecting the presence of submarines. Immediately upon the
close of hostilities this technique was adapted for mineral exploration.
It is possible with a single air-borne magnetometer to complete 500 to

[51] Henry C. Cortes and Ronald N. Gsell, "Geophysical Prospecting Over Continental
Shelves," *Proc. Fourth World Petroleum Congress* (1955), Section 1, pp. 575–604.

[55] P. A. Rodgers, "Gravity and Magnetic Effects of Subsurface Bodies," *Subsurface
Geology in Petroleum Exploration* (Colorado School of Mines, 1958), pp. 585–619; Victor
Vacquier *et al,* "Interpretation of Aeromagnetic Maps," *Geol. Soc. Am.,* Memoir 47 (1951).

[56] Sigmund I. Hammer, "Modern Methods of Gravity and Magnetic Interpretation,"
Proc. Fourth World Petrol. Congress (1955), Section 1, pp. 635–646.

1000 miles of continuously recorded profile per day. In addition the airborne instrument ignores purely local magnetic disturbances, such as magnetite-bearing erratics, which may make a surface survey completely ineffective. The aerial magnetometer is being employed today as a reconnaissance tool both in oil prospecting[57] and in the search for new deposits of iron ore. The Lund field in the northeastern corner of Travis County, Texas, is credited with being the first oil pool found (December, 1948) by the flying magnetometer. It is a serpentine accumulation.

The present (1958) status of the air-borne magnetometer appears to be in surveying virtually unexplored sedimentary basins. "Once enough wells have been drilled to define the extent and thickness of a basin, there is not much interest in further magnetic exploration."[58]

Gravimetric methods.[59] The motivation for all gravity instruments is the variation in gravitative attraction produced by different masses and different densities. As in the magnetic survey, large anomalies in the gravity picture may be due to differences in the basement rock which have nothing whatsoever to do with the structure in the overlying sedimentary rocks, but the gravimeter is also sensitive to variations in the sedimentary section.

The simplest gravity instrument, the pendulum, swings with a period inversely proportional to the square root of gravity. When the underlying rock is of greater than average density it swings more rapidly. Conversely, when there is rock of inferior density, such as salt, the pendulum swings more slowly. Accurate timing of the pendulum swings from one location to another determines the variations in gravity. The pendulum and the torsion balance were the forerunners of the modern gravity meter or gravimeter (Fig. 3–11) which is now used almost exclusively in gravity surveys.

The gravity meter is an extremely sensitive instrument which measures variations in gravitative attraction by means of a mass suspended from a balance beam. A suitable spring is attached to the other end of the beam. Readings are in milligals; a milligal is one one-thousandth of a gal, and a gal is an acceleration of 1 centimeter per second per second. The gravity meter, unlike its predecessors, has extreme portability (Fig. 3–12), and

---

[57] W. P. Jenny, "Aereal Magnetic Oil Discoveries," *World Oil* (November, 1951), pp. 85 et seq; H. Wayne Hoylman, "Airborne Magnetometer Profiles Across Cuyama Valley," *Oil and Gas Jour.* (Dec. 29, 1949), pp. 55 et seq.

[58] Milton B. Dobrin, *letter*, Oct. 30, 1958.

[59] Sigmund Hammer, *op. cit.*; W. R. Sype, "Gravity Data Indicate Structural Conditions," *World Oil* (February, 1955), pp. 82 et seq.; "Role of Gravity in an Exploration Program" (Symposium, Geophysical Society of Tulsa), *Oil and Gas Jour.* (Sept. 28, 1953), pp. 106 et seq.; Willis H. Fenwick, "A Practical Approach to Gravity Interpretation," *Mines Mag.*, Vol. 41 (October, 1951), pp. 91–95.

Fig. 3–11. Gravity meter. *Courtesy North American Geophysical Company.*

observations can be taken relatively quickly. When roads are available, the gravity meter is carried in a car or truck. At each station a tripod is lowered through the floor and the gravity meter is moved into position on the tripod from its cradle in the truck bed. Where vehicle progress is impossible, the gravity meter may be carried by hand or in a canoe and periodically set up for readings to be taken. It is also used in offshore surveys.[60] Here the gravity meter is lowered in a pressure-tight case to the sea floor. Leveling and unclamping of the meter are carried out by remote control, and the readings are taken by a camera enclosed with the gravity meter.

[60] R. D. Wyckoff, "Geophysical Exploration," *Oil and Gas Jour.*, Vol. 47 (Nov. 11, 1948), p. 344.

Fig. 3-12. Air-borne gravity meter. *Courtesy North American Geophysical Company.*

Gravity devices react sharply to a major discontinuity[61] (the contact plane between rocks of markedly different density) except when the plane of the discontinuity is horizontal. The gravity meter has been successful in the tracing of shallow faults where the density contrast is large.

Gravity methods have had their greatest success in the discovery of the piercement-type salt domes because of the marked variations in gravity between the prism of salt and the intruded country rock. The gravity survey map shows an area of low gravitative attraction above the buried salt mass. Not only have gravity surveys been successful in discovering and delineating inland, onshore, and offshore salt domes in the Gulf Coast, but they have also been successful in mapping salt-cored structures in other parts of the world, including the maritime provinces of eastern Canada and the Middle East. As one would expect, igneous intrusions behave oppositely from salt cores; they appear in the gravity survey map as areas of high gravity attraction.

Many of the Silurian-reef gas fields in southwestern Ontario were dis-

[61] Jonathan W. Phillips, "Discontinuity—Key Word in Oil Finding," *Oil and Gas Jour.* (March 17, 1958), pp. 159–173.

covered by drilling gravity anomalies.[62]   It is uncertain, however, whether it is the reefs or the associated evaporites which produce the anomalies.

It is even possible to trace the rise and fall of a gravity discontinuity across country providing that (1) the discontinuity is large enough, and (2) it does not keep to excessive depths. Some arched rocks, such as the Lost Hills anticline in the San Joaquin Valley of California, can be mapped with gravity instruments due to the presence of a lighter rock, in this case diatomaceous shale. Other anticlines are mappable because an abnormally heavy rock is arched over the top. Obviously the geologist must be well acquainted with the stratigraphic section before he can interpret the positive and negative anomalies in terms of geologic structure.

The gravity meter is perhaps most valuable in deciphering the regional tectonics. Within large basin areas the major structural axes can be traced and the tectonic history determined[63] by gravity surveys.

Seismic methods.[64]   The most elaborate, the most expensive, and the most effective geophysical instrument at the present time is the seismograph. The magnetometer and the gravity meter are generally considered to be reconnaissance instruments and the seismograph to be a device for determining the structural geology in detail. Wyckoff[65] calls attention to the fact that magnetic and gravimetric methods measure forces at a distance and consequently are indirect, whereas seismic methods record sound waves generated at or near the surface, which actually "reach down and touch or traverse the rock strata."

Seismic surveying is done by either the refraction or the reflection method. Both techniques are based on the fact that shocks created by man-made explosions travel downward into the bedrock and then are refracted and reflected back to the surface by rocks of different density. The length of time involved in that round trip is a reliable indication of the depth of the refracting or reflecting layer. Therefore, these techniques can be used in tracing the structural rise and fall of the rock formations beneath the surface.

The difference between the refraction and the reflection methods is shown in Fig. 3–13. The refraction method was first used in the United States about 1923. It was very successful in the search for new salt domes on the Gulf Coast of Texas and Louisiana. Refraction shooting has re-

---

[62]W. B. Dyer, "Gravity Survey Pays Its Way," *The Oil and Gas Jour.* (July 9, 1956), pp. 86–92.

[63]Lester L. Logue, "Gravity Anomalies of Texas, Oklahoma, and the United States," *The Oil and Gas Jour.* (April 19, 1954), pp. 132 et seq.

[64]John C. Hollister and W. P. Hasbrouck, "Seismic Prospecting," *Subsurface Geology in Petroleum Exploration* (Colorado School of Mines, 1958), pp. 555–583; Maurice Ewing and Frank Press, "Seismic Prospecting," *Handbuch der Physik*, Band XLVII, pp. 153–168; C. H. Dix, *Seismic Prospecting for Oil* (Harper and Brothers, New York, 1952).

[65]*Op. cit.*

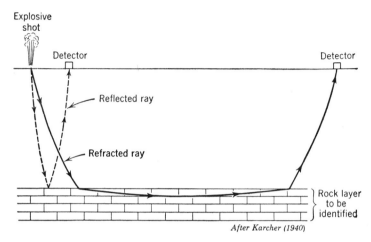

After Karcher (1940)

Fig. 3-13. Reflected and refracted seismic waves. *Courtesy American Institute of Mining and Metallurgical Engineers.*

cently become important again as a method of detailing rock structure in deep high velocity carbonate and evaporite sediments where reflection shooting fails to deliver.[66] Furthermore in some areas it is the most economical way to carry out large scale reconnaissance. Refraction has the advantage of supplying data which enable the interpreter to identify formations. For example, the Rundle limestone in the Canadian foothills belt can be mapped by refraction surveys because of the characteristic velocity of its surface.[67]

The reflection method (Fig. 3-14) was first used commercially in Texas in 1926. It has been used not only in the salt-dome search but also in the search for anticlinal structures. It has largely superseded the refraction method.

In surveying with the reflection seismograph, a shot hole is drilled and an explosive charge is placed in it. For each explosion, or synthetic earthquake, a series of geophones ("jugs") are placed on the ground along a line or according to a pattern. The vibration of the geophones converts the earth shock waves into electrical energy which is transmitted by wire to the recording truck. The time taken by the earth waves in traveling from the explosion to the reflecting surface and back to the earth's surface is converted into depths in feet by using different velocities for different types of rock. Variations in these depth figures give the interpreter a picture of the geologic structure pattern. Under favorable conditions, it is

[66]E. A. Blomerth, Jr., "Using Refraction Work in South Florida," *Oil and Gas Jour.* (Feb. 14, 1955), pp. 180–182.

[67]Milton B. Dobrin, *letter*, Oct. 30, 1958.

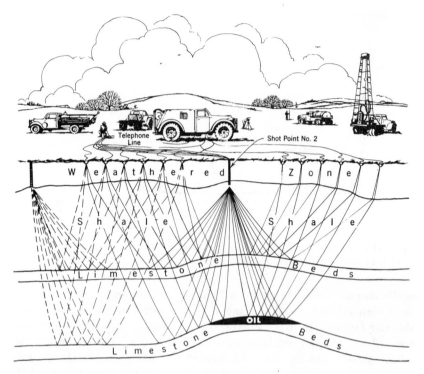

Fig. 3-14. A seismograph party "shoots for oil." Explosion waves from a dynamite charge travel down through the earth and are reflected to the surface from rock layers. At the surface geophones pick up echoes and convert them to electrical impulses. In the recording truck the impulses are traced on a strip of film, called "seismogram." Possible oil-bearing structures may be disclosed. *Courtesy The Horizons, Pan-American Oil and Gas Company.*

possible with the reflection seismograph to map the three-dimensional aspect of an anticline or a fault with considerable accuracy.

Beginning in 1944 the oil companies began to explore the geology of the floor of the Gulf of Mexico by marine seismic exploration.[68] The procedure is the same as on the land surface, except that the explosive is detonated on the sea floor, or below the surface of the water, instead of in a shot hole. The detectors are attached to a cable and let down to the sea floor in a linear arrangement. The recorders are located in a boat instead of a truck. Aside from the weather, which is a most important factor, the only unusual complication in marine work is the problem of exactly

[68]R. P. Palmer, "Techniques and Problems in Marine Seismic Exploration," *Oil and Gas Jour.*, Vol. 47, No. 8 (June 24, 1948), pp. 160 et seq.

locating the detector points. This locating is done either by triangulation or by one of several methods of electronic surveying.

Growing use is being made in seismic prospecting of the "weight-drop technique," in which a three-ton weight is hoisted and dropped onto the surface instead of exploding charges in shot holes drilled below the surface.[69]

During the last decade the reflection seismograph has increased markedly in efficiency due to progress in instrumentation[70] and in accuracy due to improvements in interpretation.[71] In areas of complex structure, since an erroneous picture results if the dips are plotted beneath the shot hole, they must be "migrated" to their true position.[72]

The ultimate objective of all search methods is to be able to pinpoint the location of actual petroleum accumulations of commercial size beneath the surface of the earth. In May, 1974 the secret of the "bright spot" leaked out to the press. This is a light (or even dark) inconspicuous anomaly that has been observed on some seismic charts. It is thought to be a wave reaction caused by the large density contrast between gas, water, and rock. This contrast is much less with most crude oils; however, an oil deposit with a gas cap should produce the bright spot pattern. The use of the bright spot resulted in some excellent discoveries, but there have also been failures, due to similar reactions from other substances in the sedimentary column. This has taken some of the glow from the bright spot, but it will continue to be a possible indicator of hydrocarbons. It is possible, if not probable, that at some future time the state of the art of oil finding will improve to the point of developing a "direct oil finder" even though we cannot expect 100 per cent success even then.

Radioactive methods. The attempted employment of radioactivity measuring devices to detect directly the presence of oil was discussed earlier in this chapter under "Soil Analysis." Considerably greater success has attended the use of radioactive measurements obtained by both

---

[69]Alan D. Waldie, "Weight-drop Technique—How It's Working Out," *World Oil* (April, 1956), pp. 148 et seq.; R. L. Palmer, "A New Approach to Seismic Exploration," *World Oil* (June, 1954), pp. 140 et seq.

[70]Geophysical Society of Tulsa, "Magnetic Recording—New Tool in Seismic Prospecting," (Symposium), *Oil and Gas Jour.* (April 4, 1955), pp. 172 et seq.; Henry Salvatori, "New Developments in Seismic Methods," *Proc. Fourth World Petrol. Congress*, Section 1, 1955, pp. 605–618; F. J. Agnich, "Geophysical Instrumentation Is Making Rapid Progress," *World Oil* (April, 1955), pp. 109 et seq.

[71]P. I. Bediz, "Interpretation of Seismic Data in Locating Reefs in Alberta," *World Oil* (Feb. 1, 1953), pp. 86 et seq.

[72]John Daly, "Third and Fourth Dimensions in Exploration," *Bull. Am. Assoc. Petrol. Geol.*, Vol. 38 (February, 1954), pp. 319–332.

surface and air-borne devices in the mapping of lithologic contacts. This method makes possible the recognition of hidden faults and the preparation of structural and areal geological maps.[73]

[73]Robert B. Guillou and Robert G. Bates, "Correlation of Airborne Radioactivity and Areal Geology," *Bull. Geol. Soc. Am.*, Vol. 68 (December, 1957), p. 1740; W. J. Williams and Phillip J. Lorenz, "Detecting Subsurface Faults By Radioactive Measurements," *World Oil* (April, 1957), pp. 126–128; Bernard Lang, "Gammatron Surveys," *World Oil* (November, 1950), pp. 86 et seq.

4

SUBSURFACE AND PRODUCTION GEOLOGY

Subsurface geology [1] is the study of geology in three dimensions through the utilization of all available samples and records. These records include those obtained by surface geology and geophysics, by core- and slim-hole ("strat") drilling, and by the drilling of water wells, wildcats, production wells, brine disposal wells, and holes for secondary recovery purposes. The motivation may be solely scientific, or it may be for the purpose of exploring for new oil fields or for the efficient development of an oil field already discovered. The application of subsurface geology to the development and possible extension of an existing field is known as *production geology.*

Because one is a part of the other, production geology cannot be divorced from subsurface geology. Both use the same sources of information, although with some differences in emphasis.

SOURCES OF DATA

The subsurface or production geologist is always collecting data. He seeks out all the surface-obtained information available, both geological and geophysical. The drilling of every well supplies him with cores, cuttings, and logs. Lastly, there are the completion, production and workover records, all of which supply reservoir information.

Surface Sources. All of the information obtainable at the surface that was available to or developed by the exploration geologists is also

[1] John W. Haun and L. W. LeRoy, editors, *Subsurface Geology in Petroleum Exploration* (Colorado School of Mines, 1958); Daniel A. Busch, "Subsurface Techniques," *Applied Sedimentation* P. D. Trask, editor, John Wiley and Sons, New York, 1950), pp. 559–578.

available to the subsurface geologist. Both surface geological reports
and the results of geophysical surveys can be collected and correlated
with the subsurface data obtained by drilling.

Geological data. The subsurface geologist has a head start in identifying the formations penetrated by the drill, if he knows the stratigraphic
position and the geologic structure of the rocks occupying the bedrock
surface at the well site. He should also know the regional stratigraphic
section. Largely through prior activity and tradition the formation names
which he uses are in most cases surface names. It is the job of the subsurface geologist to carry this nomenclature into the buried rocks where-
ever possible. Therefore, the better acquainted he is with the surface
stratigraphy the easier is his task.

The structural geology at the surface is also of considerable importance
to the geologist mapping in three dimensions. Regional and local dips
can be extrapolated downward. The location and dip of fault planes at
the surface can be used to forecast the points of interception in a well.

In most areas, government sponsored reports and geological maps have
been published prior to exploration, although all too often these are reconnaissance in nature and scale. Usually surface geological parties sent
out by one or more oil companies have added more detail to the picture,
and their reports may be in the company files.

Attention to the surface geology should not be relaxed during development, and surface maps should be kept up to date. Excavation work for
roads, well locations, tank sites, and so on, furnish new exposures which
should be examined when they are fresh. Moreover, they may be only
temporarily accessible for investigation.

The solution of local subsurface geological problems repeatedly requires consideration of the regional geology. For instance, (1) the stratigraphic section encountered in a field does not always include the complete
sequence of formations existing in the region and regional data must be
consulted to determine the possibilities of stratigraphic traps in layers
wedging out along the flanks of a producing structure; (2) the regional
information is necessary to determine the deeper possibilities in a field,
(3) regional knowledge of the depositional conditions is of importance in
the study of reservoir rocks, and (4) regional patterns of faulting and
folding may suggest analogies to help solve local patterns.

For these reasons, the acquisition of a regional geological knowledge
from published literature, unpublished reports, and frequent consultation
with exploration geologists is most essential.

Geophysical data. The chances are that discovery was preceded by
gravity, or seismograph surveys, or both. The records of these surveys
should be combined with subsurface information, particularly in the early

stages of development when few subsurface geological data are available. The geologist, however, must be cognizant of the accuracy and limitations of the data upon which the geophysical interpretation is based. He should also be able to judge where additional geophysical work will aid him when interpreting subsurface structure.

Samples. Formation samples obtained during drilling furnish factual information on the lithology and stratigraphy, textural and structural characteristics of the rocks, porosity and permeability, and indications of fluid content of the strata penetrated. They are of the utmost value to the subsurface or production geologist. The examination and description of them is, therefore, a subject to which much attention must be paid in order to derive the maximum information from the material available.

The rock samples collected during the drilling of a well are in the form of either cuttings or cores. These samples are described in the following paragraphs. Other samples, which may be collected of the water, oil, and gas struck during drilling, are discussed in Chapter 7, "Reservoir Fluids."

Cuttings.[2] Cuttings are the rock fragments broken or torn from the rock being drilled by the bit at the bottom of the hole. They are collected for lithological and paleontological correlation, determination of subsurface markers, and investigation of reservoir layers. In soft sand and shale sections electrical logging, combined with sidewall samples, has almost eliminated the use of drill cuttings, except for paleontological examination. On the other hand, in hard formations and especially in hard reservoir rocks where the high electrical resistance of the rock masks the electrical resistance of hydrocarbons, sampling and examination of drill cuttings remain important. The same holds for salt-bearing sections.

When cable tools are used, the cuttings are broken from the rock by the pounding of the sharpened edge of the bit. They are periodically removed by the bailer, and samples are caught on the derrick floor when the bailer is dumped. Drill cuttings from cable-tool wells afford a fairly good record of the penetrated layers. They are better than rotary samples for insoluble residue studies.[3]

Rotary cuttings are torn from the rock by the rotating bit and are brought to the surface by the circulation of the drilling mud. They are collected from the mud ditch or shale shaker, or are collected by diverting part of the drilling mud to a settling box. The box is equipped with a gate at the end to facilitate cleaning after each sample is taken. In earlier days of rotary drilling, the samples obtained were of questionable value

[2] Julian W. Low, "Examination of Well Cuttings," *Subsurface Geology in Petroleum Exploration* (Colorado School of Mines, 1958), pp. 17–58.

[3] H. A. Ireland, "Insoluble Residues," *Subsurface Geology in Petroleum Exploration* (Colorado School of Mines, 1958), pp. 75–94.

because of (1) contamination with cavings from higher beds, (2) pulveriza-
tion of the sample by the bit, (3) failure of the mud to carry the samples to
the surface, and (4) recirculation of the cuttings themselves. Through an
intensive study of drilling muds, the engineers have worked out the
weights, consistencies, and pressures necessary to plaster or "mud off"
the higher beds adequately and at the same time pick up the cuttings and
bring them to the surface with efficiency and dispatch. Recirculation of
rock fragments can be prevented by the use of settling pits.

The introduction of reverse circulation (sending the mud down the annular space between drilling pipe and hole wall and returning it to the
surface through the drill pipe) to rotary drilling has been an important
factor in obtaining better samples. As the cuttings are larger and there
is less chance of contamination, they are suitable for porosity and permeability analysis by newly developed methods. The shorter time lag in
reaching surface enables the true depth of the samples to be estimated
more accurately than when normal circulation is used. This method is
particularly valuable for securing samples of limestone and dolomite
reservoir rocks as conventional and wireline coring usually do not recover
the most porous parts.

Drill-cutting samples normally must be rinsed clean and placed in cloth
sacks or cans which are labeled as to depth, well name, and location.
However, cuttings from soft formations cannot be washed or even slightly
rinsed without loss of sample. Where special care is needed to prevent
such loss, for example, for certain paleontological investigations, extra
samples should be collected and canned without rinsing. Further treatment can then be carried out in the laboratory under controlled conditions.

Drill cuttings recovered from an oil or oil-base mud stream should be
rinsed with hot water or first with kerosene and then hot water. The
original hydrocarbon saturation in the formation is not removed in this
cleaning process. The samples are usually dried before they are sacked,
labeled, and shipped. Reversed-circulation cuttings for laboratory saturation tests and physical analysis are washed, and after the water has
dripped off, the cuttings are canned in order to reduce the loss of hydrocarbons.

Samples of drill cuttings are taken at intervals of 1 foot to 30 feet, depending upon the purpose for which the samples are collected. For lithological correlation and construction of lithologic logs, samples at intervals
of 5 to 10 feet are adequate, and for investigation of reservoir beds,
samples every 5 feet usually suffice. Reversed circulation samples, which,
owing to the faster return of the drilling fluid through the drill pipe, reach
the surface in good condition, are collected at intervals of 1 foot for limestone and dolomite and 2 feet for sandstone. In soft formations where

the lithologic log is less useful, sampling is sometimes confined to the zones which are subjected to paleontological investigation. The interval is usually 30 feet but may be shortened to 15 feet or even 10 feet at the paleontologist's request. It must be realized, however, that curtailment in collecting drill cuttings eliminates a source of subsurface information and the procedure should only be adopted after it is ascertained that the information is dispensable.

Rotary cuttings reach the surface some time after they were cut at the bottom of the hole. This time lag varies with the speed and size of the mud pumps and with the depth and the diameter of the hole. It is checked whenever necessary but, in particular, when pay sections are sampled. An effective check may be made by putting two handfuls of red- or yellow-colored mica into the mud stream when a pipe connection is made. The time the drilling fluid takes to move from the top to the bottom of the hole can be calculated fairly accurately, and the difference from the time taken by the mica to reappear in the ditch gives the time lag for the cuttings.

Another good indication of time lag can be obtained by noting the arrival of cuttings after starting drilling. It will also be observed then that the smaller cuttings appear first, which indicates a differential lag between larger and smaller cuttings and, consequently, a certain degree of mixing of cuttings on their way to the surface. Furthermore, breaking from hard, slower-drilling into soft, faster-drilling strata is usually accompanied by a change to a greater concentration of cuttings, sometimes of different lithological character. Finally, correction can be made by correlating the formation contacts obtained from the cuttings with the electric log or with the drilling time curve.

Cores.[4] Cores, the largest samples obtained of subsurface formations, furnish the most reliable information concerning lithology, texture, rock structure, fossil content, dips and tectonic disturbance. In addition, laboratory analysis of the cores provides valuable information on porosity, permeability, fluid content, and salinity of connate water. In hard formations, the use of diamond core bits is tending to become general practice, because virtually full recovery of cores up to sixty feet in length can be obtained at reasonable cost. Diamond coring is particularly valuable in the investigation of limestone and dolomite reservoir rocks, as the more porous parts of these rocks are generally not recovered in conventional or wireline coring. The true nature of many of these reservoir rocks could not be appreciated before diamond cores were obtainable.

[4] H. L. Landua, "Applications of Coring," *Subsurface Geology in Petroleum Exploration* (Colorado School of Mines, 1958), pp. 711–714, J. G. Crawford, "Core Analysis," *ibid*, pp. 229–249.

As coring added considerably to the cost of drilling, it stimulated the development of cheaper methods. Today, electrical and radioactivity logging in conjunction with sidewall sampling and improved methods of collection and investigation of ditch cuttings have largely replaced coring. However, the general curtailment of coring has resulted in a loss of dip information which has seriously handicapped subsurface interpretation in structurally complicated fields. It is the task of the production geologist, who deals with such fields, to press for the coring of selected intervals for dip information, if it is not possible to obtain this information by dipmeter surveys.

In general, a more liberal attitude is taken in providing for coring in exploration wells. No hard and fast rules are established, as the weight of information derived from coring varies greatly from project to project.

Upon extraction of the core from the core barrel, the top and bottom of the core and the amount of core recovery should be noted immediately. The cores are then examined geologically and samples are selected to be shipped for laboratory analysis, paleontological examination, or for other specially arranged purposes.

Cores selected for laboratory determination of porosity, permeability, and fluid content must be handled in such a way that the fluid content is disturbed as little as possible. The cores should not be removed from the core barrel by water pressure, as this practice may increase the water saturation. For the same reason, the drilling mud should be removed by scraping and not by washing. The core must be placed in a can, which should be filled to prevent breakage. The can size must, therefore, conform approximately to the diameter of the core. No cloth or paper should be enclosed if saturation tests are to be made. In unconsolidated sand, a sample of clean material free of contamination with drilling mud should be shipped. The cores must be labeled (name of company, well name and number, name of field or district, and depth of sample) and sealed as soon as possible, as a core exposed to the atmosphere rapidly loses water and light fractions of hydrocarbons.

Samples for laboratory analysis should be taken from every foot of core where the interval is less than ten feet in thickness. However, fewer samples may be taken from thicker, uniform sections except where an oil-water contact or a gas-oil contact is suspected to exist in the interval. Alternating sand and shale sections should be sampled as closely as possible, every six inches to one foot. In wildcat operations, two samples of the drilling mud should be forwarded with the cores to the laboratory.

The selection of core samples for paleontological investigation should be done according to the instructions of the paleontologist, but it is no longer considered good practice to sample only selected lithologies (for example, shales only from sand-shale sequences), as new techniques of sample examination requiring all types of material are continually being evolved. There is a general tendency to dispose of core material that is not used immediately for laboratory investigation. However, it has been the experience that old core material is repeatedly re-examined if available during later subsurface reviews. Re-examinations made to check data usually produce additional information which was overlooked in the original examination. Valuable information from re-study of core material may also be obtained for secondary recovery projects in the later stages of production. Furthermore, old core material is often requested in connection with new techniques which were not used or conceived at the time the cores were taken.

The preservation of core material is, therefore, essential. Cores should be stored so that they are readily accessible, and the storage place should not be used for any purpose except facilities for core examination. The storage facilities, core boxes, and labeling must be such that the cores and their identification will withstand the vagaries of climate and insects.

Side-wall samples and side-wall cores are taken principally to investigate possible accumulations indicated from the examination of electrical logs. Frequently, producing sections are sidewall-cored for laboratory analysis to assist in determining the completion interval. The samples, as with cores, should be handled most carefully to minimize loss of fluid content by evaporation.

Examination and description.[5] Ordinarily cores are examined with the aid of a hand lens, and drill cuttings with a binocular microscope, generally using 10- to 18-power magnification, as this is sufficient for observations of the essential characteristics. The first and sometimes only examination of samples is made by the well-site geologist. It is, therefore, of the utmost importance that his descriptions be complete, accurate, and clear, both for the benefit of his own work and that of his successors in the field and for those relying on the written records filed in offices elsewhere. For the same reasons, descriptions should employ standard terms in order to achieve uniformity.

The chief geological characteristics recorded in the examination of cores and drill cuttings are lithologic type, color, accessory minerals, fossil con-

[5] John M. Hills, "Examination of Well Cuttings," *Subsurface Geology in Petroleum Exploration* (Colorado School of Mines, 1958), pp. 17–58; Daniel A. Busch, "Subsurface Techniques," *Applied Sedimentation*, P. D. Trask, editor (John Wiley and Sons, New York, 1950), pp. 559–578; Julian W. Low, "Examination of Well Cuttings," *Quarterly Colorado School of Mines*, Vol. 46 (October, 1951), pp. 1–48; George W. Johnson, "Writing Scientific Reports," *Subsurface Geology in Petroleum Exploration* (Colorado School of Mines, 1958), pp. 779–791.

tent, rock texture and structure, porosity, and indication of fluid content. These characteristics furnish basic data for subsurface correlation and for the geological and petrophysical study of reservoir rocks. The first step is the identification of the rock or rocks represented in the sample; the second is the search for (a) lithological and paleontological characteristics which may be diagnostic, and (b) evidences of unconformities.[6] Apparently unimportant things, such as coarse sand grains or calcareous zones in shale, should not be overlooked. They may be clues that lead to nearby reservoir rocks—for example, lenses of sandstones in shale, or carbonate reefs.[7]

Rock texture and structure are examined and described in detail because they furnish distinct criteria for subsurface correlation and for the interpretation of the framework of reservoir beds. They concern rock features which are formed during sedimentation (for example, stratification, contemporaneous deformation) and after sedimentation (cementation, solution, weathering, deformation by tectonic action). Cores are an important source of textural information, as it is not ordinarily possible to observe these features when dealing with fragments. Even structural features, such as unconformities and rock dips[8] may be visible in cores, especially in the large-diameter cores now being obtained with diamond bits. Both Weaver[9] and Wagner[10] have pointed out the value of petrofabrics in determining the structural position of a single isolated well and in recognizing faults by slip planes and shattering.

Porosity in sedimentary rocks can be classified into two types: normal or primary porosity, which is related to the primary texture of the rock and which formed during sedimentation and lithifaction; and secondary porosity, which is superimposed on the primary porosity after lithifaction, mainly by the action of solution, fracturing, or weathering. The description of sandstone texture includes size of grain, sorting, and amount of cementation. These factors are intimately connected with the porosity, permeability and other properties. Porosity is estimated qualitatively as

[6] W. C. Krumbein, "Criteria for Subsurface Recognition of Unconformities," Bull. Am. Assoc. Petrol. Geol., Vol. 26 (January, 1942), pp. 36–62.

[7] Paul Weaver, "The Geological Interpretation of Exploratory Wells," Oil and Gas Jour., Vol. 47 (March 17, 1949), pp. 102–104; Bull. Am. Assoc. Petrol. Geol., Vol. 33 (July, 1949), pp. 1135–1144.

[8] Charles A. Sansom, "The Interpretation of Core Evidence," Science of Petroleum (Oxford University Press, 1938), Vol. 1, pp. 502–507.

[9] Paul Weaver, "The Geologic Interpretation of Exploratory Wells," Oil and Gas Jour., Vol. 47 (March 17, 1949), pp. 102–104; Bull. Am. Assoc. Petrol. Geol., Vol. 33 (July, 1949), pp. 1135–1144.

[10] Warren W. Wagner, "Petrofabric Analysis," Subsurface Geologic Methods (Colorado School of Mines, 1950), pp. 157–172.

poor, fair, good, and excellent, depending upon the size and frequency of visible pores.

The examination of cuttings of carbonate rocks involves three main considerations: (1) the calcite-dolomite ratio, (2) the texture of the matrix, and (3) the character of the visible pore structure. The calcite-dolomite ratio may be important in pointing out the direction of greater dolomitization, with perhaps greater porosity as well.[11] The texture of the matrix covers the character of the minute pore structure between the crystals, granules, or fossils. This minute pore structure cannot be readily seen or investigated, yet it is important in the fluid distribution within the rock. The size and distribution of the visible pores is estimated in order to obtain an approximation of the total porosity of the rock.[12]

It is impossible to place too much emphasis on the importance of detecting hydrocarbons in well samples. The success of an exploration well may depend on this detection. The discovery of new accumulations and sometimes of new fields often has been delayed because of failure to recognize hydrocarbons in drill cuttings.

Only qualitative tests of fluid content are made in the field examination of cores, side-wall samples, and drill cuttings, except when a trailer laboratory is available. The principal objectives of the field examination are, apart from the determination of the presence of oil and gas, to aid in the interpretation of the electric and radioactivity logs, to locate gas-oil and oil-water contacts,[13] and to select the intervals for drill-stem tests and for the completion of wells.

Gas is indicated by gas bubbles on fresh samples of cores and cuttings and in the mudstream. The emergence of gas bubbles can be accelerated by applying a vacuum.[14] In a trailer laboratory the gas detector records minute quantities. Dry gas has no odor, but if it is wet it has a petroleum or distillate odor and gives positive ether and fluorescence reactions. *Condensate* is indicated by odor and light ether and fluorescence reactions. *Oil* is indicated by oil saturation and stains, odor, oil color (black, brown to straw, or green to yellow), iridescence, ether reaction (also acetone, carbon tetrachloride and chloroform reactions) and by fluorescence. Acetone also gives a color when carbonaceous material is present, so that care must be exercised in its use.

[11] Kenneth K. Landes, "Porosity through Dolomitization," *Bull. Am. Assoc. Petrol. Geol.*, Vol. 30 (March, 1946), pp. 305–318.

[12] G. E. Archie, "Classification of Carbonate Reservoir Rocks and Petrophysical Considerations," *Bull. Am. Assoc. Petrol. Geol.*, Vol. 36 (February, 1952), pp. 278–298.

[13] John G. Caran, "Core Analysis-Predicting Well Behavior," *Subsurface Geologic Methods* (Colorado School of Mines, 1950), pp. 295–320.

[14] Jack E. Bliss, "Vacuum Method Detects Gas-Bearing Formations," *World Oil* (October, 1953), pp. 218 et seq.

Ether reactions or cuts should be standardized as far as possible by using a constant amount of crushed sample in a constant amount of ether. The resulting reaction or "cut" is evaluated by means of a color scale ranging from dark to light amber, through faint and very faint, to no cut. The intensity of the cut depends not only on the amount of oil, but also on its specific gravity and type.

Fluorescence tests are graded as strong or good, through medium or fair, to slight or positive. The use of a fluorescent lamp is very convenient for examining drill cuttings for hydrocarbons. If oil staining is very slight, ether cuts are recognized more easily when examined by ultraviolet light.

The acid test has been proved to be a practical way of detecting microscopic oil stains in limestone, dolomite, and calcareous sandstone. When the cuttings are immersed in diluted hydrochloric acid, 60 per cent water and 40 per cent HCl, relatively large bubbles are formed which last long enough to float the specimen. The bubbles which form on the surface of a similar unstained cutting break away before they are large enough to float it.

Formation water in samples may be recognized by salty or sour taste, odor (sulphur water), strong silver nitrate reaction when crushed sample is covered with distilled water, the spotty appearance of oil solution, grayish luster caused by water film on the rock grains, and stains of dead oil or black, tarry oil.

In addition to the routine lithologic description by the well-site geologist, samples are commonly examined by various experts in specialized techniques for improving correlation, both local between wells and regional between oil fields. The production geologist must be familiar with the bases of these specialized methods and must be able to use the results intelligently in the development of oil fields.

Micropaleontology [15] is one of the most important of these aids to correlation and the study of environment of the sediments. In such areas as the Gulf Coast and California, it is routine procedure to examine the microfossils. In recent years fossil pollen and spores have become im-

[15] H. V. Dunnington, "Close Zonation of Upper Cretaceous Globigerinal Sediments by Abundance Ratios of Globotruncana Species Groups," *Micropaleontology*, Vol. 1 (July, 1955), pp. 207–219; Martin F. Glaessner, *Principles of Micropalaeontology* (John Wiley and Sons, New York, 1949); A. Morley Davies, "Paleontology," *The Science of Petroleum* (Oxford University Press, 1938), Vol. 1, pp. 306–308; W. L. F. Nuttall, "Micro-Paleontology," *ibid.*, pp. 309–311; Hubert G. Schenck, "Applied Paleontology," *Bull. Am. Assoc. Petrol. Geol.*, Vol. 24 (October, 1940), pp. 1752–1778; Carey Croneis, "Micropaleontology—Past and Future," *Bull. Am. Assoc. Petrol. Geol.*, Vol. 25 (July, 1941), pp. 1208–1255; William S. Hoffmeister, "Micropaleontological Analysis," *Subsurface Geology in Petroleum Exploration* (Colorado School of Mines, 1958), pp. 203–228.

portant geologic age indicators.[16] When cores are available macrofossils may occur and can be used for age determination just as at the outcrop. Sedimentation, or sedimentological studies,[17] particularly of sands and incoherent sandstones, may indicate the environment of deposition of reservoir sands. The separation and examination of the heavy minerals[18] from sands and insoluble residues from carbonate rocks[19] have given valuable data for correlation. Recent developments in the examination of rocks and minerals include the use of X-ray diffraction patterns,[20] differential thermal analysis *(DTA)*, and thermoluminescence studies of carbonate rocks. Examination of thin sections of both cores and cuttings is becoming increasingly common in the study of the porosity relations of both arenaceous and carbonate rocks.

Logs.[21] In oil-field terminology, a "log" is a record made during or after the drilling of a well. It furnishes, directly or indirectly, a report of the geological formations penetrated. Some logs are relatively simple, giving the driller's identification of the rocks drilled or the time consumed in drilling each foot of hole. Other logs are more complex, made only after a thorough study of the drill cuttings or from observations taken with help of elaborate physical or chemical equipment.

[16] Ted A. Armstrong, "New Exploration Tool," *Oil and Gas Jour.* (March 9, 1953), pp. 64–65.

[17] Gordon Rittenhouse, "Detrital Mineralogy," *Subsurface Geologic Methods* (Colorado School of Mines, 1950), pp. 116–140; L. W. LeRoy, four papers on sedimentology, *ibid.*, pp. 184–202; John R. Hayes, "Miscellaneous Petrologic Analyses," *Subsurface Geology in Petroleum Exploration* (Colorado School of Mines, 1958), pp. 95–118.

[18] R. D. Reed and J. P. Bailey, "Subsurface Correlation by Means of Heavy Minearls," *Bull. Am. Assoc. Petrol. Geol.*, Vol. 11 (April, 1927), pp. 359–368; A. Brammall, "The Correlation of Sediments by Mineral Criteria," *Science of Petroleum* (Oxford University Press, 1938), Vol. 1, pp. 312–314.

[19] Raymond Sidwell, "Aid of Sedimentary Petrology to the Discovery of Oil," *Jour. Sediment. Petrol.*, Vol. 13 (December, 1943), pp. 112–116; H. A. Ireland, "Insoluble Residues," *Subsurface Geology in Petroleum Exploration* (Colorado School of Mines, 1958), pp. 75–94.

[20] Ralph E. Grim, "Clay Mineralogy and the Petroleum Industry," *World Oil* (March, 1951), pp. 61 et seq.; D. H. Reynolds, Eldon A. Means, Lindsey G. Morgan, "Application of X-Ray Crystal Analysis to a Problem of Petroleum Geology," *Bull. Am. Assoc. Petrol. Geol.*, Vol. 21 (October, 1937), pp. 1333–1339; N. C. Schieltz, "X-Ray Analysis," *Subsurface Geology in Petroleum Exploration* (Colorado School of Mines, 1958), pp. 149–177.

[21] Harold Vance, *Elements of Petroleum Subsurface Engineering* (St. Louis, 1950), Chapter 3, pp. 10–57; E. DeGolyer and Harold Vance, *Bibliography of the Petroleum Industry, Bull. Agr. and Mech. Coll. Texas,* 83 (1944), pp. 369–370 and 405; Hubert Guyod et al., "Well Logging Methods Conference," *Bull. Agr. and Mech. Coll. Texas,* 93 (1946), pp. 166–171 (6 pages of references); Hubert Guyod, "Well Logging Methods Studied at Texas A. and M.," *Petrol Engineer,* Vol. 17 (April, 1946), p. 218; *ibid.* (August, 1946), pp. 62–66; Carl A. Moore, editor, *A Symposium on Subsurface Logging Techniques* (University Book Exchange, Norman, Oklahoma, 1950).

Logs can be classified into two types: (1) those which are based upon actual rock samples, and (2) those that are reaction patterns, based upon the effect of the wall rock upon a tool lowered into the hole. In the first category are driller's logs and all types of sample logs. Drilling time logs, electrical and radioactivity logs, and caliper logs are reaction patterns. These may permit accurate and efficient identification of formations *after* the initial determination of the formations in a subsurface section has been made from samples. Mechanical logging always has to start in new territory from a "type" formation log based upon sample studies.

It would be difficult to overemphasize the importance of logs in oil searching. By means of the subsurface geological information available in logs, it is possible to determine the structural pattern of the rocks and to recognize lateral changes in facies. Both of these are important in discovering oil- or gas-bearing traps.

Driller's logs.[22] The first well logs were lithologic identifications and depth figures stored in the mind of the driller, and some water-well drillers still operate on a memory basis. The next step was to write the data down in a "log book" kept by the driller. For thousands of wells drilled during the earlier decades of the present century, the driller's log is the only record available. Therefore the geologist operating in such areas must learn to interpret the terminology used. It should be remembered that the cable-tool driller makes his rock identifications largely by *feel*—the feel of the drilling line while the formation is being drilled, the feel of the cuttings washed out of the bailer, and the feel of the cutting edge of the bit when withdrawn from the hole.[23] The experienced driller is quite capable of distinguishing in this manner the three most commonplace sedimentary rocks, shale ("soapstone"), sandstone ("sand"), and limestone ("lime").

The driller's log of a rotary-drilled hole is much less reliable than that of a cable-tool hole. This fact was one of the main reasons for the development of accurate sample logging and the various indirect methods of determining the characteristics of the strata penetrated.

When only driller's logs are available, the usual procedure, for purposes of correlation and subsurface geologic study, is to plot these on specially prepared paper log strips at a scale of 100 feet to the inch. Colored pencils are used for the different types of rock, usually yellow for sandstone and blue for limestone, with shale denoted by the omission of any color pattern. These strips can be "slipped" up and down on the drafting

[22] L. W. LeRoy, "Driller's Logging," *Subsurface Geologic Methods* (Colorado School of Mines, 1950), pp. 475–478.

[23] James F. Swain, "Interpretation of Cable Tool Drilling Logs," *Bull. Am. Assoc. Petrol. Geol.*, Vol. 27 (July, 1943), pp. 997–1000.

table in respect to each other until the most likely "match" is obtained. In this way formations are correlated in the subsurface from well to well and eventually from one side of a sedimentary basin to the other side. Although they are of less relative importance in correlation today, because of the availability of many other data, driller's logs are still kept, and in some companies a two-column strip log is plotted, one column containing the sample log, the description of which follows, and the other column the driller's log.

An additional and important feature of the driller's log, especially in cable-tool wells, is its notation of "shows" of oil and gas and of the presence of water-yielding formations. Some oil shows consist only of a "rainbow" of colors on the surface of the water bailed out of the well. A gas show may be merely a slight bubbling or effervescence noted when the bailer is dumped. These occurrences can be observed only by the men working on the derrick floor. They leave no record in the samples for subsequent consideration by the geologist, and yet even such minor shows of oil or gas may be very significant. Many an oil field has been discovered through the encouragement given by oil or gas shows in otherwise unsuccessful wildcat tests.

In the same way, the notation in the log by the driller of "HFW" (hole full of water), or other reference to the presence of water in fair abundance in a subsurface formation, may be of great value to the geologist. Such information indicates a porous and permeable stratum which under a more favorable environment may be just as full of oil as it is here full of water. Even the casing depths entered in the driller's log may be interpreted in terms of lithology and water content.

Sample or lithologic logs.[24] Sample logging is one of the most important methods of logging a well, but the accuracy of the resulting log depends upon many factors, the most important of which is probably the experience of the geologist examining and interpreting these samples.

Sample logs are plotted strip logs which show the identifications made by a geologist who examines the well cuttings with a binocular microscope. Although basically similar to the plotted driller's log in showing rock types by color patterns, the sample log includes infinitely more detail. The less common, or less easily distinguished rock types, such as dolomite and anhydrite, are included. Furthermore, approximate percentages of chert in limestone, or anhydrite in dolomite, or other ingredients in a rock mixture can be shown by lateral thickness of the color pattern in the vertical-log column. Because of variations in the occurrence of minor constituents which are ordinarily missed by the driller, it may be possible

[24] Bob Greider, "Lithologic Logging," *Subsurface Geologic Methods* (Colorado School of Mines, 1949), pp. 297–302.

to obtain more precise correlations with sample logs, and it is certainly true that facies changes can be studied in much more detail. Inasmuch as a considerable proportion of each sample obtained during the drilling of a rotary well consists of rock sloughed off from a higher level, the percentage type of sample log does not give a true picture of each individual stratum penetrated. Therefore, in some areas the percentage description of rotary samples has been superseded by *interpretive logs*, in the construction of which the geologist attempts to describe only the new material cut by the bit. Still another type of sample log is the *particle-type log,*[25] in which the cuttings, particles, or grains are classified and described as minerals instead of as rock types.

The complete sample log regardless of type, also contains all available information on oil and gas shows.

In almost all subsurface geologic work, the study of samples and the plotting of sample logs is routine procedure. In order to obtain consistency in textural descriptions from day to day and from person to person, the use of a "textural standard"[26] has been advocated.

Mounted logs. A variation in plotting sample logs is to mount the samples themselves on log strips at a scale of 20 feet to the inch. (Fig. 4–1). First glue is spread over that part of the strip which represents the interval of the sample, and then the cuttings are sprinkled over the glue until the surface is covered. Breaks in the lithology are just as obvious on the mounted as on the plotted logs, and color variations are much more noticeable. Furthermore, the original material rather than an interpretation of the original material is used in slipping the logs for correlation and in tracing facies. Because of the labor of preparation and because of their bulk, mounted logs are not widely used.

Drilling-time logs.[27] Drilling time or rate-of-penetration logs consist of a curve plotted on a time-depth basis. The slope of the curve designates the speed of penetration. Any abrupt breaks show contacts

25 J. E. Banks, "Particle-Type Well Logging," *Bull. Am. Assoc. Petrol. Geol.*, Vol. 34 (August, 1950), pp. 1729–1736.

26 Gordon Rittenhouse, "Textural Standard for Sample Log Work," *Bull. Am. Assoc. Petrol. Geol.*, Vol. 29 (August, 1945), p. 1195.

27 W. B. Mather, "Rock Penetrability in Making Hole," *Oil and Gas Jour.* (Oct. 25, 1951), pp. 98 et seq.; P. B. Nichols, "Mechanical Well Logging," *Bull. Agr. and Mech. Coll. Texas*, 93 (1946), pp. 105–118 (15 references); G. Frederick Shepherd and Gordon I. Atwater, "Geologic Use of Drilling Time Data," *Oil Weekly*, Vol. 114 (July 3, 1944), pp. 17 et seq.; *ibid.* (July 10, 1944), pp. 38 et seq.; Lester C. Uren, "Recent Developments in Formation Logging," *Petrol. Engineer*, Vol. 14 (February, 1943), pp. 63 et seq. (bibliography with 10 references on p. 70); Robin Willis and R. S. Ballantyne, Jr., "Drilling Time Logs and Their Uses," *Bull. Am. Assoc. Petrol. Geol.*, Vol. 26 (July, 1942), pp. 1279–1283; G. Frederick Shepherd, "Drilling-Time Logging," *Subsurface Geology in Petroleum Exploration* (Colorado School of Mines, 1958), pp. 367–387.

Fig. 4-1. Research geologist uses mounted samples with electric logs in sub-surface correlation and facies studies. *Courtesy R. Dana Russell and the Ohio Oil Company.*

between rocks of unequal penetrability. The major factors in controlling the speed of drilling are the toughness and strength of the rock being drilled. The toughness is dependent upon the mineral content, the kind of cement and the degree of cementation, the texture, and the porosity, whereas the strength varies with the pressure of the fluid in the pore space. It is obvious that penetration speed is also a function of various equipment and operational factors, such as design and sharpness of bit, the pressure on the bit, the characteristics and velocity of the drilling mud, the rotation speed of the bit, and the skill of the driller. In spite of these variables the drilling-time log is ordinarily a reliable index to the nature of the formations penetrated. The log is commonly made auto-matically by means of a device such as the "Geolograph" or "Log-O-

Graph", although in their absence the rate of penetration can be recorded manually.

Drilling-time logs can be used in correcting the lag in sample return and in correlating from well to well. They are especially valuable during the drilling of a well because of their immediate availability. The passing of a formation contact may be recognized by the time log before the circulation of the drilling mud brings the samples to the surface. Coring points may be selected, and the time log may be used to interpret gaps in the core record due to incomplete recovery. This type of log has been of value in locating porous and permeable intervals during the drilling of thick carbonate rock sections.

Mud-analysis logs.[28] Detailed examination of the drilling fluid returning from the bottom of a hole is an important service, especially in wildcat well drilling. The service company supplies a completely equipped mud-logging truck or trailer as well as trained operators. Part of the mud stream coming out of the hole is diverted through this mobile laboratory. Mud analysis allows the detection of minute amounts of oil and gas in the drilling mud and in the drill cuttings. The log constructed from the data consists of three curves and is usually combined with the lithologic and/or drilling-time log. One curve shows the oil content, another the content of hydrocarbon gases other than methane (and ethane), and a third the total gas content.

The chief value of mud-analysis logging is the indication it gives of fluid content of the reservoir sands penetrated. Although only qualitative, this indication allows the amount of coring to be reduced without too much risk of missing a potentially productive zone.

Temperature logs.[29] The earliest physical test to be made in boreholes was the insertion of thermometers to determine earth temperatures at the depths reached. More recently, the development of sensitive recording thermometers has made it possible to obtain a continuous record

[28] Robert O. Patterson, "Logging From Drill Returns," *Mines Mag.* Vol. 43 (October, 1953), pp. 127–132; Jack E. Bliss, "Hydrocarbon Well Logging," *Oil and Gas Jour.* (Aug. 9, 1951), pp. 76 et seq.; R. W. Wilson, "Methods and Applications of Mud-Analysis and Cuttings Analysis Well Logging," *Symposium On Subsurface Logging Techniques at University of Oklahoma,* (April 5–6, 1949), pp. 81–87; B. Otto Pixler, "Mud Analysis Logging," *Bull. Agr. and Mech. Coll. Texas,* 93, (1946), pp. 14–28 (9 references); Robert E. Souther, "Application of Mud Analysis Logging," *Geophysics,* Vol. 10 (January, 1945), pp. 76–90; J. T. Hayward, "Continuous Logging of Rotary Drilled Wells," *Oil and Gas Jour.,* Vol. 39 (Nov. 14, 1940), pp. 100–110; W. H. Russell, "Mud and Cuttings Logging," *Subsurface Geology in Petroleum Exploration* (Colorado School of Mines, 1958), pp. 357–366.

[29] Hubert Guyod, "Temperature Logging," *Bull. Agr. and Mech. Coll. Texas,* No. 93 (1946), pp. 146–165; Wilfred Tapper, "Caliper and Temperature Logging," *Subsurface Geology in Petroleum Exploration* (Colorado School of Mines, 1958), pp. 345–355.

of the temperatures in a borehole. This record, however, normally does not represent the true formation temperature, as a borehole and the contained drilling fluid are rarely in thermal equilibrium. The drilling fluid has a temperature greater than that of the shallower formations and less than that in the deeper part of the hole. After circulation ceases, the temperature difference begins to equalize by conduction of heat at rates which vary in different formations. For example, shale conducts heat at a lesser rate than sand. The most pronounced anomalies occur about 24–36 hours after circulation has ceased, and logs taken at this time reflect changes in lithology that to some extent are correlatable from well to well. However, the principal use of temperature logs is to determine the top of cement fill behind casing. The heat generated by setting cement has a pronounced effect on the temperature of the borehole—an effect which persists for several days after the cement is in place.

The recording of detailed temperature logs by means of a special high-sensitivity thermometer gives a method of locating zones of gas entry in wells.[30] A small decrease in temperature is caused by the expansion of gas into the well. Similarly, a temperature log under favorable conditions may indicate fluid entry by a small increase in temperature.

Electric logs.[31] Electric logging is, essentially, the recording of *resistivities* of the subsurface formations and *spontaneous potentials*

[30] K. S. Kunz and M. P. Tixier, "Temperature Surveys in Gas Producing Wells," *Jour. Petrol. Technol.*, Vol. 7 (July, 1955), pp. 111–119.

[31] Schlumberger Well Surveying Corporation, "Introduction to Schlumberger Well Logging," *Schlumberger Document*, No. 8 (1958); Hubert Guyod, "Applicability of Electric Logging Devices," *World Oil* (July, 1957), pp. 107–118; Dr. R. G. Hamilton and Paul Charrin, "How to Interpret Electric Logs . . . Quantitative Log Analysis," *Oil and Gas Jour.* (Nov. 12, 1956–June 3, 1957); Ernest E. Finklea, "Formation Evaluation of Some Limestone Reservoirs with Particular Reference to Well Logging Techniques," *Jour. Petrol. Technol.* Vol. 8 (August, 1956), pp. 25–31; "Fundamentals of Logging," *Univ. of Kansas Petrol. Eng. Conference*, (April 2–3, 1956), pp. 1–169; Maurice Martin, "With the Microlog . . . You Can Be Sure," *Oil and Gas Jour.* (Sept. 19, 1955), pp. 108–111, and (Oct. 24, 1955), pp. 106–109; D. Taylor Smith, "Well-Logging Methods," *Petroleum* (August, 1955), pp. 285–289; M. R. J. Wyllie, *The Fundamentals of Electric Log Interpretation* (Academic Press, Inc., New York, 1954), pp. 1–107; Leendert de Witte, "Resistivity Logging in Thin Beds," *Jour. Petrol. Technol.*, T.P. 3803, Vol. 6 (July, 1954), pp. 29–35; A. Poupon, M. E. Loy, M. P. Tixier, "A Contribution to Electrical Log Interpretation in Shaly Sands," *Jour. Petrol. Technol.* (T.P. 3800), Vol. 6 (June, 1954), pp. 27–34; L. A. Puzin, "New Well Logging Developments," *World Oil* (December, 1953, and January, 1954); L. W. Storm, "Symposium on Well Logs—Some Notes on Well Logs of Various Types with a Resume of Opinions About Them," *The Mines Mag.*, Vol. 43 (October, 1953), pp. 87–107; H. G. Doll, "Two Decades of Electrical Logging," *Jour. of Petrol. Technol.*, Vol. 5 (September, 1953), pp. 33–41; H. G. Doll, "The Microlaterlog," *Jour. of Petrol. Technol.* (January, 1953), pp. 17–31; J. E. Walstrom, "The Quantitative Aspects of Electric Log Interpretation," *Jour. Petrol. Technol.*, T.P. 3280, Vol. 4 (February, 1952), pp. 47 et seq.

generated in the borehole. Since the advent of electric logging in the late 1920's, it has become general practice when a hole has been drilled, or at intervals during the drilling, to run an electrical survey for the purpose of obtaining quickly a complete record of the formations penetrated. This recording is of immediate value for the geological correlation of the strata and for the detection and evaluation of possibly productive horizons.

Electric logs have reduced but not eliminated the need for sample logs and cores. In addition to the speed and economy of formation logging by this method, there are extra advantages in that the log is mechanically recorded and is uninfluenced by personal factors, the depth measurements are accurate and continuously printed, and the electrical measurements are made in the borehole with the formations in their natural state, except for some mud invasion and the effect of the borehole itself on reservoir pressures.

Along with the continued improvement of conventional electric-logging instruments, advances have been made in the art of interpreting the available data in terms of essential geological and engineering parameters. As the efficiency of interpretation improved it was evident that special auxiliary devices were needed for the measurement of additional parameters and for operation in extreme mud types. In the broad sense these auxiliary or supplementary logs are considered part of electric logging and will be individually discussed following a discussion of the conventional electric log.

These various electrical logs are made in a truck-mounted instrument cab (Fig. 4–2). The recording instruments are connected to the downhole device by means of a cable, containing insulated conductors, which is spooled on a drum on the rear of the truck (Fig. 4–3). As the downhole device is raised or lowered in the well, a continuous record of the readings and their corresponding depths is printed on film. The depths at each 100-foot level are printed in a special column on the log and intermediate levels are identified by the depth grid.

The electrical log, as delivered in the field, normally consists of two depth scales — one for detailed study and the other for correlation. The multi-galvanometer recorders that are presently used make it possible to present, on each of these two films, a complete complement of curves simultaneously recorded. Special dark rooms are provided in the instrument cab so that the film recording can be developed immediately following the survey. Field prints or copies of the survey are then run in a special printing machine. The complete logging operation, starting when the drill stem has been removed from the well and ending with the delivery of the field print, normally takes only a few hours.

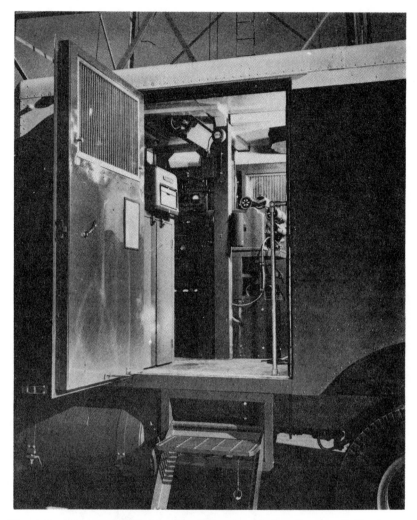

Fig. 4–2. Instrument cab of an electric logging truck. *Courtesy Schlumberger Well Surveying Corporation.*

It is possible to use electric logs as a subsurface-correlation tool with little or no concept of the meaning of the erratic pattern of the curves. Electric logs can be "slipped" alongside of each other in the same way as plotted or mounted logs, and correlations can be made from well to well by the pattern of the "kicks" on the curves. If the formations have been identified on one electric log, based on sample studies, it may be possible to carry these formations, with their separate behavior patterns, for considerable distances laterally by electric logs alone. However, many pitfalls can be avoided if the user knows something about the possible reasons for the "kicks" on the curves, and much valuable knowledge of lithology and fluid content is available to those who can interpret electric logs.

Fig. 4–3. Electrical logging truck. This rear view shows the surveying cable spooled on the drum. The striped doors on each side open equipment racks in which the various down-hole instruments are carried. On the sides are sheave wheels that are hung in the derrick and through which the cable passes in the performance of the logging operation. The braces with the serrated edges, next to the sheave wheels, are dropped to the ground and anchor the truck against the pull of the cable during the operation. *Courtesy Schlumberger Well Surveying Corporation.*

The resistivities of the formations are important clues to their lithology and fluid content. Formations conduct electric current only by means of the mineralized water that they contain.. The minerals that constitute the solid parts of the strata are insulators when absolutely dry. The few exceptions are metallic sulphides, graphite, etc., which conduct electricity like metals. In a similar manner, any pure oil or gas in the formations is electrically non-conductive. Other important factors in formation resistivity are the shape and the interconnection of the pore spaces occupied by the water. These depend upon the lithology and, in reservoir rocks, on the presence of non-conductive oil and gas.

Electrical resistivity is that property which tends to impede the flow of electricity through a substance. It is a specific property of the substance and is independent of the character or the shape in which the substance occurs. The unit of measurement of resistivity in well logging is the ohmmeter.

As previously mentioned, formations conduct electric current only by means of the mineralized water that they contain. Therefore, the resistivity of a formation depends directly upon the amount and the resistivity of the water that is contained. In electric logging the resistivity of a formation water quite often has to be deduced from a knowledge of its salt content and temperature. The more concentrated the salt solution becomes, the lower is the water resistivity; the higher the temperature of the solution, the lower is the water resistivity.

The spontaneous potential (SP) curve of the electrical log (Fig. 4–4) provides information from which the water salinity, and resulting resistivity, can be determined. The SP log is a record of variations in natural potentials along the borehole. Differences in potential are recorded when the electrode passes from a shale to a sand or lime formation. The base line of the curve corresponds, in most cases, to shale and impervious beds, whereas the peaks are generally opposite permeable beds. This variation with lithology has long made the SP a valuable correlation curve. The magnitude of the SP deflection is closely related to the relative salinities of the mud and formation water. This relation has been widely studied and is a fundamental factor in qualitative and quantitative interpretation.

Once the formation water resistivity has been deduced from SP curve it is possible to determine, from the resistivity curves, the amount of water present in the formation. Two primary factors govern the amount of water that can be contained in a formation: first is the porosity of the formation, and the second is the percentage of that porosity that is filled with hydrocarbons. Other factors remaining constant, the lower the formation resistivity, the higher is the porosity. A zone of a given poros-

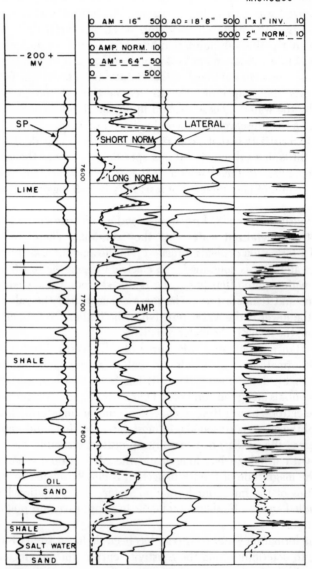

Fig. 4–4. Spontaneous potential log on left. The center two columns show the three standard resistivity curves. The right column is a microlog. *Courtesy Schlumberger Well Surveying Corporation.*

ity and formation-water resistivity would indicate a higher resistivity when part of the pore space is occupied by hydrocarbons than it would when completely saturated with water. This, essentially, is the basis upon which electrical logs are interpreted to provide the valuable knowledge of reservoir and fluid content data that is required by geologist and engineer.

Accurate evaluation of a reservoir depends, to a major extent, upon the accuracy with which the true formation resistivity is determined. The vast scope of information required in modern well logging, together with the restrictions imposed by the borehole, the heterogeneity in thickness, fluid content, and physical characteristics of the formations rules out the possibility of completely fulfilling the requirements of well logging by any single survey. These variations require different logs in different areas and often even in different sections in the same well.

The *conventional electrical log* is a combination of spontaneous potential measurement and formation-resistivity measurements using spaced electrodes. It is at its best when used in fresh mud, formations of low to moderate resistivity, and fairly thick beds. Under these conditions the influences of the mud in the borehole as a shunting conductor across the electrodes, and of beds adjacent to the formation at the point of measurement, are at a minimum. As with other resistivity measuring devices, the use of the electrical log is restricted to uncased wells, since the metal of casing short-circuits the electrodes. Often, corrections are applied to recorded readings to obtain true resistivities.

The resistivity log usually includes three curves (Fig. 4–4) chosen by experience to give the maximum information. Multi-curve resistivity recording is essential to proper interpretation of formation characteristics. Short-spacing or shallow-penetration curves are necessary to define bed boundaries accurately and to help determine the resistivity of the mud-invaded zone close to the borehole. Long-spacing or deep-penetration resistivity curves are equally essential to help determine the true resistivity of the formation, which is the key to locating and evaluating possible productive zones. The arrangement of the electrodes and their spacings determines the response and the depth of investigation under any particular set of conditions. Actual resistivity measurements with this conventional system are accomplished by causing a known electric current to flow from a set of electrodes into the formation and measuring the resulting potentials between another set of electrodes. One of the major problems with this system is that the currents are not focused; they are free to follow the path of least resistance. Since the system depends upon the flow of current from the electrodes into the formation, it is obvious that the device cannot be employed when the borehole is filled with a non-conductive medium such as oil, oil-base mud, gas, or air.

The *induction log* is a focused logging method which requires no current flow from the tool into the formation. Instead, the formations are energized by an alternating magnetic field created by the tool. Since no contact through the mud is required, the induction log makes it possible to record conductivity (reciprocal of resistivity) and resistivity curves in empty holes or non-conductive oil-base muds, as well as in normal water-base muds. The curve characteristics under these different conditions are almost identical. When the use of induction logs is routine, logs across a field may be compared directly even though the drilling and completion program may require changes to oil-base mud, or the use of gas instead of drilling mud.

The recorded signal of the induction log is less affected by resistive materials within the volume of measurement than by conductive materials. Therefore its recorded values closely approach true formation resistivity in resistive muds (fresh-water or oil-base), and it is able to detect conductive zones, that is, salt-water sands, even behind a resistive invaded zone.

A system of focusing permits the induction log to determine more accurately the true resistivity of relatively thin beds. Even in normal drilling muds the greater depth of investigation and the reduction of effects from surrounding beds by focusing have made the induction log an outstanding tool. The conventional electrical log is commonly affected by various reversals, blind zones, reflection peaks, and lags that confuse interpretation of the deeper investigating curves. There are no such effects on the induction log.

The advantages of the induction log have led to its combination with the short normal and the SP curve to form the induction-electrical log. (Fig. 4–5). This combination is replacing, to an important degree, the conventional electrical log as a basic surveying tool. It performs the same basic functions of correlation for geological study and the detection and evaluation of productive formations, but the sharpness of detail, deeper investigation, and high accuracy in thin beds permits recording a more efficient log. The induction-electrical log is broadly applicable under the same conditions as the electrical log, that is, in normal fresh drilling muds.

In salt muds both the conventional electric log and the induction log are affected by the low-resistivity mud column. For these conditions, where the mud is considerably more conductive than the formations, the *laterolog* was developed. As in the electric log, current is sent into the formation from electrodes on the down-hole tool. However, by means of

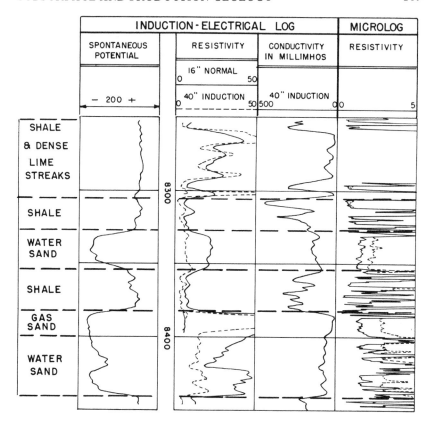

Fig. 4–5. Electrical log and microlog, with induction curves added in two middle columns. *Courtesy Schlumberger Well Surveying Corporation.*

an appropriate electrode arrangement and an automatic control system, the current is sharply focused into a sheet of predetermined thickness (Fig. 4–6). This focusing permits the measurement of a portion of the ground of limited vertical extent and virtually eliminates the shorting effect of the mud column.

The laterolog has its primary application in wells drilled with salt-saturated muds. Under these conditions the recorded values are close to true resistivity in salt-water bearing and impervious formations. In oil-bearing formations the recorded values are also close to true resistivity if invasion is not deep. In fresh muds the recorded values in salt-water bearing formations are too high because of invasion of the resistive mud filtrate. With fresh muds, in general, the laterolog is not used for true resistivity determination unless the formation resistivity is large.

Microdevices are used to measure the resistivities of small volumes of formation just behind the borehole wall. They are of two general types— non-focused, such as the microlog, and focused, such as the microlaterolog. They provide such important information as the detection and sharp boundary definition of permeable beds, thereby permitting an accurate sand count, and the degree of porosity of the permeable formations.

As previously pointed out, the true formation resistivity, for a given formation water resistivity, will be determined by the amount of water

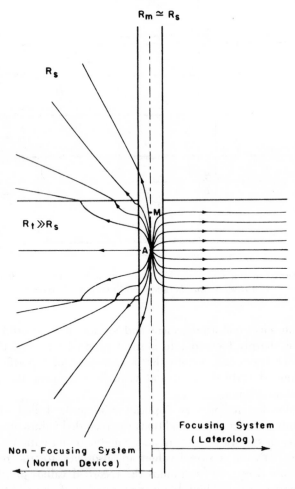

Fig. 4–6. Schematic drawing showing comparative distribution of current lines for normal device on left and laterolog on right opposite a thin resistive bed. *Courtesy Schlumberger Well Surveying Corporation.*

that it contains. Since a decrease in porosity would affect the formation resistivity in the same manner as an increase in hydrocarbon saturation, it is necessary to have a means of evaluating porosity that is independent of the true resistivity measuring devices. The microdevices, by measuring

Fig. 4-7. Modern powered micro-sonde. The pads are mounted on parallelogram arms and held against the hole wall by coil springs. For the descent into the well, and for more rapid recovery of the tool after completion of the survey, the arms are collapsed against the sonde body by means of an electric motor. Note the three small, circular electrodes on the face of the pad. *Courtesy Schlumberger Well Surveying Corporation.*

the resistivity of the zone that is flushed by mud filtrate, provide this independent evaluation of porosity.

The two general types of microdevices are each measured with closely spaced electrodes that are mounted in an insulating rubber pad which is held, by spring action, against the wall of the hole (Fig. 4–7). In this way the shorting effect of the mud column is minimized. With the non-focused type, as in the case of the electrical log, the electric current is allowed to take the least resistant path, whereas the focused device, as in the laterolog, beams the current into the formations. As might be expected, the focused device is primarily used in salt muds where the shorting effect of the mud is most severe. Logs made with both types usually include a micro-caliper curve, which is a diameter measurement between the face of the insulating pad and the face of a back-up pad on the opposite side of the down-hole tool.

The *dipmeter*[32] is a survey instrument run in the hole to determine the amount and direction of the dip of various formations penetrated by the drill. Any one of three types of electrical curves—SP, resistivity, or microresistivity—is recorded by means of electrodes imbedded in the arms of a mandrel held against the wall of the hole. The arms are 120° apart and the electrodes are positioned in a plane perpendicular to the axis of the instrument. The curves register sharply all formational changes, and if a bedding plane crosses the borehole at an angle, a vertical displacement between curves results. Correlation of the three curves establishes the amount of displacement (Fig. 4–8).

The *continuous dipmeter*, as the name implies, makes a continuous recording of three microfocused resistivity curves. At the same time directional surveys are made and recorded by a photoclinometer, and a microcaliper log is run. As a result, one film, produced in the logging truck at the surface, contains data from which the strike and dip of the rock layers penetrated can be determined, with corrections made for any crookedness in the borehole, and correlation can be made with other wells. The older *interval dipmeter* takes a series of readings, either SP or resistivity, through 40 to 60-foot intervals where prior electric logs show

[32] Jack Grynberg and M. I. Ettinger, "The Continuous Dipmeter . . . Its Use—and Abuse," *Oil and Gas Jour.*, Part 1 (April 1, 1957), pp. 166–178; B. Osborne Prescott, "Calculating Dip and Strike From Continuous Dipmeters," *Oil and Gas Jour.*, (March 7, 1955), pp. 118–126; Pierre de Chambrier, "The Microlog Continuous Dipmeter," *Geophysics*, Vol. 18 (October, 1953), pp. 929–951; F. G. Boucher, A. B. Hildebrandt, and H. B. Hogen, "New Dip-Logging Method," *Bull. Am. Assoc. Petrol. Geol.*, Vol. 34 (October, 1950), pp. 2007–2026; E. F. Stratton and R. G. Hamilton, "Application of Dipmeter Surveys," *Subsurface Geology in Petroleum Exploration* (Colorado School of Mines, 1958), pp. 389–393; H. G. Doll, "The S. P. Dipmeter," *Am. Inst. Mining Met. Engrs.*, Tech. Publ. No. 1547, (1943), pp. 1–10.

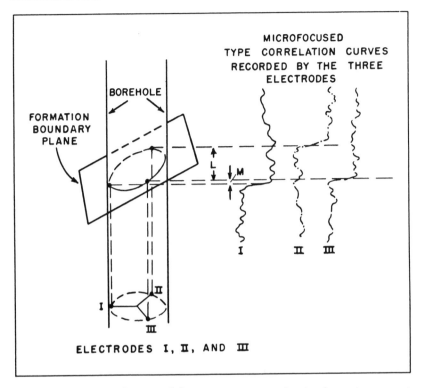

Fig. 4–8. Schematic drawing of dipmeter curves at a dipping formation contact. The sharp break in each curve occurs as its corresponding measuring electrode crosses the pictured formation boundary. The discrepancies "L" between the curves recorded by electrodes I and II, and "M" between those recorded by electrodes I and III are the basic elements for the computation of the formation boundary's dip. *Courtesy Schlumberger Well Surveying Corporation.*

sharp contacts. Directional pictures can also be taken on the same run, and as many as 12 intervals can be logged on one trip into the hole.

The dipmeter is one of the most useful aids a subsurface geologist has for determining the structural picture. In the absence of good correlation it may enable a realistic interpretation to be deduced. The interpretation of dipmeter records is a job for a specialist familiar with the area concerned, and not just a computer. The data can also be invaluable in preparing paleogeological maps, as the strike and dip may be quite vague without them. The dipmeter also gives information from a single well unobtainable any other way and permits a better planning of future development from a discovery well, or from an outstepping well.

Sonic logging[33] (also known as continuous velocity logging, or CVL) fundamentally involves the recording of the time required for a sound wave to travel through a definite length of formation. These travel times are recorded continuously, versus depth, as the sonic sonde is pulled up the borehole, and are inversely proportional to the speed of sound in the various formations.

Table 1 gives the sonic velocities and the corresponding travel times (in microseconds per foot) for various fluids and solid materials. Some of the values are averages, and may vary with depth in subsurface formations.

TABLE 1

Material	Sonic Velocity, V (feet per sec)	Travel Time, Δt ($= 10^6/V$ in μ sec/ft)
Air (S.T.P.)	1088	919
Methane (S.T.P.)	1417	706
Oil	4300	232
Water (mud)	5000 – 5300	200
Neoprene (typical)	5300	189
Shales	6000 – 16,000	167 – 62.5
Rock Salt	15,000	66.7
Sandstones	up to 18,000	55.6
Anhydrite	20,000	50.0
Limestones	up to 23,000	43.5
Dolomite	24,500	40.8

Speed of sound in subsurface formations depends upon the elastic properties of the rock matrix, the porosity of the formations, and their fluid content and pressure. Below the "weathered" or low-velocity layer, which may extend from the surface down to 50 or perhaps 100 feet, the sound velocities may range from about 6000 feet per second for shallow shales, to as much as 26,500 feet per second for dolomites. The sonic log is quite detailed and reflects so faithfully changes in lithology that even in

[33] A. A. Stripling, "Velocity Log Characteristics," *Jour. Petrol. Technol.*, Vol. 10 (September, 1958), pp. 207–212; H. M. Breck, S. W. Schoellhorn, and R. B. Baum, "Continuous Velocity Logging," *Subsurface Geology in Petroleum Exploration* (Colorado School of Mines, 1958), pp. 409–426; John L. P. Campbell, "Density Logging in the Gulf Coast Area," *Jour. Petrol. Technol.*, Vol. 10 (July, 1958), pp. 21–25; H. R. Breck, S. W. Schoellhorn, and R. B. Baum, "Velocity Logging and Its Geophysical and Geological Applications," *Bull. Am. Assoc. Petrol. Geol.*, Vol. 41 (August, 1957), pp. 1667–1682; E. M. Denton, "Continuous Velocity Log," *Oil and Gas Jour.* (Nov. 19, 1956), pp. 224–234; W. G. Hicks and J. E. Berry, "Application of Continuous Velocity Logs to Determination of Fluid Saturation of Reservoir Rocks," *Geophysics*, Vol. 21 (July, 1956), pp. 739–745; G. C. Summers and R. A. Broding, "Continuous Velocity Logging," *Geophysics*, Vol. 17 (July, 1952), pp. 598–614.

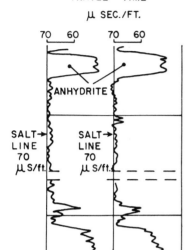

WELL C   WELL D

TRAVEL   TIME
μ SEC./FT.

Fig. 4–9. Sonic logs of two wells, showing how anhydrite and salt are readily differentiated, and how accurate correlation between wells is possible. *Courtesy Schlumberger Well Surveying Corporation.*

its present early stage of development it is an excellent log for correlation purposes (Fig. 4–9).

In moderate to hard formations, where porosities are low, the sonic log is considerably affected by the amount of fluid in the formations; hence, the sonic log gives reliable indications of their porosity. In all types of formations, this device is of value to the geophysicist in interpreting seismic charts.

Much of the success of the sonic log can be traced to the excellence of the two-receiver type device (Fig. 4–10) which virtually eliminates the effect of the borehole. In this system a short pulse of sound energy is emitted by a transmitter every tenth of a second. Energy from this pulse travels through the mud and formation and is detected by each of two receivers. The *difference* in time between reception of the energy at the two receivers is recorded. This travel time, Δt, is the time required for sound to traverse one foot of formation. Hence, it is the inverse of the velocity of sound through that foot of formation. Only that portion of formation situated between the receivers affects the reading. Used in conjunction with other electrical logs the sonic log has proven itself a valuable aid to formation and reservoir analysis.

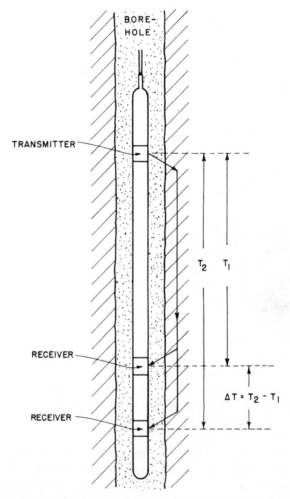

Fig. 4–10. Schematic drawing of two-receiver sonic logging instrument. *Courtesy Schlumberger Well Surveying Corporation.*

Radioactivity logs.[34] Radiation logs are of two very general types: those which measure the natural radioactivity of formations (*gamma-ray logs*), and those which indicate radiation reflected from or induced in the formations as a consequence of bombarding the formations with neutrons emitted from a source contained in the down-hole tool (*neutron logs*). Gamma-ray logging began in 1940 and neutron logging about a year later. A major advantage of radioactivity logs is that they may be obtained through several strings of casing and even through cement. Radioactivity logs are better considered an auxiliary to sample logs than an end in themselves. The curves obtained are subject to various interpretations, and sample logs are necessary to the correct solution of lithology.

The theory of the gamma-ray curve is simple. All rocks contain radioactive material but the amount is highly variable. The radioactive substances are undergoing constant disintegration, in the course of which rays are emitted. The most penetrating of these rays are the gamma rays. As a general rule, since shales contain more radioactive material than sandstones and limestones, shale produces a prominent "kick" in the gamma-ray curve. To this extent, gamma-ray logging is also lithologic logging.

The neutron curve, produced by an artificially induced neutron bombardment of the wall rocks, is an inverse measurement of the amount of

[34] Schlumberger Well Surveying Corporation, "Introduction to Schlumberger Well Logging," *Schlumberger Document*, No. 8 (1958); Richard L. Caldwell and Robert F. Sippel, "New Developments in Radioactive Well-Logging Research," *Bull. Am. Assoc. Petrol. Geol.*, Vol. 42 (January, 1958), pp. 159–172; Gilbert Swift and Russell G. Norelius, "New Nuclear Radiation Logging Method . . . Employing a Pair of Neutron Curves," *Oil and Gas Jour.* (Oct. 15, 1956), pp. 109–114; H. R. Brannon, Jr. and J. S. Osoba, "Spectral Gamma-Ray Logging," *Petroleum Transactions, Jour. of Petrol. Technol.*, Vol. 8 (February, 1956), pp. 30–35; Herman E. Schaller, "New Radiation Counter Solves Many Logging Problems," *The Mines Mag.*, Vol. 43 (October, 1953), p. 86; J. T. Dewan and L. A. Allaud, "Experimental Basis For Neutron Logging Interpretation," and A. Blanchard and J. T. Dewan, "The Calibration of Gamma Ray Logs," *Petrol. Engineer*, (August and September, 1953); M. R. J. Wyllie, "Procedures for Direct Employment of Neutron Log Data in Electric Log Interpretation," *Geophysics*, Vol. 17 (October, 1952), pp. 790–805; W. L. Russell, "Interpretation of Neutron Well Logs," *Bull. Am. Assoc. Petrol. Geol.*, Vol. 36 (February, 1952), pp. 312–341; C. W. Tittle, H. Faul, and C. Goodman, "Neutron Logging of Drill Holes: The Neutron-Neutron Method," *Geophysics*, Vol. 16 (October, 1951), pp. 626–638; R. E. Bush and E. S. Mardock, "The Quantitative Application of Radioactivity Logs," *Jour. Petrol. Technol.*, Vol. 3 (July, 1951), pp. 191–198; F. P. Kokesh, "Gamma Ray Logging," *Oil and Gas Jour.* (July 26, 1951); H. Faul and C. W. Tittle, "Logging of Drill Holes by the Neutron-Gamma Method, and Gamma Ray Scattering," *Geophysics*, Vol. 16 (April, 1951), pp. 260–276; Gilbert Swift and Arthur Youmans, "Radioactivity Well Logging," *Subsurface Geology in Petroleum Exploration* (Colorado School of Mines, 1958), pp. 329–343; Warren J. Jackson and John L. P. Campbell, "Some Practical Aspects of Radioactivity Well Logging," *Am. Inst. Mining Met. Engrs. Tech. Publ.* 1923 (September, 1945), 27 pp. (30 references); Robert E. Fearon, "Gamma Ray Measurements," *Oil Weekly*, Vol. 118 (June 4, 1945), pp. 33–41 (66 references).

hydrogen present. Because most hydrogen in rock is in either interstitial water or oil, the neutron curve indicates (by subnormal readings) the presence of fluids and hence is a porosity log. It is not possible to distinguish oil from water with the neutron log. However, this curve does complement the gamma-ray log in that it makes possible the distinction between porous and non-porous formations. The accuracy of the identifications is increased further by using the radioactive logs in conjunction with electric logs (Fig. 4–11).

Although the curves may be singly recorded, the gamma-ray and neutron curves are generally obtained simultaneously during a single traverse of the well by means of modern dual logging equipment (Fig. 4–12). As in electric logging, an insulated cable transmits down-hole measurements to the surface instruments and recorder which are mounted in the instrument cab of a truck. Various types of gamma-ray detecting devices are used. The Geiger-Mueller counters or ionization chambers that are sometimes used become electrically conductive when penetrated by gamma rays. The scintillation detector employs a sodium iodide crystal that glows when struck by gamma rays. The more concentrated the gamma rays, the greater is the brightness of the glow. A photoelectric cell and multiplier are used to transform these light variations into measurable electrical quantities.

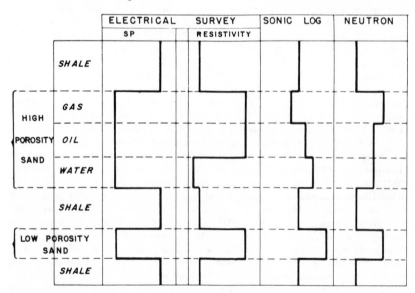

Fig. 4–11. Example of typical reactions of sonic log, neutron and electrical survey. Since each log measures a different parameter, accurate qualitative interpretation is possible. *Courtesy Schlumberger Well Surveying Corporation.*

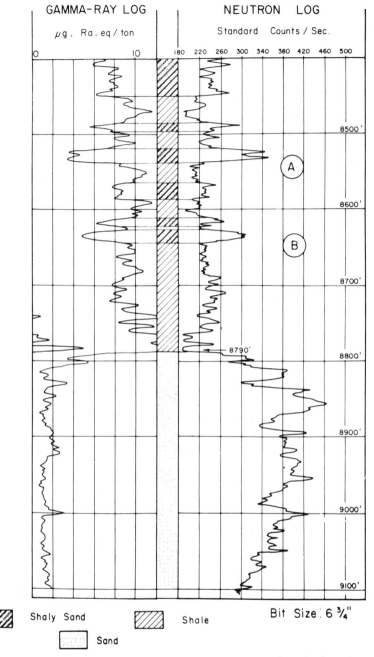

Fig. 4-12. Gamma-ray and neutron logs of shaly (upper) and sandy (lower) section taken in open hole with water-base mud. Reaction of thin shaly sands A and B is intermediate between shales and sands. *Courtesy Schlumberger Well Surveying Corporation.*

119

Neutron-logging detectors may respond to either induced gamma radiation or to neutrons themselves. Referred to as a neutron-gamma process, a properly shielded ionization chamber responds to the high-energy gamma rays of capture that occur when a neutron is captured by an atom. In the neutron-neutron process the detector responds directly to neutrons. The results of the two methods are nearly identical. Logging speeds are commonly in the range of 20 to 50 feet per minute.

The ability of radioactivity devices to log through casing (they are the only tools that can do so at present), causes them to be used widely in workover and completion programs. Many old wells were drilled and cased prior to the increased efficiency and widespread application of electrical logs. There are many such wells that have been recompleted as commercially productive wells based upon the information gained from radioactivity logs. Subsurface geology and mapping have frequently been improved by information obtained from old cased wells that were strategically located. In cased wells it is customary to record the position of each collar on the radioactivity log by means of a magnetic casing-collar locator that is incorporated in the logging instrument. Since the casing collars are thus tied into the formations, any chosen depth for perforation may be accurately located by means of a casing-collar locator on the perforating gun, and it is unnecessary to rely upon the precise measurement of a great length of cable.

In addition to assisting the production engineer, radioactivity logs can be and are used by geologists in subsurface structure mapping. Some formations produce a highly distinctive radioactivity pattern which remains fairly constant over great distances. A notable example of this pattern is the Chattanooga shale, which is easily recognized by the distinctive gamma-ray "kicks" from the Appalachians to west Texas, and from Canada to Oklahoma.

Caliper logs.[35] The borehole caliper is a device used to determine the variations in well diameter from bottom to top. Because of differences in brittleness, cohesion, and solubility between different rock layers, the borehole does not have smooth vertical walls. Some formations are especially likely to cave, or to erode by the jetting action of the mud fluid emerging from the bit, or to wear by the subsequent circulation of the drilling mud. Other rocks, especially salt, are soluble in the water of the circulating fluid, unless previously saturated, so lateral leaching may

[35] Schlumberger Well Surveying Corporation, "Introduction to Schlumberger Well Logging," *Schlumberger Document*, No. 8 (1958); Wilfred Tapper, "Caliper and Tempera-ture Logging," *Subsurface Geology in Petroleum Exploration* (Colorado School of Mines, 1958), pp. 345–355; Hubert Guyod, "Caliper Well Logging," *Oil Weekly* (Aug. 27, 1945), pp. 32–35; (Sept. 3, 1945), pp. 57–61; (Sept. 10, 1945), pp. 65–69; (Sept. 17, 1945), pp. 52–54.

take place. It is obvious that considerable control on variations in bore-hole diameter can be exercised by drilling procedures, but the control is mainly in the magnitude rather than in the presence of the variations. As a general rule, where boreholes pass through massive limestones and indurated sandstones they are but slightly greater in diameter than the diameter of the bit. In soft, unconsolidated or poorly cemented rock, such as shale, the hole diameter may be considerably enlarged.

The caliper consists of three or four collapsible arms evenly spaced around a steel shank. The tool is lowered to the bottom of the hole, where the arms are released. It is then raised to the surface at a rate of about 100 feet per minute. During the trip, individual springs keep each arm pressed against the sides of the hole, and the deviations from the bit diameter are automatically and continuously plotted with the depths by an electrically operated recorder mounted in the service truck.

Caliper logging was first conceived and developed for the purpose of assisting in solving certain engineering problems such as the amount of cement needed to plug a hole or the amount of material needed to gravel-pack a well. It is also used in surveying the effects of shooting and acidizing. The caliper log is a help in sample examination, giving clues as to both the amount and the source of cavings in the samples. In addition, the peaks and valleys on the caliper profile of a well represent a stratigraphic succession, so the caliper log can be used in well-to-well correlation. It is still essential to have sample logs available for calibrating the caliper logs, however.

Logs recorded with modern microresistivity devices include a caliper curve which is the distance between the faces of the two pads that are in contact with opposite sides of the borehole. This microcaliper is very helpful in that the presence of filter cake, indicated by an under-gauge hole, readily indicates those formations that are sufficiently permeable to permit invasion. Caved zones are detected as in the case of the regular caliper device.

USES OF DATA

The mere collection of subsurface data is a necessary chore which is of no geological value until it has been studied, classified, recorded, and interpreted. The primary objectives of the exploration subsurface geologist are to find hitherto hidden traps, whether structural or stratigraphic; the production geologist searches for information that may lead to an increased yield of hydrocarbons within a producing area. But neither one can begin to obtain his objective until he has worked out the stratigraphic section in minute detail. This project includes the correlation of sedimentary units and the recognition of unconformities, wherever they may occur. The primary use made of both logs and samples is in the correlation [42] of the sedimentary layers from well to well. Samples are essential in the correlation from surface to subsurface. Correlations in the subsurface are made by comparing lithologies of samples, by means of fossils in cuttings and cores, and by comparing reaction patterns such as those obtained in electric and radioactive logs. When the well data are contained on log forms it is customary to slide these up and down in order to find the level at which the best match between wells takes place.

Differences in thickness of stratigraphic sections between wells may be the result of original differences in depth of deposits, differences in degree of compaction, intersection by faults, or the presence of *unconformities*. Although unconformities are rarely actually seen by the subsurface geologist (and when they are seen it is only in cores), he is in a better position to find them than is the surface geologist because of his three-dimensional regional viewpoint. Most unconformities are identified by the shortened or lengthened sections caused by missing stratigraphic units in one instance and additional strata in the other.

The data obtained by the subsurface geologist are invaluable in evaluating the possibilities of different formations functioning in the area investigated as source, reservoir, or seal rocks. The physical descriptions and the mechanical measurements include such pertinent information as organic content, porosity, permeability, and thickness.

After the stratigraphic spade work has been completed (or while it is under way) the data obtained are entered on maps and sections. Then it is time for geological interpretation, conclusions, and recommendations.

Subsurface Maps and Sections. [43] The logs and rock samples obtained during the drilling of individual wells may be of considerable value when considered alone, [44] but when compiled with similar information obtained from other wells, the resulting maps and cross sections are of inestimable value in oil finding. It is the task of the subsurface geologist to assemble the individual well records into geologic illustrations which

[42] L. W. LeRoy, "Stratigraphic Correlation," *Subsurface Geology in Petroleum Exploration* (Colorado School of Mines, 1958), pp. 439–452; Carl A. Moore, "Some Aspects of Subsurface Geologic Correlations," *The Shale Shaker Digest*, Vol. 1–5 (1950–1955), pp. 167–172.

[43] F. H. Lahee, "Maps," *Science of Petroleum* (Oxford University Press, 1938), Vol. 1, pp. 276–283; Julian Low, "Subsurface Maps and Illustrations," *Subsurface Geology in Petroleum Exploration* (Colorado School of Mines, 1958), pp. 453–530; Hugh McClellan, "Stereographic Problems," *ibid*, pp. 531–551.

[44] Paul Weaver, "The Geologic Interpretation of Exploratory Wells," *Bull. Am. Assoc. Petrol. Geol.*, Vol. 33 (July, 1949), pp. 1135–1144.

guide the administrative officers of the company in developing their exploration program. Descriptions of the various types of subsurface compilations follow.

Geologic maps. In many areas, owing to thick veneers of glacial drift or other types of mantle, the geologic formations of the bedrock surface can be mapped only by means of well data. This was probably the earliest application of subsurface geology. Water wells, salt wells, and oil wells furnished the information that was needed in compiling the geologic

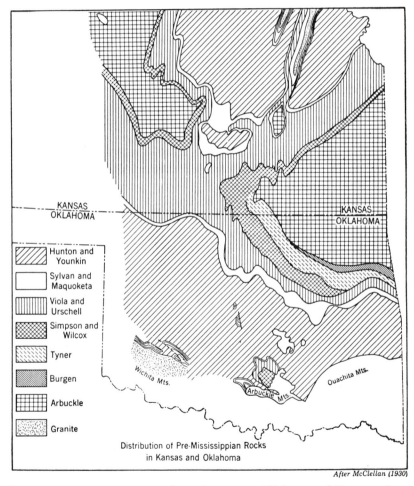

KANSAS
OKLAHOMA

KANSAS
OKLAHOMA

Hunton and Younkin

Sylvan and Maquoketa

Viola and Urschell

Simpson and Wilcox

Tyner

Burgen

Arbuckle

Granite

Wichita Mts.

Arbuckle Mts.

Ouachita Mts.

Distribution of Pre-Mississippian Rocks
in Kansas and Oklahoma

After McClellan (1930)

Fig. 4–13. Pre-Mississippian paleogeologic map, Oklahoma and Kansas. *Courtesy American Association of Petroleum Geologists.*

map of the southern peninsula of Michigan, where there are entire blocks of counties without a single outcrop.

The utilization of areal geologic maps in the search for oil was discussed in an earlier section. Such structural features as domes, anticlines, and noses may be discernible on the geologic map by the presence of inliers and other significant outcrop patterns. Faults may be of utmost importance in oil finding.

Paleogeologic maps.[45] A paleogeologic map is a map showing the areal geology at a given period in the past. It is, in effect, the result that would be obtained by peeling off the layers of younger rock above an unconformity.[46] Paleogeologic maps are constructed by plotting on a map the formations shown by well data to underlie the plane of the unconformity and then sketching in the formation boundaries after all the well data are plotted.

The paleogeologic map is of immeasurable value to the petroleum geologist. It reveals not only many structural features masked by the unconformity but also much information of the geologic past that is pertinent to the search for sedimentary or stratigraphic traps. "Oil and gas are found as high in their containing reservoir as it is possible for them to move. If true of the present, it must have been true in the past, and a study of the changing areal and structural geology during geologic time becomes of prime importance."[47]

One of the earliest uses of the paleogeologic map was in the Mid-Continent, where a map of the pre-Mississippian formations of Oklahoma and Kansas[48] (Fig. 4–13) has been widely used in guiding exploration.

An interesting type of paleogeologic map is the *worm's eye*[49] or *vermioculus*[50] map (Fig. 4–14). This map shows the distribution of geological formations immediately above an unconformity, as though the unconformity were a window through which one was looking from beneath. No other type of map does as good a job of illustrating the successive wedge-out of overlapping sedimentary formations. These wedge-outs

[45] A. I. Levorsen, "Application of Paleogeology to Petroleum Geology," *Science of Petroleum* (Oxford University Press, 1938), Vol. 1, pp. 300–303.

[46] W. C. Krumbein, "Criteria for Subsurface Recognition of Unconformities," *Bull. Am. Assoc. Petrol. Geol.* Vol. 26 (January, 1942), pp. 36–62.

[47] A. I. Levorsen, "Studies in Paleogeology," *Bull. Am. Assoc. Petrol. Geol.*, Vol. 17 (September, 1933), pp. 1107–1132.

[48] Hugh W. McClellan, "Subsurface Distribution of Pre-Mississippian Rocks of Kansas and Oklahoma," *Bull. Am. Assoc. Petrol. Geol.*, Vol. 14 (December, 1930), pp. 1535–1566.

[49] L. L. Sloss, W. C. Krumbein and E. C. Dapples, "Integrated Facies Analyses," *Geol. Soc. Am.*, Memoir 39, 1949, p. 115; W. C. Krumbein and L. L. Sloss, *Stratigraphy and Sedimentation* (W. H. Freeman and Co., San Francisco, 1951), p. 421.

[50] James K. Chrow, "A Paleogeologic Study of the Michigan Basin," *unpublished master's thesis*, University of Michigan (1958).

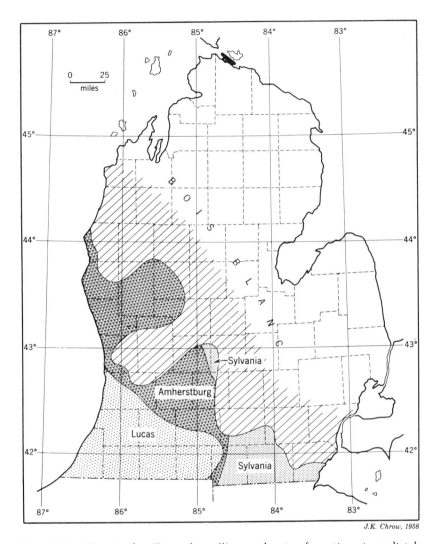

Fig. 4-14. Vermioculus ("worm's eye") map showing formations immediately overlying the Devonian-Silurian unconformity in the Southern Peninsula of Michigan. The regional dip is to the northeast; the Lucas formation is the youngest shown and the Bois Blanc ("Bob Low") the oldest. By J. K. Chrow, 1958.

may be due to original deposition (shore lines), or they may be due to emergence and truncation after more widespread deposition.

Paleogeographic maps.[51] The paleogeographic map purports to show the submergent and emergent areas at a given time in the geologic past. Most often the shore lines shown are pure guesswork. The data available for the construction of a paleogeographic map for a period of past time are confined to the erosional remnants of the sediments deposited during that time. These residua can be areally mapped in both the surface outcrop and in the subsurface, but the boundaries are almost always the result of erosion, and the original shore line was an entirely unknown distance out beyond the present boundaries.

True paleogeographic maps can be constructed only where the actual strand lines have been submerged and covered with younger sediments before erosion has destroyed all evidence of the former land-water contact. This ideal situation is rare. However, better paleogeographic maps than those extant today could be made by more thorough studies of the paleoecology of the sediments and contained faunas. These studies would at least result in a more intelligent guess as to the distance from outermost remaining sedimentary deposit to original land area.

Structure contour maps.[52] Subsurface-structure contour maps are prepared by first plotting and then contouring the elevations of a buried datum surface (usually a formation top). These elevations may be obtained, with the help of surface elevations, from core-drill information, logs of all types, and other well records, sample studies, and geophysical (especially seismic) surveys. Contour maps may be made for any traceable datum. Many oil fields have been discovered by subsurface-structure maps contoured on a near-surface formation penetrated by water wells, core tests, and shallow exploratory wells.

Obviously a structure contour map based on the reservoir rock (Fig. 3–5) is much more pertinent in a study of oil accumulation than one based on an outcropping formation. Usually the subsurface-structure map shows more relief and greater closure[53] than the surface-structure map. There are two principal reasons for increase in closure with depth. One is that many anticlines have been folded more than once. If the older

[51] John Emery Adams, "Paleogeography and Petroleum Exploration," *Jour. Sed. Petrol.* Vol. 13 (December, 1943), pp. 108–111.

[52] Louie Sebring, Jr., "Chief Tool of the Petroleum Exploration Geologist: The Subsurface Structural Map," *Bull. Am. Assoc. Petrol. Geol.*, Vol. 42 (March, 1958), pp. 561–587; Russell F. Klinger, "Regional Subsurface Maps Reveal the High and Low of It," *Oil and Gas Jour.* (April 4, 1955), pp. 168 et seq.; Daniel F. Merriam, "Kansas Structure Indicator," *World Oil* (September, 1955), pp. 84 et seq.

[53] The amount of closure is the vertical distance between the lowest possible closed structure contour line and the highest point on the structural feature.

(deeper) formations went through more periods of compression than the surface formations, obviously the intensity of the folding would be greater. The second reason for greater closure with depth is the dampening upward that takes place when the underlying rock is pushed upward (relatively) from one cause or another or when the sediments settle over buried hill, reef, or sand bar. The effect is similar to that of poking a finger into a pillow or mattress.

There are also examples of closure decreasing with depth. This situation may take place where the regional dip steepens downward. The closure of an anticline of the same magnitude is greater when it is superimposed upon a nearly flat surface than when it occurs on a dipping plane, as the flank of a basin.

The position of an anticlinal apex may shift laterally with depth owing to asymmetry, convergence, or unconformity. The axial plane of an asymmetrical anticline dips away from the steep flank, so the top of the structure at depth is beneath the more gently dipping flank. Convergence of the strata between surface and reservoir causes the top of the anticline to migrate in the direction of that convergence. A shifting of the axis of most intense uplift after emergence and resubmergence produces an offset in the top of the fold above the unconformity.

The oil and water in a reservoir are, of course, in adjustment with the structure of that reservoir and not with the structure of overlying formations.

The structure contour map also reveals the location of faults[54] cutting through the formation mapped. Due to upward dampening, or the presence of unconformities, the faults may not be present in the outcropping formations, and yet they may have played an important role in oil accumulation. As will be shown in Chapter 13, the intersection of a plunging anticline by a tight fault plane creates a hydrocarbon trap. Subsurface-structure contour maps of reservoir formations give the location of such traps. Another aid to trap hunting is a contour map of the fault plane itself. Mapping faults in three dimensions has demonstrated that fault surfaces are more often curved than flat.

Both the local and the regional subsurface structure contour map divulge geological information of value. The local map pinpoints possible oil or gas accumulations in structural (anticlinal) traps. It also shows the location of anticlinal noses which are essential to completing the closure in most types of varying permeability ("stratigraphic") traps which are described in Chapter 14. Regional structural maps show the depths (with the help of surface contour maps) to possible reservoirs. They also show

[54] L. W. LeRoy, "Stratigraphic, Structural, and Correlation Considerations," *Subsurface Geologic Methods* (Colorado School of Mines, 1950), pp. 51–55.

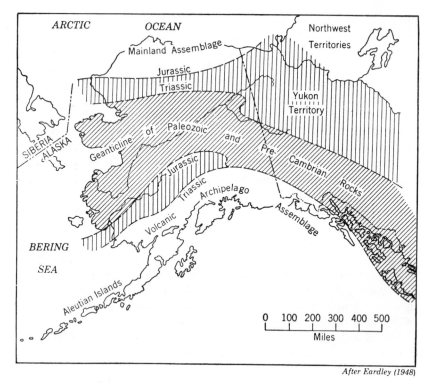

After Eardley (1948)

Fig. 4–15. Paleotectonic map of Alaska, showing regional structure during the Triassic and Jurassic periods. *Courtesy Journal of Geology.*

the size of the current potential gathering area through which oil may have moved before becoming trapped.

Paleostructure, paleotectonic,[55] *or palinspastic maps.*[56] Palinspastic maps show the structural or tectonic history of a region (Fig. 4–15). As will be seen later, under "Time of Accumulation" (Chapter 10), the dating of the earth movements is of utmost importance. Obviously a trap in which oil has accumulated must have been formed *before* the present oil deposit was caught. Woodward[57] describes three periods of folding in the Appalachians, in the pre-Silurian, Late Devonian, and

[55] A. J. Eardley, "Paleotectonic and Paleogeologic Maps of Central and Western North America," *Bull. Am. Assoc. Petrol. Geol.,* Vol. 33 (May, 1949), pp. 655–682.

[56] George Marshall Kay, "Paleogeographic and Palinspastic Maps," *Bull. Am. Assoc. Petrol. Geol.,* Vol. 29 (April, 1945), pp. 426–450.

[57] Herbert P. Woodward, "Multiple Folding in the Appalachian Basin," *World Oil* (December, 1957), pp. 110–113.

Late Paleozoic.   Each folding was adequate to produce its own set of potential structural traps.

The simplest type of paleostructure map is one constructed by eliminating the regional dip.[58]   This map shows the character of the folding before the regional tilting took place.   If that tilting was quite recent, the oil may not yet have become completely adjusted to the new structural environment, so a picture of the older structural situation may serve as a guide to oil.

Isopach maps.[59]   An isopach is a line drawn through points of equal thickness.   The subsurface geologist determines the thickness of a rock unit and enters these data on a map.   Isopach lines can then be drawn in the same manner as contours.

If the upper surface of the section included in a thickness or isopach map represents an originally horizontal surface of deposition rather than an erosion surface, then the isopach lines portray a *negative* picture of the sea floor at the start of deposition of the section mapped.   This negative picture shows hills or anticlines as hollows, and valleys or synclines are ridge-like.   In other words the negative portrays either a surface of erosion or a structure surface.   If the latter, the *next lower* stratigraphic surface should be parallel.

If the bottom and top surfaces of a section included in an isopach map are erosional unconformities and were once flat or nearly so, the map portrays the structure of the lower surface at the time of deposition of the upper surface.   If, however, either surface was eroded irregularly prior to deposition, the structural picture may be masked, owing to the presence of local channels in the sedimentary section.   As a general rule, isopach maps are reliable structural indicators in areas of folded rocks where erosional irregularities are relatively insignificant.   They are also useful in determining regional paleostructure in plains regions where the folding has been on a subdued scale.   However, the geologist must know when interpreting thickness maps covering local plains areas the degree to which both surfaces were dissected prior to deposition.

The sedimentary basins, wherein lies most of the world's oil, are identified and studied by means of isopach maps (Fig. 4–16).   The num-

[58] John Rich, "Graphical Method of Eliminating Regional Dip," and "Fault-Block Nature of Kansas Structures Suggested by Elimination of Regional Dip," *Bull. Am. Assoc. Petrol. Geol.*, Vol. 19 (October, 1935), pp. 1538–1543; G. D. Hobson, "The Application of Tilt to a Fold," *Proc. Geol. Assoc.*, Vol. 55 (Jan. 26, 1945), pp. 216–221.

[59] Wallace Lee, "Thickness Maps as Criteria of Regional Structural Movement," *Kansas Geol. Survey*, Bull. 109, Part 5 (1954), pp. 65–80; "Thickness Maps Can Reveal Mid-Continent Structures," *World Oil* (Aug. 1, 1955), pp. 77 et seq.; Jay B. Wharton, Jr., "Isopachous Maps of Sand Reservoirs," *Bull. Am. Assoc. Petrol. Geol.*, Vol. 32 (July, 1948), pp. 1331–1339.

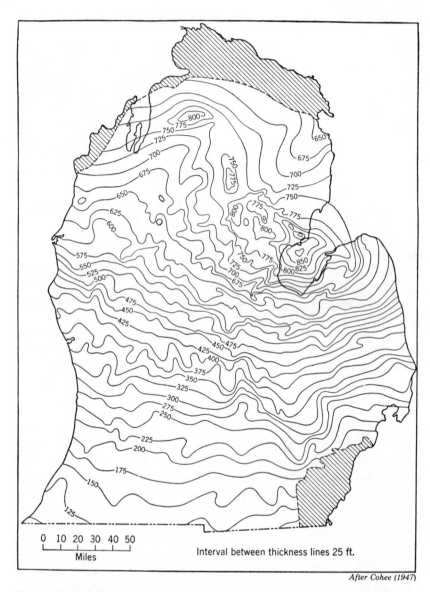

0 10 20 30 40 50
Miles

Interval between thickness lines 25 ft.

After Cohee (1947)

Fig. 4–16.   Isopach map, Traverse (Devonian) group, Michigan.   Thickness decreases southward from the center of the Michigan basin toward axes of Kankakee and Findlay arches and northward toward Precambrian shield.   Ruled areas show where Traverse rocks have been eroded.   *United States Geological Survey* (revised).

ber and the times of successive downwarps can be determined by isopaching different sedimentary intervals. These maps also delineate the areas of "pinch-out" of possible reservoir rocks. Such thinning-to-extinction is due in most places to planation and truncation prior to the deposition of the sediment overlying the upper surface of the isopach interval, but in some instances wedge-outs are actually original shore lines.

Isopach maps are very commonly employed in the search for oil. For example, a local reduction in thickness on an isopach map, if not caused by the presence of a hill on the lower surface or a channel into the upper surface, is in all probability an anticline or dome which was formed and truncated before the rock that overlies the zone covered by the isopach map was desposited.[60]

A special type of thickness map is the *convergence map*[61] which shows the direction and degree of the thinning (convergence) of a stratigraphic section. If this thinning is of considerable magnitude, the structure (and especially the position of the anticlinal axes) of the older rocks is quite different from that shown by the rocks overlying the converging unit. In order to determine the structural shift between the surface and the rocks underlying the converging section, it is necessary to have a surface structure map as well as a convergence map. The construction of the preconvergence-structure contour map is relatively simple and is based on the calculation and plotting of the deeper rock elevations where the higher structure contour lines intersect the isopach or convergence lines.

Isopach maps are also useful in determining the volume of a reservoir for the purpose of calculating petroleum reserves.

It has been pointed out[62] that an isopach map only shows the total thickness for whatever interval is mapped, and that in order to obtain the lateral changes for different geological times it is necessary to have a series of such maps. This necessity can be avoided by determining the percentage of thinning or thickening for different intervals between pairs of wells and plotting these data on a single sheet.

Facies maps.[63] Differences from place to place in the character of the physical sediment (lithofacies) or in the organic sediment (biofacies)

[60] Wallace Lee, "Relation of Thickness of Mississippian Limestones in Central and Eastern Kansas to Oil and Gas Deposits," *Kansas Geol. Survey Bull.* 26 (1939), 42 pp.

[61] Robert R. Wheeler and Robert M. Swesnik, "Stratigraphic Convergence Problems," *World Oil*, Vol. 130 (April, 1950), pp. 57–61; Robert M. Swesnik and Robert R. Wheeler, "Stratigraphic Convergence Problems in Oil Finding," *Bull. Am. Assoc. Petrol. Geol.*, Vol. 31 (November, 1947), pp. 2021–2029; A. I. Levorsen, "Convergence Studies in the Mid-Continent Region," *Bull. Am. Assoc. Petrol. Geol.*, Vol. 11 (July, 1927), pp. 657–682.

[62] A. H. Wadsworth, Jr., "The Percentage-of-Thinning Chart," *Oil and Gas Jour.* (March 2, 1953), pp. 72–73.

[63] W. C. Krumbein, "Regional and Local Components in Facies Maps," *Bull. Am. Assoc. Petrol. Geol.*, Vol. 40 (September, 1946), pp. 2163–2194; E. C. Dapples, "General Litho-

can be shown by means of facies maps (Fig. 4–17). The usual procedure is to convert the data into numbers, such as percentage figures, and then construct contour-like ("isopleth") maps by drawing lines through points of equal value. The initial data are obtained from cores or cuttings from wells, or from outcrop samples, or from both.

Krumbein[64] has listed a considerable number of sedimentary rock characteristics that have been or can be used in the construction of facies maps. Examples of lithofacies maps include those showing variations in grain size, porosity, permeability, heavy-mineral content, clay mineral content[65] sphericity of grains, magnesium content, insoluble residues, and degree of cementation. Maps which show lateral changes in sand or shale content are especially valuable. Variations in the character of contained fluids, such as the salt content of the connate water, and in the degree of saturation are readily shown by isopleth maps.

Among types of biofacies maps are those showing variations in percentage of a particular type of organism or in overall organic content.

Facies maps can be three-dimensional or two-dimensional. Figure 4–17 is three-dimensional; it shows the dominant lithology throughout the Cambrian section at any one place. Two-dimensional facies maps show the rock lithofacies or biofacies either immediately below or immediately above a formation contact. The latter may be called a worm's eye facies map (Fig. 4–18).

Facies are of great importance in problems of the origin and accumulation of oil. By means of facies studies the paleogeographic and paleo-

---

facies Relationship of St. Peter Sandstone and Simpson Group," *Bull. Am. Assoc. Petrol. Geol.*, Vol. 39 (April, 1955), pp. 444–467; T. H. Philpott, "Paleofacies—the Geologists New Tool," *Oil and Gas Jour.* (March 24, 1952), pp. 164 et seq.; Parke A. Dickey and Richard E. Rohn, "Facies Control of Oil Occurrence," *Tulsa Geol. Soc. Digest*, Vol. 23 (1955), pp. 227–232; L. L. Sloss, "Stratigraphic Analysis," *Oil and Gas Jour.* (Sept. 13, 1951), pp. 102 et seq.; L. W. LeRoy, "Stratigraphic, Structural, and Correlation Considerations," *Subsurface Geologic Methods* (Colorado School of Mines, 1950), pp. 22–32; R. C. Moore, E. D. McKee, S. W. Muller, E. M. Spieker, E. Wood, II, L. L. Sloss, W. C. Krumbein, and E. C. Dapples, "Sedimentary Facies in Geologic History," *Geol. Soc. Amer. Memoir 39* (June 17, 1949). W. C. Krumbein, "Sedimentary Maps and Oil Exploration," *N. Y. Acad. Sci. Trans.*, Ser. 2, Vol. 7 (May, 1945), pp. 159–166; W. C. Krumbein, "Lithofacies Maps and Regional Sedimentary-Stratigraphic Analysis," *Bull. Am. Assoc. Petrol. Geol.*, Vol. 32 (October, 1948), pp. 1909–1923; E. C. Dapples, W. C. Krumbein, and L. L. Sloss, "Tectonic Control of Lithologic Associations," *Bull. Am. Assoc. Petrol. Geol.*, Vol. 32 (October, 1948), pp. 1924–1947; W. C. Krumbein, L. L. Sloss, and E. C. Dapples, "Sedimentary Tectonics and Sedimentary Environments," *Bull. Am. Assoc. Petrol. Geol.*, Vol. 33 (November, 1949), pp. 1859–1891.

[64] W. C. Krumbein, "Sedimentary Maps and Oil Exploration," *N. Y. Acad. Sci. Trans.*, Ser. 2, Vol. 7 (May, 1945), pp. 159–166.

[65] Ralph E. Grim, "Clay Mineralogy and the Petroleum Industry," *World Oil* (March, 1951), pp. 61 et seq.

ecologic[66] picture can be brought into focus. On the regional scale the emergent and submergent areas at any one time in the geologic past can be identified; the type of emergent terrain and its relief estimated, the paleoclimatology determined, and the submergent areas classified as to water depth, nearness to shore, and sea floor topography. All of this information is grist for the mill in the study of possible source, reservoir, and seal rocks, and in the search for favorable trap areas, including especially permeability wedge-outs.

[66] H. S. Ladd, editor, "Marine Ecology and Paleoecology," *Geol. Soc. Am. Memoir* 67, Vol. 2 (1957).

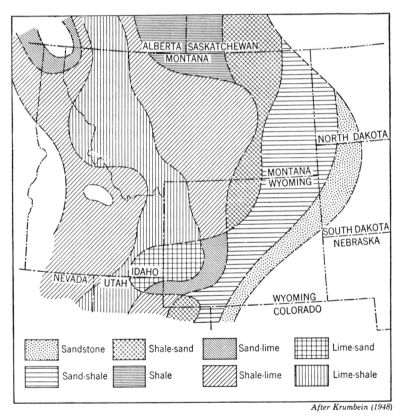

After Krumbein (1948)

Fig. 4-17. Lithofacies map showing character of Cambrian deposits in northern Rockies. *Courtesy American Association of Petroleum Geologists.*

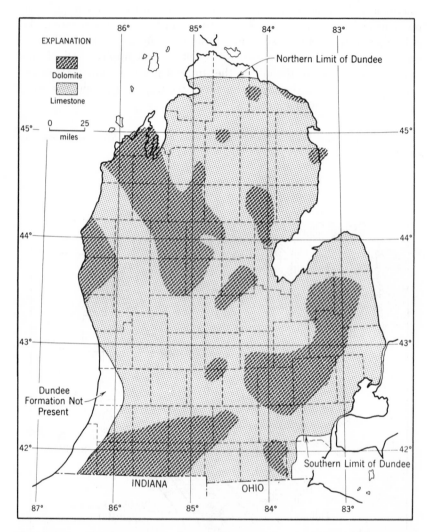

Fig. 4-18. Worm's eye map showing facies of basal Dundee formation in Southern Peninsula of Michigan. The basal Dundee rock immediately overlying the pre-Dundee unconformity is either dolomitic limestone or high-calcium limestone. *After Landes, 1951.*

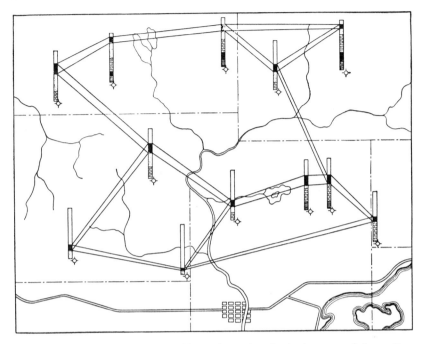

Fig. 4–19.   Log map.   The plotted log is drawn beside the location of the well on the map.   The relationship between convergence and other lateral changes and geographic position can be seen at a glance.   *Drawing by John Jesse Hayes.*

On the local scale facies studies may point out the direction to hitherto undiscovered sand bars, reefs, and porous dolomite[67] zones.   The facies changes in a shallow formation may even be a reflection of deeper anticlines and synclines.[68]

Cross sections.   By means of cross sections, the subsurface geologist can assemble in one drawing information which is spread over several maps.   A single cross section can be used to show surface topography, overburden thickness, the stratigraphic column, facies changes, convergence, unconformities, overlap, and the structure at various levels.

[67] Kenneth K. Landes, "Porosity Through Dolomitization," *Bull. Am. Assoc. Petrol. Geol.*, Vol. 30 (March, 1946), pp. 305–318; Richard Louis Jodry, "Rapid Method For Determining Magnesium-Calcium Ratios of Well Samples and Its Use in Predicting Structure and Secondary Porosity in Calcareous Formations," *Bull. Am. Assoc. Petrol. Geol.*, Vol. 39 (April, 1955), pp. 493–511.

[68] Richard L. Jodry, "Reflection of Possible Deep Structures by Traverse Group Facies Changes in Western Michigan," *Bull. Am. Assoc. Petrol. Geol.*, Vol. 41 (December, 1957), pp. 2677–2694.

The disadvantage of the cross section lies in its limitation to two dimensions. Therefore cross sections should always be used in conjunction with maps, and vice versa.

Cross sections are made by plotting the information obtained from wells and interpolating the formation boundaries and other data between wells. They can be plotted with sea level as a base, and then the topography and structure can be shown as they are today, or a formation boundary can be taken as a base in order to emphasize convergence or facies changes.

The usual American custom of exaggerating the vertical scale with respect to the horizontal scale of the cross section has been criticized, and with justification, by Suter.[69]   A true dip of 5° becomes 46° if the vertical scale is exaggerated 12 times, which is by no means unusual.   The only excuse for the increased vertical scale is to show stratigraphic detail which would be too compressed for visibility if true scale were used.

[69] H. H. Suter, "Exaggeration of Vertical Scale of Geologic Sections," *Bull. Am. Assoc. Petrol. Geol.*, Vol. 31 (February, 1947), pp. 318–339.

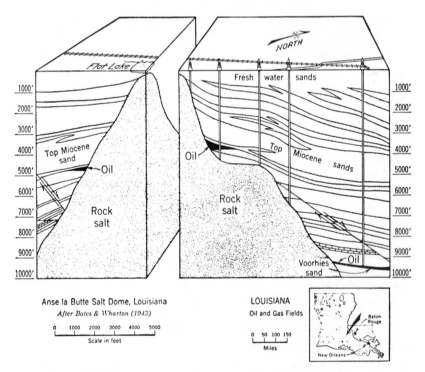

Anse la Butte Salt Dome, Louisiana
After Bates & Wharton (1943)

0   1000   2000   3000   4000   5000
Scale in feet

LOUISIANA
Oil and Gas Fields

0   50 100 150
Miles

Baton Rouge

New Orleans

Fig. 4–20. Block diagram.   Anse la Butte Salt Dome, Louisiana.   *Courtesy Colorado School of Mines*

Miscellaneous methods of geologic illustration include the log map[70] (Fig. 4–19), the block diagram (Fig. 4–20), the structure diorama,[71] and the peg model, which has a wooden rod on which the formations are marked off in color, for each well. The block diagram was used most effectively by Bass[72] to show reservoir sand thickness. Each well is the intersection of two sand "fences" running normal to each other (Fig. 14–10).

Trap Finding. The principal reason for the employment of petroleum geologists is their ability to find deposits of oil. Inasmuch as no direct method of oil finding has been perfected as yet, petroleum geologists concentrate mainly on the hunt for buried traps in which oil or gas may be hiding. In Chapter 3 we reviewed the methods, geological and geophysical, by means of which traps are sought through data obtained at the surface without recourse to drilled wells. Those methods have been most successful in discovering anticlinal and other types of structural traps, but they have not been consistently successful in the search for varying permeability ("stratigraphic") traps.

Subsurface geology increases each year in relative value as a trap-finding method as (1) surface developed prospects are used up, and (2) more subsurface information becomes available through the continued drilling of exploration wells. The successes of subsurface geology can be considered dividends resulting from the expenditures made on dry holes. Although applications of subsurface geology have resulted in the discovery of many structural traps, its greatest promise, perhaps because it is the only consistently proved method at the present time, is in the search for traps of the stratigraphic type.

The major handicap of trap hunting by subsurface geology is that it is unable to pinpoint the location, as surface methods can do for anticlines. Therefore oil hunting by subsurface geology calls for the drilling of more exploratory wells to probe the rock section, and hence greater expense per barrel of oil discovered thereby.

Trap searching by means of subsurface geology involves first the collection of all available data and second the digestion of those data and their projection on maps, charts, diagrams and sections. It has already been pointed out, under each individual heading, how these illustrations point

[70] T. H. Bower, "Log Map, New Type of Subsurface Map," *Bull. Am. Assoc. Petrol. Geol.,* Vol. 31 (February, 1947), pp. 340–349.

[71] Frederick Squires, "The Structure Diorama (Lawrence County, Illinois)," *Oil and Gas Jour.,* Vol. 43 (Aug. 19, 1944), pp. 86–93.

[72] N. W. Bass, "Origin of the Shoestring Sands of Greenwood and Butler Counties, Kansas," *State Geol. Survey of Kansas Bull.* 23 (September, 1936), p. 76.

the finger at the general vicinity of possible oil-bearing traps, both structural and stratigraphic. The superposition of the pertinent data on a single map may help in delimiting the areas of most promise.[73]

After the trap has been found, and oil discovered therein, production geology plays an important role in the subsequent development and secondary recovery activities.

[73] C. W. Shenkel, "Superposed Geologic Data as an Exploration Tool (abstract)", *Bull. Am. Assoc. Petrol. Geol.*, Vol. 40 (February, 1956), p. 428.

5

ORIGIN AND EVOLUTION
OF OIL AND GAS[1]

Petroleum geologists have written more on the origin of petroleum than on any other subject.[2] Speculation in this field was initiated in the 1860's, and the problem of origin is still with us. "Each year the crop of new papers adds to the fearsome pile threatening to bury the subject by the very dead weight of diversified opinion, and the earnest student, anxiously searching for enlightenment, gets lost in the maze of conflicting opinions."[3] This prodigious amount of research and discussion has not, however, been entirely in vain. Many theories have been tested and found wanting. Although many more theories have appeared to replace those discarded, it is but a matter of time before these too are tested and the inadequate ones rejected, until a fairly clear picture is obtained of the origin and evolution of oil and gas. "The 'mysteries' relative to petro-

---

[1] William B. Heroy, Jr. "Petroleum Geology," *Geol. Soc. Am. 50th Anniversary Vol.* (1941), pp. 512–548; V. C. Illing, "Origin of Petroleum, *Science of Petroleum* (Oxford University Press, 1938), Vol. 1, pp. 32–38; U. S. Geol. Survey, "Geology and Occurrences of Petroleum in the United States," *Petroleum Investigation,* Hearings, 73d Congress, H.R. 441 (1934), pp. 895–910; E. DeGolyer and Harold Vance, "Petroleum Bibliography, *Bull. Agr. and Mech. Coll. Texas,* 83 (1944), pp. 330–338; A. G. Skelton and M. Skelton, "A Selected Bibliography of the Theories of the Origin of Petroleum," *Oklahoma Geol. Survey, Mineral Rept.* 7 (1942); G. D. Hobson, "Petroleum Geology," *Rev. of Petrol. Technol.* (Institute of Petroleum, London, 1947), Vol. 7, pp. 1–16; F. Morton, "Origin of Petroleum," *Petrol. Engineer* (January, 1953), pp. A-60 et seq; Nelson P. Stevens, "Origin of Petroleum—a Review," *Bull. Am. Assoc. Petrol. Geol.,* Vol. 40 (January, 1956), pp. 51–61.

[2] G. M. Knebel, "Progress Report of API Research Project 43, The Transformation of Organic Material into Petroleum," *Bull. Am. Assoc. Petrol. Geol.,* Vol. 30 (November, 1946), p. 1945.

[3] V. C. Illing, "Geology of Petroleum," *Jour. Inst. Petrol. Technologists,* Vol. 21 (1935), p. 491.

leum formation are largely disappearing in the face of more and better chemical, geological, and bacteriological information." [4]

The last decade has seen considerable advance in our understanding of petroleum genesis. This fact has been due to (1) developments in techniques, and (2) adequately financed industry-sponsored research. One relatively new technique is the immediate freezing of samples of recent sediment. This practice prevents destruction of organic matter by bacteria prior to the sample analysis. Chromatography permits more accurate determination of minute percentages of hydrocarbons than has been possible heretofore. Lastly, the age in years of geologically recent carbon compounds can be found by means of carbon 14 determinations. As a result it is now possible to analyze sediments and rocks for their hydrocarbon content more accurately than ever before, and to distinguish older petroleum from recent organic matter.

Industry has subsidized investigations in petroleum genesis through individual company research laboratory activities (which are referred to subsequently) and through research projects established by trade associations such as the American Petroleum Institute (API). Especially noteworthy in the latter category are API Research Project 6, "The Composition of Petroleum," [5] of which further mention is made in Chapter 7, API Project 43, "Transformation of Organic Material into Petroleum," [6] and API Project 51, "Study of Near-Shore Recent Sediments and Their Environments in the Northern Gulf of Mexico." [7]

Investigation of petroleum genesis is not purely academic exercise. When we can determine exactly under what conditions commercial deposits of oil are formed and the date of their formation, we can confine our explorations for oil to (1) places that meet those conditions, and (2) to "traps" of oil that were in existence at the proper time. [8]

PROBLEMS AND STAGES IN THE
NATURAL HISTORY OF OIL AND GAS

The problems in analyzing the birth and evolution of oil and gas, and the stages involved in that evolution, are several. The more important ones are:

[4] Benjamin T. Brooks, letter addressed to Dr. Robert W. Schiessler, July 7, 1955.

[5] Frederick D. Rossini, "Hydrocarbons from Petroleum," *Jour. Inst. Petrol.*, Vol. 44 (April, 1958), pp. 97–107.

[6] "Fundamental Research on Occurrence and Recovery of Petroleum, 1950–1951," *Am. Petrol. Inst.* (1952), pp. 98–225; 1952–53 (1954), pp. 151–220.

[7] *Ibid*, 1950–51 (1952), p. 256; 1952–53 (1954), pp. 241–371; 1954–55 (1956), pp. 91–340.

[8] Harold W. Hoots, "Origin, Migration and Accumulation of Oil in California," *Calif. Div. Mines Bull.* 118 (August, 1941), p. 259.

1. The source of the elements (hydrogen and carbon).
2. The accumulation of the source materials.
3. The burial of the source materials.
4. The transformation from solid to liquid (or gas).
5. Evolution of petroleum.

These five topics cannot be given in their exact chronological order, for some overlap may occur between stages, and no one knows for sure just when the transformation from solid to liquid does take place.

The problem of petroleum genesis is made difficult by the very complexity of petroleum itself (Chapter 7). Crude oil "contains molecules ranging from one to fifty or more carbon atoms per molecule in size, comprising a diversity of types."[9] By mid-1957, one hundred eight hydrocarbon compounds had been isolated from the gas and gasoline fraction alone of a crude oil. Furthermore it is believed that "all petroleums contain substantially the same hydrocarbon compounds" and within any one of five main classes of hydrocarbons "the individual compounds occur in proportions which are usually of the same order of magnitude for different petroleums."[10] Any theory of petroleum origin has to be compatible with this *uniformity of complexity* of natural petroleums.

The solution of the problem is definitely in the bailiwick of the chemist according to Brooks: "I am firmly convinced that the formation of petroleum is essentially a *chemical problem* and that geologists have already determined the conditions (such as the framework of environment) *within* which the chemical processes have to be explained."[11]

SOURCE OF HYDROGEN AND CARBON[12]

Primary Source. There can be but little doubt that the carbon and hydrogen in petroleum and natural gas were, in their initial stages, like everything else on this earth in solution in hot liquid rock or magma. Methane, CH_4, the most stable hydrocarbon, has been detected in volcanic gas. Meteorites, samples of space rock furnished us by Mother Nature, may contain hydrocarbons.[13] The spectra of Neptune, Jupiter, Saturn, and Uranus show the presence of methane in (fittingly) astronomic

---

[9] Frederick D. Rossini, *op. cit.*, p. 97.

[10] *Op. cit.*, p. 103.

[11] Benjamin T. Brooks, *letter* dated Aug. 12, 1958.

[12] F. W. Clarke, "Data of Geochemistry," *U. S. Geol. Survey Bull.* 770 (1924), pp. 744–755.

[13] Theodore A. Link, "Whence Came the Hydrocarbons?", *Bull. Am. Assoc. Petrol. Geol.*, Vol. 41 (July, 1957), pp. 1387–1402.

quantities.[14] Within the igneous rocks of the earth's crust hydrogen and carbon occur combined in mineral form, but not in the quantity and concentration needed for commerical exploitation. Examples are the black hydrocarbons (referred to by mineralogists as "thucolite") that have been found in pegmatites in minute amounts.

The question facing the petroleum geologist is whether or not originally magmatic (inorganic) hydrogen and carbon were removed by plant or animal before conversion into petroleum. In other words, did the carbon and hydrogen pass through an organic phase in the natural history of crude oil? The following sections describe five lines of evidence which should be considered before reaching an answer to this question.

Synthesis of Hydrocarbons from Inorganic Raw Materials. It has been frequently demonstrated in the chemical laboratory that hydrocarbons may be obtained from (1) carbon dioxide, free alkali metal, and water and (2) metallic carbides and water. The carbon dioxide, alkali, and water method was first demonstrated by Berthelot in 1866. Mendeleef is credited with obtaining hydrocarbons from water and carbides in 1877. Since then both these syntheses have been repeated many times. These are the two principal methods of obtaining hydrocarbons from inorganic materials. It should be noted that both procedures require substances (free alkali metal and metallic carbides) that are unknown or at least as yet undiscovered in nature. It is true that the earth does have a heavy center, but studies of the earth's magnetism and of meteorites lead us to believe that this higher specific gravity is due to the concentration of iron and nickel in the core of the earth rather than a free alkali metal or iron carbide. Furthermore, the rock pressures are of such magnitude below the earth's crust that permeability disappears with depth. For this reason, it is impossible to visualize any transfer of materials between the earth's center and the upper parts of the earth's crust.

Natural Synthesis of Hydrocarbons Within Organisms.[15] Hydrocarbons can be and are produced by normal life processes on the surface of the earth. The isolation in minute amounts of various hydrocarbons from lichens, algae, higher plants, barnacles, corals, insects, worms, fishes, and higher animals including man has been reported.[16] In fact fatty oils are produced commercially in quantity from a number of

[14] Wallace E. Pratt, "Discussion: Whence Came the Hydrocarbons?", Am. Assoc. Petrol. Geol., Vol. 41 (November, 1957), p. 2584; C. E. Van Orstrand, "Cosmic Origin of Oil and Gas," World Oil, Vol. 128 (November, 1948), pp. 150–158.

[15] F. C. Whitmore, Review of API Research Project 43B, Research on Occurrence and Recovery of Petroleum (American Petroleum Institute, 1943) p. 124; Werner Bergmann, "Coral Reefs May Add to Petroleum Supply Some Day," Science News Letter (Oct. 30, 1948), p. 281.

[16] P. V. Smith, Jr., "Status of Our Present Information On the Origin and Accumulation of Oil," Proc. Fourth World Petrol. Congress (1955), Section 1, pp. 359–376.

plants and animals. Examples are whale oil, cod liver oil, castor and soy bean oil, and palm oil.

Photosynthesis. Perhaps the most prolific organisms in internal hydrocarbon production are the green plants because of their ability to photosynthesize when exposed to light. Light causes green plants to take up carbon dioxide and water to produce organic carbon compounds plus oxygen. "In the leaves of some dicotyledons the organic compound can be detected as starch; in other leaves sugar accumulates; in some algae fats are formed."[17] The possibility that at least some algae can produce fats by photosynthesis may be extremely important, for these plants are most abundant in plankton.

Artificial Synthesis of Hydrocarbons from Organic Matter. In addition to the previously mentioned plants and animals that can be tapped or squeezed to yield natural oils, hydrocarbons can be produced by synthesis from any carbonaceous matter including fish, farm waste, peat, coal, and oil shale. Petroleum products are being produced commercially today from coal and oil shale in various parts of the world.

The fact that oil *can be* produced from inorganic materials is therefore not a valid argument that it *has come* from inorganic sources, for it can also be synthesized from organic remains.

Testimony of Oil and Gas. Chemical analyses of oil and gas and optical studies of oil have yielded considerable information, some of which appears to have a bearing on the origin of oil. A few natural gases contain helium, a product of radioactive disintegration. It is easier to explain this element as a wanderer from disintegrating radioactive minerals caught in the same trap with natural gas, than to tie its occurrence in with any organic cycle. On the other hand, the presence of nickel and vanadium in some petroleum cokes can probably be best explained by an organic background, for analyses of the ashes left after some types of plants have been burned show the presence of these metals. Such plants appear to remove the metals from the surrounding soil, thereby concentrating them. Other indications of plant origin include the presence of nitrogen compounds in some oils and the optical activity of petroleum hydrocarbons.

Sanders[18] and Waldschmidt[19] have described such microscopic objects as Foraminifera, diatoms, plant remains, insect scales, and spines, as well as fragments of other materials, present in crude oils. These dis-

---

[17] C. P. Whittingham, "Photosynthesis," *Endeavor*, Vol. 14 (October, 1955), pp. 173–180.

[18] J. McConnell Sanders, "The Microscopical Examination of Crude Petroleum," *Jour. Inst. Petrol. Technologists*, Vol. 23 (1937), pp. 525–573.

[19] W. A. Waldschmidt, "Progress Report on Microscopic Examination of Permian Crude Oils," *Program 26th Annual Convention, Am. Assoc. Petrol. Geol.* (1941), p. 23.

coveries may or may not be significant in terms of the origin of oil. It may be that the petroleum in the course of its travels through sedimentary rocks rich in organic materials picked up the microscopic objects.

Of greatest importance to petroleum genesis was the announcement by Treibs[20] in 1934 of the discovery of chlorophyll porphyrins in pigments in petroleum. Chlorophyll porphyrins are also found in bituminous shales, coal, and asphalt. This "is conclusive evidence that the original organic source material of petroleum was associated with green algae or other marine plant forms, that the organic debris was rather quickly protected from oxidation and maintained under anaerobic conditions, and that the persistence of these chlorophyll porphyrins proved a relatively low-temperature history throughout."[21] The importance of low temperature history of crude oil will be discussed later in this chapter.

Testimony of Associated Rocks. The geological associations of oil and gas have led practically all geologists to reject the inorganic theories as entirely inadequate. Over 99 per cent of the world's oil and gas so far produced has come from sedimentary rocks. Furthermore, in every oil-producing region the sedimentary rock section includes beds which either contain or have contained considerable organic material. In fact, Heald[22] expresses the belief that the distribution of oil within the sedimentary series is dependent upon the localization of source material within the rock column. It is his belief that the oil occurring in the beds lying above and below widespread water-bearing rocks must have had its source in that interior zone.

One would expect the crystalline rocks, when porosity exists, to contain more oil than the sedimentary rocks if the source of the oil is inorganic. As a matter of fact, there are instances of oil occurring in commercial quantities in crystalline rocks. "More than 15 million barrels of oil have been produced from igneous and metamorphic rocks; one gas field produces from a basalt flow; and millions of tons of asphalt . . . are known in serpentine in one area . . ."[23] Oil is being produced commercially today

[20] Alfred Treibs, "Chlorophyll und Haminderivate in bituminosen Gesteinen, Erdolen, Erdwachsen und Asphalten; Ein Beitrag zur Enstehung des Erdols," *Annalen der Chemie,* Band 510 (1934), pp. 42–62; Alfred Treibs, "Chlorophyll- und Hamin-Derivate in bituminosen Gesteinen, Erdolen, Kohlen, Phosphoriten," *Annalen der Chemie,* Band 517 (1935), pp. 172–196; Alfred Treibs, "Chlorophyll- und Haminderivate in organischen Mineralstoffen," *Angewandte Chemie,* Band 49 (1936), pp. 682–686.

[21] Benjamin T. Brooks, "Origin of Petroleums," *The Chemistry of Petroleum Hydrocarbons,* (Reinhold, New York, 1954), Chapter 6, pp. 83–102.

[22] K. C. Heald, "Essentials for Oil Pools," *Elements of the Petroleum Industry* (American Institute of Mining and Metallurgical Engineers, 1940), p. 26.

[23] Sidney Powers and F. G. Clapp, "Nature and Origin of Occurrences of Oil, Gas and Bitumen in Igneous and Metamorphic Rocks," *Bull. Am. Assoc. Petrol. Geol.,* Vol. 16 (August, 1932), p. 719.

from fissures in Precambrian granite in Kansas[24] and pre-Cretaceous schists in California.[25] Many other examples could be cited,[26] but in every instance organic sedimentary rocks which could have been the source of the bituminous material occur nearby, frequently, at a lower elevation.

Conclusions. To a geologist the conclusion is inescapable that the petroleum hydrocarbons passed through an organic stage. That the literature on the inorganic origin of petroleum has grown to such proportions is apparently due to the failure of geologists to emphasize to the lay public the geological environment of petroleum and to the ignoring by the chemists of this geological environment. "If . . . it had been laid down, as it might have been, that no theory on the origin of petroleum was admissible unless it explained the close association between oil and sediments most of the chemical theories on its origin would not have been put forward."[27] Chemist Brooks concurs in this statement, but adds: "Some of the theories proposed in the past by geologists without chemical knowledge are just as bizarre."[28] Touché!

POSSIBLE SOURCE ORGANISMS

It is not possible to name, with any certainty, the particular organisms that produce the oil in any given oil reservoir. We do suspect that any type of organism containing fatty oils or fatty acids is potential source material. However, between the time of the death of the organism and its transformation to oil many changes take place, during which the organic material tends to lose its identity.

We are able, by following two lines of reasoning, to eliminate from further consideration as possible source material some rather broad classes of plant and animal life. One of these directions of reasoning involves the timing factor. It is obviously impossible for ancient oils to have been

---

[24] Robert F. Walters, "Oil Production from Fractured Precambrian Basement Rocks in Central Kansas," *Bull. Am. Assoc. Petrol. Geol.*, Vol. 37 (1953), pp. 300–313.

[25] John H. Beach, "Geology of Edison Oil Field, Kern County, California," *Structure of Typical American Oil Fields* (American Association of Petroleum Geologists), Vol. 3 (1948), pp. 58–85.

[26] Sidney Powers et al., "Symposium on the Occurrence of Petroleum in Igneous and Metamorphic Rocks," *Bull. Am. Assoc. Petrol. Geol.*, Vol. 16 (August, 1932); F. M. Van Tuyl and Ben H. Parker, "The Time and Origin of Petroleum," *Quarterly Colo. School of Mines*, Vol. 36 No. 2 (April, 1941), pp. 145–146.

[27] V. C. Illing, "Geology Applied to Petroleum," *Oil Weekly*, Vol. 122 (July 15, 1946), p. 36.

[28] Benjamin T. Brooks, letter dated July 28, 1955.

derived from modern organisms. "Inasmuch as petroleum is found in rocks that were formed millions of years before there were such complex forms of life as trees, land animals, or even fishes, it is apparent that very simple forms of life sufficed to provide materials which might be transformed into petroleum . . ."[29] Another logical factor for consideration is that of opportunities for survival. If an organism lived under such environmental conditions that the chances of its existing long enough after death to become buried in sediment were very slight, then the chances of that organism's being source material for petroleum also would be very slight.

These limitations immediately rule out non-aqueous land plants and animals from further consideration. Most of them did not exist at the time the Ordovician oil, for example, was formed. For organisms raised on the land, the chances of escaping destruction by putrefaction or devourment by scavengers are very slight. Possible exceptions are land organisms that can be readily transported by wind, like spores, pollen, and leaves. If these materials settle into bodies of water, they may join the muck on the sea or lake floor and thus become preserved. Organisms living in fresh waters, especially lakes, are of course another story. If they become preserved in sediment, they have just as much chance of becoming crude oil as do aqueous marine forms. Studies of carbon isotopic compositions of petroleums have shown that crude oils from marine organisms can be distinguished from those derived from freshwater source materials. This fact in turn is evidence that petroleum can come from non-marine organic matter.[30]

Marine life, except the vertebrates, was abundant and fairly well advanced by early Paleozoic time, and it is in these phyla that we suspect the greater part of the source organisms of petroleum belong.

The one-celled plants and animals, such as the Algae and Foraminifera, were abundant in the seas from the earliest Paleozoic to the present and could have contributed to the oil deposits of all ages. The more advanced types of marine life, besides being less abundant in the more remote geologic periods, also had lesser opportunity for survival because of the presence and vigor of sea-floor scavengers. It is difficult to visualize the accumulation of the soft body parts of fish (fairly common from the Devonian on), shellfish, or even corals in enough abundance on the sea floor to provide a quantitatively sufficient source of oil.

[29] F. C. Whitmore, Review of API Research Project 43B, *Research on Occurrence and Recovery of Petroleum* (American Petroleum Institute, 1943).

[30] Sol R. Silverman and Samuel Epstein, "Carbon Isotopic Compositions of Petroleums and Other Sedimentary Organic Materials," *Bull. Am. Assoc. Petrol. Geol.*, Vol. 42 (May, 1958), pp. 998–1012.

More and more investigators of petroleum source materials are concentrating on marine phytoplankton (floating plants) as the probable mother substance for most of the world's oil. This material has many points in its favor:

1. Plants, as the primary source of food for all animals, are more abundant than animals.[31]

2. Phytoplankton is the most abundant life in the oceans. It is also most abundant in lakes, but the total quantity there is, and probably always has been, relatively insignificant compared with marine life.

3. Phytoplankton is most prolific in the near shore areas where river-brought nutrients are available; this is also the area of greatest sedimentation.

4. Due to photosynthesis at least some of the plants in phytoplankton have produced fatty oils prior to deposition.

5. Phytoplankton consists predominantly of microscopic one-celled forms which have been on earth since earliest Paleozoic and probably before that.

Nothing in this concept rules out intermixture of other life types. The sea-floor muck includes not only the remains of planktonic material (including animal forms) which settle to the bottom, but also other "waste products of the biota inhabiting the environment in which the source beds accumulate."[32]

It has long been the custom to ascribe differences in petroleums to differences in the relative abundance of the various types of source organisms.[33] Thus organisms rich in fats would be expected to produce a paraffinic oil, whereas protein bearing organisms would yield an asphaltic oil. Actually, the present trend in thinking is contrary to any concept of greatly dissimilar raw material. Although it may be that very young oils owe some of their differences in chemical composition and physical properties to differences in the predominant type of source material, it is also probable that differences in nature of mature oils are due more to time and environment than to initial differences in source material.[34] In other

---

[31] Nickola Propopovich, "Primary Sources of Petroleum and Their Accumulation," *Bull. Am. Assoc. Petrol. Geol.*, Vol. 36 (May, 1952), pp. 878–883.

[32] Parker D. Trask et al., *Origin and Environment of Source Sediments of Petroleum* (Gulf Publishing Co., Houston, Texas, 1932), p. 234.

[33] Joseph A. Taff, "Physical Properties of Petroleum in California," *Problems of Petroleum Geology* (American Association Petroleum Geologists, 1934), p. 177; David B. Reger, "Gravity of Oils in the Appalachian Province," *Problems of Petroleum Geology* (American Association of Petroleum Geologists, 1934), p. 107.

[34] This conclusion is contrary to that reached by L. C. Snider, "Current Ideas Regarding Source Beds," *Problems of Petroleum Geology* (American Association Petroleum Geology, 1934), p. 63.

words, two oils from dissimilar source materials, having similar post-generation histories, would tend to become similar in nature.

TIME OF OIL GENERATION

The student of petroleum genesis finds himself dealing with at least five different ages: (1) the age of the source materials; (2) the age of the oil itself; (3) the age of the trap in which the oil accumulates; (4) the age of the actual accumulation; and (5) the age of the reservoir rock. We know least about the first four of these ages, and yet they are of the utmost importance in oil finding. Ordinarily the age of the producing formation is known, and for the sake of convenience that date is commonly accepted for the oil itself, but actually there is not much justification for this practice.

The following paragraphs contain a discussion of the times at which conversion from solid organic matter to liquid petroleum may take place; the other four ages are considered in subsequent sections.

Geologists differ widely in their opinions regarding time of conversion. Evidence is growing that hydrocarbon compounds produced by organisms, especially plankton, are carried down to the seafloor when the minute plants or animals die and become incorporated within the fine sediment accumulating there. This is the syngenetic ("born with") theory. Next are those who believe that oil is epigenetic, formed later than the enclosing sediments but before lithifaction took place. Last is the postlithifaction school, some members of which believe that oil is being formed in buried rock strata today.

The determination of age in years by the carbon 14 method *can be applied only to geologically recent material.* It has been determined that two California crude oils from Upper Pliocene reservoirs are in one instance "older than 27,780 years" and in the other "older than 24,000 years."[35] Hydrocarbons extracted from sediments collected off the Louisiana coast proved to be in the 12,000–13,000 year age range.[36] A separate Gulf sediment study found that the samples examined ranged in age from about 3000 to 10,000 years, whereas similar hydrocarbons found in soils proved to be less than 500 years old.[37] A 5000-year age has been reported from an oil-bearing sand in northeastern Venezuela.[38] All of

[35] W. F. Lilley, "Chicago Radiocarbon Data," *Science*, Vol. 116 (Dec. 19, 1952), p. 677.

[36] P. V. Smith, Jr., "Studies on Origin of Petroleum: Occurrence of Hydrocarbons in Recent Sediments," *Bull. Am. Assoc. Petrol. Geol.*, Vol. 38 (March, 1954), p. 383.

[37] Nelson P. Stevens, Ellis E. Bray, and Ernest D. Evans, "Hydrocarbons in Sediments of Gulf of Mexico," *Bull. Am. Assoc. Petrol. Geol.*, Vol. 40 (May, 1956), pp. 975–983; Reprinted in *Habitat of Oil* (American Association of Petroleum Geologists Special Vol., 1958), pp. 779–789.

[38] "How Long Does it Take to Form Crude Oil?", *Oil and Gas Jour.* (March 11, 1957), p. 263.

this proves that some hydrocarbons are very young; it does not disprove that some petroleums may be very old.

The possible causes of transformation will be considered later. Only the dating of the oil generation will be discussed in this section.

Syngenetic Oil. If oil is syngenetic with the sediments, the liquid hydrocarbons must have been formed (or have been already present) while the sediments were accumulating on the sea floor. It can be readily demonstrated by experimental methods that petroleum dispersed through fine sediment is both carried to the sea floor and retained there by that associated sediment.[39] "Placer" deposits of oil are therefore a possibility. Corbett[40] believes that the heavy oil ("tar") of the Athabaska, Alberta, area was deposited with the enclosing sand.

The use of dry ice to keep samples of sea-floor sediment frozen between time of collection and analysis has led to the discovery of widespread hydrocarbons in such sediment. Samples collected prior to the use of this technique may have lost their original hydrocarbons because of the activity of hydrocarbon oxidizing bacteria.[41]

By means of modern analytical procedures, including chromatographic, infra-red, and mass-spectra methods, investigators working with samples obtained from beneath the sea floor in the Gulf of Mexico,[42] California,[43] and Russia,[44] as well as from Minnesota[45] and Wisconsin[46] lake sedi-

---

[39] F. M. Van Tuyl and Ben H. Parker, "The Time of Origin and Accumulation of Petroleum," *Quarterly, Colo. School of Mines*, Vol. 36 No. 2 (April, 1941), p. 134, includes reference to: (1) Murray Stuart, "Sedimentary Deposition of Oil," *Indian Geol. Survey Records*, Vol. 40 (1910), pp. 320–333; and (2) F. M. Van Tuyl and C. F. Barb, "Notes on the Sedimentary Deposition of Petroleum," *Mines Mag.*, Vol. 24 (October, 1934), pp. 19–20; O. A. Poirier and George A. Thiel, "Deposition of Free Oil by Sediments Settling in Sea Water," *Bull. Am. Assoc. Petrol. Geol.*, Vol. 25 (1941), pp. 2170–2180.

[40] C. S. Corbett, "In Situ Origin of McMurray Oil of Northeastern Alberta and Its Relevance to General Problem of Origin of Oil," *Bull. Am. Assoc. Petrol. Geol.*, Vol. 39 (August, 1955), pp. 1601–1620.

[41] Claude E. ZoBell, C. W. Grant, and H. F. Haas, "Marine Microorganisms Which Oxidize Petroleum Hydrocarbons," *Bull. Am. Assoc. Petrol. Geol.*, Vol. 27 (September, 1943), p. 1189.

[42] P. V. Smith, Jr., "Studies on Origin of Petroleum: Occurrence of Hydrocarbons in Recent Sediments," *Bull. Am. Assoc. Petrol. Geol.*, Vol. 38 (March, 1954), pp. 377–404; P. V. Smith, Jr., "Status of Our Present Information on the Origin and Accumulation of Oil," *Proc. Fourth World Petrol. Congress* (1955), Section 1, pp. 359–376; Nelson P. Stevens, Ellis E. Bray, and Ernest D. Evans, "Hydrocarbons in Sediments of Gulf of Mexico," *Bull. Am. Assoc. Petrol. Geol.*, Vol. 40 (May, 1956), pp. 975–983.

[43] Wilson L. Orr and K. O. Emery, "Composition of Organic Matter in Marine Sediments: Preliminary Data on Hydrocarbon Distribution in Basins off Southern California," *Bull. Geol. Soc. Am.*, Vol. 67 (September, 1956), pp. 1247–1258.

[44] IU. N. Petrova and I. P. Karpova, "Chemical Composition of Dispersed Organic

ments, have shown the consistent presence of minute but measurable percentages of liquid hydrocarbons. The same has been true of soils.[47] Furthermore the percentages, when multiplied by cubic miles of sediment, result in an enormous volume of liquid hydrocarbons.

Thus Smith[48] has found paraffinic, napthenic, and various aromatic components in cores taken from the floor of the Louisiana Gulf Coast in proportions ranging up to a maximum of 11,700 parts per million of dried sediment. When translated to barrels of hydrocarbons per cubic mile of sediment the maximum figure becomes 10,400,000, or slightly over 3 barrels per acre foot.

Somewhat similar calculations were made by Orr and Emery[49] from samples collected in three sea floor basins off southern California. Hydrocarbons made up 2 to 19 per cent of the extractable organic matter, or 0.003 to 0.038 per cent of the total dry sediment. Even these small percentages indicate a potential liquid hydrocarbon yield from these basins equal to 10 times the anticipated eventual petroleum recovery from the oil fields of the Los Angeles basin. However, if the same off shore sediments were subjected to pyrolysis (high temperature decomposition) the yield would be greater by a factor of 10.

Stevens, Bray, and Evans[50] point out that the complex hydrocarbon-asphaltic mixtures which they found in soils and recent marine sediments in the Gulf of Mexico "differ significantly from those comprising crude oils." Therefore if these hydrocarbons are pre-petroleums they are subject to later changes.

Orr, Emery, and Grady[51] investigated the loss en route of chlorophyll derivatives between the live phytoplankton in the sea and the sediments on the sea floor. The object was to account for the porphyrins found in petroleum. "If as much as 1 per cent of the pigments preserved in the surface zones of basin sediments are eventually converted to the porphyrins

---

Matter in Sedimentary Rocks," Summary by George V. Chilingar, *Bull. Am. Assoc. Petrol. Geol.*, Vol. 39 (July, 1955), pp. 1417–1419.

[45] Frederick M. Swain, "Stratigraphy of Lake Deposits in Central and Northern Minnesota," *Bull. Am. Assoc. Petrol. Geol.*, Vol. 40 (April, 1956), pp. 600–653.

[46] Sheldon Judson and Raymond C. Murray, "Modern Hydrocarbons in Two Wisconsin Lakes," *Bull. Am. Assoc. Petrol. Geol.*, Vol. 40 (April, 1956), pp. 747–761.

[47] Nelson P. Stevens et al, *op. cit.*

[48] P. V. Smith, Jr., *op. cit.*

[49] Wilson L. Orr and K. O. Emery, *op. cit.*

[50] Nelson P. Stevens, Ellis E. Bray, and Ernest D. Evans, *op. cit.*

[51] Wilson L. Orr, K. O. Emery, and John R. Grady, "Preservation of Chlorophyll Derivatives in Sediments Off Southern California," *Bull. Am. Assoc. Petrol. Geol.*, Vol. 42 (March, 1958), pp. 925–962.

found in petroleum, the amounts are ample to furnish crude oils with a normal porphyrin content."

We can conclude that there are disseminated liquid hydrocarbons in sea and lake sediment which are not reworked (placer) deposits of petroleum. Whether these hydrocarbons are the result of metabolism of the organisms, and are therefore truly syngenetic, or are the result of diagenetic processes, which would place them in the next (prelithifaction oil) category, we do not know. We do know that the hydrocarbons in the recent sediments are *not of the petroleum type,* as yet at least.

There appear to be three alternatives possible for petroleum genesis at this stage: (1) the hydrocarbons in the sediment will become the petroleum of the future through qualitative changes, but without quantitative enrichment; (2) further liquid hydrocarbon generation will take place from the organic matter entombed in the sediment which will mix with the indigenous liquid hydrocarbons to form petroleum; and (3) the syngenetic liquid hydrocarbons will disappear through either dissipation or destruction, in which case petroleum will result solely from subsequent generation.

Prelithifaction Oil. The case for the generation of oil after deposition but before lithifaction has been stated well by Hoots:

> Burial and its accompanying expulsion of fluids from compressible sediments is an experience necessary to the formation of fine-grained sedimentary rocks. No shales have avoided it. In addition to the effect from burial, lateral compression may cause additional compaction. The resulting movement of large quantities of water offers the most positive opportunity for oil to move along and across the bedding of fine-grained source beds into adjoining permeable beds. Because of this fact most theorists agree that petroleum generation and its migration to the reservoir takes place during compaction.[52]

Although Hoots was primarily concerned with the origin of oil in the California oil fields where the source rock is without doubt shale, it is also possible that a calcareous ooze undergoing compaction might likewise be the source of liquid hydrocarbons. Trask[53] found calcareous deposits (as well as siliceous muds) containing relatively high percentages of organic materials. He has observed also that "the period of time shortly after the deposition of sediments is probably of major importance in the generation of petroleum."[54]

[52] H. W. Hoots, "Origin, Migration, and Accumulation of Oil in California," *Calif. Div. Mines, Bull.* 118 (August, 1941), p. 260.

[53] Parker D. Trask et al., "Origin and Environment of Source Beds of Petroleum," (Gulf Publishing Co., Houston, Texas, 1932).

[54] Parker D. Trask, "Some Studies of Source Beds of Petroleum," *International Geol. Congress, Report of the 16th Session,* 1933, Vol. 2 (1936), p. 1011.

An argument opposed to prelithifaction oil is the presence in some oil fields of evidence of more than one time of accumulation. Several examples are given by Van Tuyl and Parker.[55] The best-known case is that at Oklahoma City, where drilling has shown the existence of asphaltic residues at the unconformable contact between the Pennsylvanian sediments and the Ordovician reservoir rocks. The natural interpretation of this situation is that oil was migrating up the dipping Ordovician formations and escaping at the truncated surface exposed in pre-Pennsylvanian time. Subsequently, this escapeway was closed by the deposition of the Pennsylvanian sediments and the great oil deposits of the Oklahoma City field accumulated beneath the unconformity. A few other examples of oil residues at unconformities are cited by Van Tuyl and Parker, but Rich[56] notes that in proportion to the total number of unconformity accumulations residual asphalt is decidedly rare.

Actually, of course, heavy oil residues at unconformities which are underlain by deposits of "live" oil are not in themselves evidence of multiple or continuous generation of oil. They can be interpreted only as evidence that oil was in motion in the reservoir rock at two widely different periods of time. It is possible that regional tilting during the time that truncated Ordovician formations were exposed at the surface at Oklahoma City permitted some spillage of the oil from a basinward anticline. After the relatively impervious Pennsylvanian sediments had been deposited across the truncated layers, further tilting might have released from below much more oil that accumulated beneath the unconformity, or the new oil might have been driven there by the overlying compacting sediments. Hiestand[57] has observed that oil is adjusted to present structural traps and concludes as a corollary that the oil and associated water "have made migratory adjustments in various directions according as a given trap was destroyed or preserved by structural modifications." Weeks[58] describes "secondary tectonic structures" which cause "secondary redistribution or localization of oil." Similar ideas are expressed by Van Tuyl, Parker, and Skeeters: "It does not necessarily follow that these substances (oil and gas) are generated in the mother rocks continuously or as successive 'crops.' It is possible that they may be stored for considerable periods after generation, either in a disseminated condition or as

[55] F. M. Van Tuyl and Ben H. Parker, "The Time of Origin and Accumulation of Petroleum," Quarterly Colo. School of Mines, Vol. 36, No. 2 (April, 1941), pp. 84–98.

[56] F. M. Van Tuyl and Ben H. Parker, op. cit., p. 96.

[57] T. C. Hiestand, "Regional Investigations, Oklahoma and Kansas," Bull. Am. Assoc. Petrol. Geol., Vol. 19 (July, 1935), p. 965.

[58] L. C. Weeks, in F. M. Van Tuyl, Ben H. Parker, and H. W. Skeeters, "The Migration and Accumulation of Petroleum and Natural Gas," Quarterly Colo. School of Mines, Vol. 40, No. 1 (January, 1945), p. 53.

pools, which may undergo renewed migration as a result of changes in geologic structure . . . In some instances there may have been renewed migration of oil and gas in pools previously formed as a result of changes in direction or amount of dip, faulting, increase in the vigor of artesian circulation or other causes."[59]

Postlithifaction Oil. Several of the older theories of oil generation are based on the assumption that conversion from solid carbonaceous matter to liquid hydrocarbons does not take place until the sediments have been deeply buried and presumably lithified. However, there is little actual evidence that oil generation takes place after lithifaction. In many oil reservoirs, especially limestone reservoirs, the age of the local accumulation is obviously postlithifaction, but that is still not evidence that the oil was generated from source materials in lithified sediments. Stadnichenko[60] determined experimentally that source materials "have different temperature points or zones at which oil and gas are formed, indicating that the petroleum found in our oil fields may contain products generated at several stages in the long course of the devolatization of the organic matter in the sediments." Furthermore, the later the oil generation in a series of generations, the higher the fixed-carbon content of the source rock and the better the quality of the petroleum. When the carbon content rises above 63 per cent, natural gas only is given off.[61] The source materials used in these experiments were oil shales, cannel coals, and other already lithified rocks. But Hoots[62] pointed out that, if the oil is not produced until after lithifaction takes place, one must accept the thesis that large quantities of oil are generated and migrate through rocks that have become compact and relatively impermeable. One naturally wonders why compact shales as a source rock could permit oil to escape outwards but as a cap rock would not allow oil to pass through.

SEDIMENTATION AND BIOCHEMICAL ACTIVITY

The first phase in the formation of oil is the deposition of the sediment, both physical and organic, which is destined to become the source rock. Concurrent with this sea-floor accumulation are certain chemical changes

[59] *Op. cit.*, p. 56.

[60] Taisia Stadnichenko, "Experimental Studies Bearing on the Origin of Petroleum," *International Geological Congress, Report of the 16th Session,* 1933, Vol. 2 (1936), p. 1009.

[61] C. David White, "The Origin of Petroleum," *Petroleum Investigation,* Hearings before a Subcommittee of the Committee of Interstate and Foreign Commerce, House of Representatives, 73rd Congress, on H. R. 441 (1934), pp. 908–909.

[62] H. W. Hoots, "Origin, Migration, and Accumulation of Oil in California," *Calif. Div. Mines, Bull.* 118 (August, 1941).

that take place in the organic sediment as the result of environment and the activity of bacteria.

Deposition of Sediment. The ecology of source rocks has received considerable attention. The petroleum industry and later the U. S. Geological Survey subsidized a long-continuing program of studies by Parker D. Trask and associates. This research resulted in the publication of numerous papers and two monographs: Parker D. Trask, *Origin and Environment of Source Sediments of Petroleum* (Gulf Publishing Company, Houston, 1932; and Parker D. Trask and H. W. Patnode, *Source Beds of Petroleum*, American Association of Petroleum Geologists, Tulsa, 1942). Another publication in this field is the symposium, *Recent Marine Sediments* (American Association of Petroleum Geologists, Tulsa, 1934). Recent research in the hydrocarbon content of sea- and lake-floor sediment was discussed in the preceding section.

Environment of deposition. One of the major factors controlling the size of an oil accumulation is the quantity of source material originally deposited within sea- or lake-floor sediment. This quantity in turn is a function of environment, involving climate, water depths, water composition, topography both onshore and offshore, rivers and river mouths, currents, and character and sources of sediment.

It is essential that inorganic sediment be deposited concurrently with the organic matter as a preserving agent. For maximum deposition of source materials (1) adequate life, especially phytoplankton, must be present, and (2) conditions for the entombment of this material must be optimum. It is possible to have abundant life without any preservation, but the reverse is of course impossible.

Plankton is most abundant where there is a constant supply of the mineral salts ("nutrients") used for food. The primary source of nutrients is the land, so plankton growth is greatest near the mouths of rivers, including deltaic environments. However, currents may carry this plankton to more distant areas.

Secondary nutrients are also available. Planktonic forms sink when they die. During this slow downward movement decomposition takes place and nutrient salts are released. These salts become reavailable for plankton food where upwelling of the water takes place as in coastal shoal areas.[63] Therefore life is most prolific in and near river mouths, in coastal zones, and in shoal-water areas.

It is axiomatic that no matter how abundant life may be in the sea, only

[63] Nickola Propopovich, "Primary Sources of Petroleum and Their Accumulation," *Bull. Am. Assoc. Petrol..Geol.*, Vol. 36 (May, 1952), pp. 878–883; Margaretha Brongersma-Sanders, "On Conditions Favouring the Preservation of Chlorophyll in Marine Sediments," *Third World Petrol. Congress*, 1951, Section I, pp. 400–411.

that which becomes entombed in the sediment is potential petroleum. If the water is deep, *all* of the sinking plankton may decompose before reaching the sea floor. Emery,[64] and Orr, Emery, and Grady[65] in investigating basin sediments off the California coast found that the degree of decomposition is controlled by water depth, oxygen content of the water, and bottom topography. In their area, the loss due to oxidation of phytoplankton while sinking to the bottom was in the neighborhood of 93 per cent. Furthermore, about half of the 7 per cent of organic debris that does reach the bottom becomes oxidized before being buried to the zero oxygen level.

Of great importance in the accumulation of source materials is the configuration of the sea floor. Trask[66] noted that the organic content of modern sediment is lowest on submarine ridges and highest in sea-floor basins. The reason is that the sediment in the basins is protected from currents; where water is in motion the light organic material cannot come to rest.

At one time it was advocated that in order to preserve deposited organic matter for eventual conversion into petroleum, the water overlying the sea floor had to be either supersaline, or so foul with hydrogen sulfide that life could not live at that level.[67] The latter condition is referred to as "euxenic" after the Euxene or Black Sea, which is notable for both its hydrogen sulfide content at depth and for the accumulation of organic debris on its floor. It is now known that even in the Black Sea the effect of hydrogen sulfide poisoning on the accumulation of organic matter has been exaggerated.[68]

The current trend in thinking is not to seek the unusual, but to consider source-bed deposition as taking place in waters of normal composition. Most source material was probably deposited in salt water, some (in deltaic areas and Maracaibo-type "lakes") in brackish water, and a relatively small amount in fresh water. The proportion of source beds deposited under each of these three environments is probably not far different from the proportions of inorganic sediment of similar texture in the same

[64] Kenneth O. Emery, "Southern California Basins," *Habitat of Oil* (American Association of Petroleum Geologists, Special Vol., 1958), pp. 955–967.

[65] Wilson L. Orr, K. O. Emery, and John R. Grady, "Preservation of Chlorophyll Derivatives in Sediments Off Southern California," *Bull. Am. Assoc. Petrol. Geol.*, Vol. 42 (March, 1958), pp. 925–962.

[66] Parker D. Trask, "Organic Content of Recent Marine Sediments," *Recent Marine Sediments* (American Association of Petroleum Geologists, Tulsa, Oklahoma, 1939) p. 428.

[67] W. A. J. M. van Waterschoot van der Gracht, "The Stratigraphical Distribution of Petroleum" *Science of Petroleum* (Oxford University Press, 1938), Vol. 1, p. 58.

[68] Leonid P. Smirnow, "Black Sea Basin," *Habitat of Oil* (American Association of Petroleum Geologists, Special Vol., 1958), pp. 982–994.

environments. In other words, there is no magic, other than quantity, to salt water. Confirmation of this statement is the fact that brackish Lake Maracaibo is reputed to be one of the most productive bodies of water, in terms of aquatic life, in the world today.[69]

Because of the environmental conditions, the inorganic or physical sediment accompanying, and far exceeding in volume, the organic sediment is usually mud consisting of clay, silt, or calcareous ooze. More rarely, fine-grained sands may contain sufficient organic matter to be a potential source of oil.[70] Smith[71] noted in a Gulf Coast core that a sandy layer contained approximately eight times as much hydrocarbon as was in an immediately overlying silty clay layer. This situation could be due to a higher original depositional content, but it is more likely the result of early compaction of the finer grained sediment, with squeezing out of contained liquids. This "primary migration" is discussed in Chapter 9.

As would be expected, the percentage of organic matter within the sediment varies widely from place to place. The average recent sediment runs about 2.5 per cent by weight organic matter. The figure is lower in deltaic and continental shelf areas where currents are relatively strong; it is higher in closed basins. In bodies of stagnant water, such as the Black Sea, the percentage may be as high as 35; in small lakes the figure may even reach 40 per cent. The average in the open ocean far from land is less than 1 per cent of organic matter.[72]

In addition to classifying source sediments into marine, brackish, and fresh-water types, there are differences in the locale of the marine deposits. Most important in the geological past have been the basins, in which periodic sinkings permitted both the entry of sea water and the deposition of sediment. These basins contained shelf areas, hingebelts, deeper parts, and more disturbed zones. The control of basins on the general distribution of oil, and the distribution of oil within basins, is discussed in Chapter 15, "Regional Aspects of Accumulation." However, it should be noted here that the locale of deposition of the source material, and the type of the enclosing inorganic sediment, is another possible factor in creating differences in the character of crude oils.

[69] Alfred C. Redfield, "Preludes to the Entrapment of Organic Matter in the Sediments of Lake Maracaibo," *Habitat of Oil* (American Association of Petroleum Geologists, Special Vol., 1958), pp. 968–981.

[70] Parker D. Trask, "Some Studies of Source Beds of Petroleum," *Proc. XVI International Geological Congress* (1933), Vol. 2 (1936), p. 1011.

[71] P. V. Smith, Jr., "Status of Our Present Information on the Origin and Accumulation of Oil," *Proc. Fourth World Petrol. Congress* (1955), Section 1, pp. 395–396.

[72] Parker D. Trask, "Organic Content of Recent Marine Sediments," *Recent Marine Sediments* (American Association of Petroleum Geologists, Tulsa, Oklahoma, 1939) p. 428.

Biochemical Processes. During the time that the organic matter is accumulating on the sea floor, it is subject to attack and chemical transformation by bacteria. Opinions vary widely as to the relative importance of the biochemical stage in the genesis of petroleum. It is certain that during this period the organic matter becomes *more* petroleum-like chemically;[73] there are those who believe that petroleum itself may be formed.[74]

The role of bacteria on the sea floor has been a subject of active investigation for some years. A paper by Hammer[75] on this subject, published in 1934, lists 81 references. ZoBell[76] includes 200 references in a paper published in 1947.

Distribution of bacteria. Bacteria have been found in recent sediments up to the depth sampled (25 feet). They are most abundant in the top few inches, below which they gradually decrease in numbers.[77] It has been claimed that samples of sedimentary rock collected aseptically from a depth of 1560 feet contained indigenous bacteria,[78] but this conclusion has been questioned. In samples of oil sand from an oil "mine" in Pennsylvania bacteria were found where the mine was exposed to the air but not in the centers of the chunks.[79] It is probable that the bacteria found in samples of ancient sediments, although abundant, are adventitious. It would be extremely difficult, if not impossible, to so control conditions on the drill floor that an uncontaminated sample could be obtained.

Functions of bacteria in oil formation. Where aerobic conditions exist, not only petroleum source materials but petroleum itself is

[73] Claude E. ZoBell, "Biennial Report for 1945–1947 on API Research Project 43A-Bacteriological and Sedimentation Phases of the Transformation of Organic Material into Petroleum," *Rept. of Progress-Fundamental Research on Occurrence and Recovery of Petroleum* (American Petroleum Institute, 1949), pp. 100–106; "Bacterial Activities and the Origin of Oil," *World Oil*, Vol. 130 (June, 1950) pp. 128–138.

[74] G. D. Hobson, "Biochemical Aspects of the Origin of Oil," *Science of Petroleum* (Oxford University Press, 1938), Vol. 1, pp. 54–56.

[75] Harold E. Hammer, "Relation of Micro-organisms to Generation of Petroleum," *Problems of Petroleum Geology* (American Association of Petroleum Geologists, Tulsa, 1934), pp. 35–49.

[76] Claude E. ZoBell, "Microbial Transformation of Molecular Hydrogen in Marine Sediments, with Particular References to Petroleum," *Bull. Am. Assoc. Petrol. Geol.*, Vol. 31 (October, 1947), pp. 1709–1751; Claude E. ZoBell, "Bacterial Activities and the Origin of Oil," *World Oil*, Vol. 130 (June, 1950), pp. 128 et seq.

[77] Claude E. ZoBell, "Occurrence and Activity of Bacteria in Marine Sediments," *Recent Marine Sediments* (American Association of Petroleum Geologists, Tulsa, Oklahoma, 1939) p. 416.

[78] Claude E. ZoBell, "The Role of Bacteria in the Formation and Transformation of Petroleum Hydrocarbons," *Science*, Vol. 102 (Oct. 12, 1945), p. 365.

[79] G. M. Knebel, "Progress Report on API Research Project 43, The Transformation of Organic Material into Petroleum," *Bull. Am. Assoc. Petrol. Geol.*, Vol. 30 (November, 1946), p. 1943.

destroyed. We have already seen how phytoplankton sinking through sea water off the California coast loses 93 per cent through decomposition; another 3½ per cent of the original is lost in the top layers of sediment before the zero oxygen level is reached. There is also ample evidence of the activity of hydrocarbon-oxidizing bacteria above water, such as the relatively rapid disappearance of oil spilled upon the ground. Also hydrocarbon-oxidizing bacteria may alter the composition and characteristics of oil by destroying some hydrocarbon compounds ahead of others. Therefore, if the process of destruction is arrested at any time, the residual oil may be quite different from the initial oil.

Reducing (anaerobic) bacteria are abundant in stagnant waters, but they are of much greater geological importance in sea- or lake-floor sediment below the zero oxygen level. These bacteria liberate oxygen, nitrogen, phosphorus, and sulfur from organic material. As a result, the percentage of hydrogen and carbon is increased, making the material more like petroleum in composition.

It is common knowledge that bacteria can produce methane from organic debris. This fact not only is subject to laboratory demonstration but also is illustrated in nature by the production of marsh gas. Whether liquid hydrocarbons are produced by these bacteria is still a matter of conjecture. ZoBell[80] has found that such hydrocarbons can be produced in the laboratory: "It is especially noteworthy that certain anaerobic bacteria, isolated from marine sediments, have been shown to produce ether-soluble, non-saponifiable, oil-like extracts from naturally occurring lipides or fatty acids."

Another effect of anaerobic bacteria in sea-floor sediment is to reduce sulfate to sulfide (pyrite); Emery and Rittenberg[81] report that in some cores taken off the California coast the sulfate totally disappeared below 7 feet.

Hydrogen-producing bacteria might conceivably contribute to the evolution of hydrocarbons by increasing the hydrogen percentage. This "biochemical hydrogenation" can probably take place with methane, CH_4, as well as with straight hydrogen.

Miscellaneous activities of bacteria which may be significant in petroleum accumulation and exploitation include the production of acids that can dissolve carbonate rocks, releasing at the same time carbon dioxide, which adds to the pressure, and the liberation of oil from the surfaces of the grains in the reservoir rock.

[80] Claude E. ZoBell, "Transformation of Organic Material into Petroleum-Bacteriological a. d Sedimentation Phases," *Fundamental Research on Occurrence and Recovery of Petroleum* (American Petroleum Institute, New York, 1943), p. 104.

[81] K. O. Emery and S. C. Rittenberg, "Early Diagenesis of California Basin Sediments in Relation to Origin of Oil," *Bull. Am. Assoc. Petrol. Geol.*, Vol. 36 (May, 1952), pp. 735–805.

Conclusions. The principal function of bacteria is to remove the elements other than hydrogen and carbon, especially oxygen and nitrogen, thereby bringing the organic matter closer to petroleum in composition. Methane can be produced in this way, but whether the liquid hydrocarbons can be so generated is an open question.

Differences in the degree and character of the biochemical attack upon the organic debris accumulating upon the sea floor may be another explanation for differences in the composition and properties of petroleums.

BURIAL AND DYNAMOCHEMICAL ACTIVITY; EVOLUTION OF PETROLEUM

Burial of source material begins with its deposition but may continue long after the sea-floor accumulation of organic matter has ceased. We have already seen that within the source material-containing sediment are hydrocarbons, which are not, however, of petroleum character. We do not know at this time whether these hydrocarbons are the result of the metabolism of live organisms, or the product of bacterial attack. We suspect the former, for although methane has been produced in the laboratory from acetates by bacterial action, attempts to do the same with the higher fatty acids have not been successful.[82] The conversion of fatty acids to hydrocarbons by loss of CO_2 is still unexplained.[83]

Also present in the source material sediment are organic solids. In potential hydrocarbon volume these exceed the syngenetic hydrocarbons by ten to one in modern sediments off the California coast.[84] It is the conversion of organic solids to liquid hydrocarbons that is most bothersome. Much has been written regarding the conditions and processes which might bring about the conversion within the earth's crust of organic debris into petroleum, but in spite of this mental productivity we still do not know with certainty how, when, or where oil is formed.

Perhaps the key to the mystery is the presence of syngenetic liquid hydrocarbons. Trask[85] believes that only a small amount of "ancestral

---

[82]Robert W. Stone and Claude E. ZoBell, "Origin of Petroleum; Bacterial Aspects," *Jour. Ind. and Eng. Chem.* (*Industrial Edition*), Vol. 44, (1952), p. 2567.

[83]Benjamin T. Brooks, letter, (July 12, 1955).

[84]Wilson L. Orr and K. O. Emery, "Composition of Organic Matter in Marine Sediments: Preliminary Data on Hydrocarbon Distribution in Basins of Southern California," *Bull. Geol. Soc. Am.*, Vol. 67 (September, 1956), pp. 1247–1258.

[85]Parker D. Trask, "Inferences about the Origin of Oil as Indicated by the Composition of the Organic Constituents of Sediments," U.S. Geol. Survey, *Professional Paper* 186-H (1937), p. 156.

petroleum" need be formed as a starter and that the volume would increase greatly by the solution of organic solids such as pigments, waxes,
and fatty acids, which the liquid met in its travels. With the solution of
this organic matter, the ancestral petroleum would evolve into true
petroleum.

Regardless of whether petroleum is exclusively syngenetic, exclusively
epigenetic, or a mixture of the two, it is highly probable that the earliest
oil is semisolid asphalt or at best a sluggish liquid composed largely of
asphaltic hydrocarbons. These hydrocarbons contain oxygen, nitrogen,
and sulfur carried over from the original organic matter. Possible examples of such "neo-petroleum" are the Athabaska heavy oil ("tar") deposits, and the asphaltic oölitic Gasper (Mississippian) limestone of
northern Alabama and the asphaltic fragmental Anacacho (Cretaceous)
limestone of Uvalde County, Texas.[86] On the other hand, these asphaltic occurrences may be the result of devolatilization of a petroleum, as in
the case of seep asphalts.

There is no doubt but that petroleum, once formed, may undergo progressive change in nature's reservoirs from a crude oil consisting largely
of asphaltic hydrocarbons to one containing paraffinic compounds.[87]
Barton, after an intensive study of Gulf Coast crude oils, concluded that:

> Temperature, pressure, and perhaps other factors act on the petroleum to pro-
> duce with the passage of time a slow evolutionary transformation of the petroleum;
> from heavy crude oil, which would consist mainly of residuum in the terms of the
> United States Bureau of Mines analyses, through a progressively lighter series of
> oils, into paraffine-base crude oils composed predominantly of gasoline and kero-
> sene; and to ultimate extinctive dispersion as the volatile end members of the
> paraffine series.[88]

More recently, McNab, Smith, and Betts have reached a similar conclusion: ". . . there is considerable evidence of a progressive evolution in
crude oils, from the heavy cyclic crudes first formed and found in young
sedimentary rocks to the lighter, more paraffinic crudes containing larger
proportions of low molecular weight components and commonly found in
producing horizons of greater age or depth."[89]

[86] D. D. Utterback, "New Questions On Origin of Oil," Oil and Gas Jour. (Aug. 23,
1954), pp. 121–124.

[87] Benjamin T. Brooks, "Origin of Petroleums," The Chemistry of Petroleum Hydro-
carbons (Reinhold, New York, 1954), Vol. I, Chapter 6, pp. 83–102; H. N. Dunning and
J. W. Moore, "Porphyrin Research and Origin of Petroleum," Bull. Am. Assoc. Petrol.
Geol., Vol. 41 (November, 1957), pp. 2403–2413.

[88] Donald C. Barton, "Natural History of the Gulf Coast Crude Oil," Problems of Petroleum Geology (American Association of Petroleum Geologists, 1934), p. 149.

[89] T. G. McNab, P. V. Smith, R. L. Betts, "The Evolution of Petroleum" Ind. and Eng.
Chemistry (Industrial Edition), Vol. 44 (1952), pp. 2556–2563.

Burial of Source Material. There are two stages in the burial of the organic sediment. The first is the concurrent deposition of the much more abundant physical sediment. In the preceding section it was noted that this material is usually a clay or fine silt because the light organic matter could come to rest only where the water was so quiet that the non-organic sediment would settle out of suspension. Entombment is essential in order to preserve the organic matter under anaerobic conditions; water almost always contains enough oxygen to support life.

The second phase in the burial of the organic sediment is the deposition of a non-organic overlying rock, followed by layer upon layer of other rocks of various types. The eventual thickness of this cover varies between wide extremes.[90] It is not possible to state with any certainty the minimum cover which existed at the time transformation took place. The oil fields of eastern Kansas are often cited as examples of oil occurring in rocks which were never very deeply buried. It is true that oil occurs here in Upper Pennsylvanian rocks at a present depth of only a few hundred feet, and the addition of higher Pennsylvanian and Permian formations which crop out to the westward would make the total cover less than 2000 feet, but still younger rocks, especially the Cretaceous, may have also covered this area.

Postburial Physical Changes. At least two changes in physical environment take place with burial. One is an increase in hydrostatic pressure with each foot of cover added. It has been determined that the pressure at a depth of 13,000 feet is over 6000 pounds per square inch.[91] Another change is caused by the rise of the isogeotherms, or levels of equal earth temperature. Sediments buried to a depth of 15,000 feet may be subjected to temperatures as high as 300° F.

The first physical effect on the organic muds to result from burial and loading is compaction. Because of the nature of the material, the sediment may in time become compressed to 50 per cent or more of its original thickness, and the specific gravity of the sediment may increase from an initial 1.3 to more than 2.[92] This compaction is accompanied by the expulsion of a like volume of water or other fluid. The importance of this

[90] F. M. Van Tuyl and Ben H. Parker, "The Time of Origin and Accumulation of Petroleum," *Quarterly Colo. School of Mines*, Vol. 36 (April, 1941), pp. 79–82.

[91] William B. Heroy, "Petroleum Geology," *Bull. Geol. Soc. Am. 50th Anniversary Vol.* (1941), pp. 512–548.

[92] G. M. Knebel, Progress report on API Research Project 43, "The Transformation of Organic Material into Petroleum," *Bull. Am. Assoc. Petrol. Geol.*, Vol. 30 (November, 1946), pp. 1948–1949; H. D. Hedberg, "Gravitational Compaction of Clays and Shales," *Am. Jour. Science*, 5th Series, Vol. 31 (April, 1936), pp. 241–287.

process in the migration of oil from source rock to carrier or reservoir bed will be considered later in Chapter 9.

Compaction and induration transform the incoherent sediment into sedimentary rock. Further dehydration and minor earth movements create jointing and cleavage in the rock, and if diastrophism or static metamorphism becomes intense, recrystallization and even the development of schistosity take place. However, hydrocarbons will be driven out before this stage is reached, as will be shown subsequently.

Eometamorphism. Definite limitations are placed on the occurrence of oil by the intolerance of petroleum in the earth's crust to elevated temperatures and pressures. These create a twilight zone where oil and gas give way to gas only, both laterally in basins bordered by tectonically disturbed belts, and downward in the deeper basins. The hydrocarbons are vulnerable to the very beginnings of metamorphism, or eometamorphism (literally "early" metamorphism).

The causes of eometamorphism are increase of heat and pressure. The recorders of eometamorphism are coal, shale, reservoir rocks, and hydrocarbons in the rocks. Both the chemical laboratory and the petroleum refinery have demonstrated that hydrocarbons are extremely susceptible to changes in character with increase in temperature and pressure. Countless borehole and mine measurements attest to the increase of both with depth. Subsurface temperatures in non-magmatic areas range from surface air temperature to a maximum, as of November, 1973, of 555°F., at a total depth of 23,837 ft., in a dry hole drilled in Webb County, Southwest Texas.[94]

Mainly because of increased heat with greater depth, petroleum becomes lighter in specific gravity, more and more condensate occurs with the oil, and condensate in turn gives way to dry gas. In any particular area, the relationship between degree of eometamorphism and depth depends upon the thermal gradient, or rate of temperature increase downward. Although there may local divergences most gradients for deep wells range from 1° to 2°F per hundred feet. Figure 6-1 shows thermal gradients of 1.0° to 1.6° as vertical lines, and depths (left column) as horizontal lines. The right column gives the temperature; the curved lines starting here cross depth lines with decreasing thermal gradients. The bold line is the oil "floor," so far as present information indicates. The gas "floor" is an undefined economic plane; it is the level below which gas in any given area cannot be produced at a profit, due to lesser porosity and permeability and to greater drilling and production

[93]Kenneth K. Landes, "Eometamorphism, and Oil and Gas in Time and Space," *Bull. Am. Assoc. Petrol. Geol.,* Vol. 51, No. 6, Part I, June, 1967, p. 826–841. Also "Oil and Gas in Three Dimensions," Exploration and Economics of the Petroleum Industry, *Southwestern Legal Foundation,* Vol. 6, 1968, p. 17–24.

[94]George Ives, "How Shell Drilled a Super-Hot, Super-Deep Well," *Petroleum Engineer,* March, 1974, p. 70–78.

costs. The metamorphic end point for dry gas (methane, CH_4) is presumably graphite, which is pure carbon.

The average temperature increase with depth is about $1.3°F$. This places the average depth of the oil floor at 21,500 feet. However, not every area is "average." Where a sedimentary basin has been deeply buried, and subsequently uplifted and eroded, the hydrocarbons in the sediments have been subjected to much higher temperatures than will be recorded in drill holes, because of the adjustment in geologic time to the lessened depth brought about by offloading. However, the chemical changes that took place during deep burial are not reversible. Therefore the depths of the petroleum floor shown in Figure 6-1 are *maximum*.

It should also be noted that the oil floor calculations are based upon experience across the years. Future discoveries may necessitate some revisions to the chart, but the *concept* of a floor appears to be valid.

Coal has been a metamorphic indicator for oil and gas occurrence for years.[94a] Where coal and oil and gas reservoirs are not too far apart stratigraphically, as in the Appalachian, Arkoma (Arkansas-Oklahoma), and western Alberta coal districts one finds the lighter oils, plus condensate and much gas, toward the mountain front where the coal has been metamorphosed into the middle bituminous rank; farther into the disturbed area is the gas only zone. The use of coal "carbon ratios" in delineating oil and gas provinces is discussed later.

Older views; the transformation of solid organic matter into liquid petroleum. Among the several agents and processes that have been credited with bringing about this change are pressure, bacteria, catalysts, heat with "cracking," natural distillation, and radioactive minerals. Geologic time has also been considered an important contributing factor.

Pressure, to be effective, would have to be of shearing magnitude in order to transform organic solid into oil. Experiments carried on at room temperatures on the application of high-shearing pressures to oil shales and cannel coals resulted in some insignificant devolatilization but no oil.[95] Actually, few if any source rocks of petroleum experience pressures of this magnitude. "Skin frictional heat"[96] has been advocated as a factor in the generation of petroleum. This heat is supposed to be generated during the compaction of the source rock, but the speed of movement, which is quite important in frictional heat, is so slight in compaction

94aM. R. Campbell, "Coal as a Recorder of Incipient Rock Metamorphism," *Econ. Geology,* Vol. 25, No. 7, Nov. 1930, p. 675–696.

[95] J. E. Hawley, "Generation of Oil in Rocks by Shearing Pressures," I, *Bull. Am. Assoc. Petrol. Geol.,* Vol. 13 (April, 1929), pp. 303–328; II, pp. 329–365; III, Vol. 14 (April, 1930), pp. 451–481.

[96] Ralph H. Fash, "Theory of Origin and Accumulation of Petroleum," *Bull. Am. Assoc. Petrol. Geol.,* Vol. 28 (October, 1944), pp. 1510–1518.

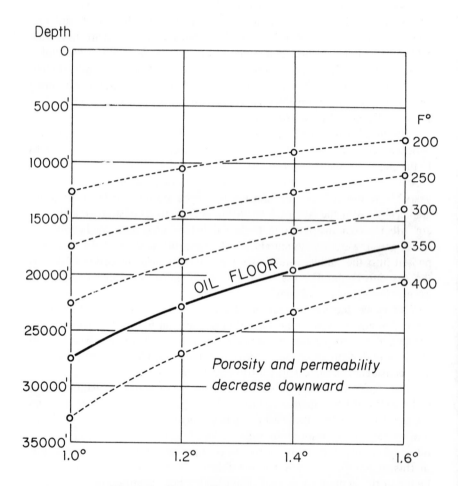

Fig. 6-1. Degrees Fahrenheit/Hundred Feet

that it is difficult to visualize a significant temperature increase from this cause.

Just when in the burial process bacteria cease to be active is not known. There are those who believe that bacteria may be responsible for the generation of liquid petroleum. "Conclusive proof is still lacking that bacteria are physiologically active in subterranean deposits of brine or oil, but bacteria are unquestionably functional in recent marine sediments to the greatest depths sampled, around 25 feet."[97] But Illing[98] questions that bacteria, even if present, would produce petroleum: ". . . all studies of such decay have tended to confirm the impression that the ultimate end product of all such decomposition tends to be methane, and until it has been proved conclusively that bacterial decay can produce the higher-molecular weight hydrocarbons, the bacterial theory can only be regarded as a tentative solution."

High temperatures not involved. The presence of chlorophyl porphyrins in crude oils is positive evidence that that oil could never have been at high temperature, for chorophyl porphyrin compounds cannot exist above about 200°C. (392°F). There is also much chemical evidence denying a high temperature history. Better yet is the fact that the deeply buried crude oils in the undisturbed Gulf Coast sediments are not at high temperature today and there is no geological evidence that they ever were. This fact makes untenable theories of petroleum genesis involving natural cracking and hydrogenation, for these reactions require temperatures in excess of 400°C. (752°F).

Also rendered untenable by the low temperature history of petroleum is the once popular theory that crude oil results from a natural distillation process. Although the most volatile hydrocarbons begin to move out at temperatures as low as 40°C (104°F) and even lower, other petroleum hydrocarbons do not vaporize until temperatures considerably above the porphyrin limit of 200° C have been reached. Furthermore, ". . . all destructive distillation at whatever temperature must leave behind a residue or spent. Such a material is readily identifiable and it is unlikely that material of this type would have escaped recognition in all the oil fields which have been investigated."[99]

Petroleum generation by radioactivity improbable. The possibility that radioactive emanations may have converted organic matter,

[97] Claude E. ZoBell, "The Role of Bacteria in the Formation and Transformation of Petroleum Hydrocarbons," *Science*, Vol. 102 (Oct. 12, 1945), p. 365.

[98] V. C. Illing, "The Origin of Petroleum," *Science of Petroleum*, (Oxford University Press, 1938), Vol. 1, p. 37.

[99] V. C. Illing, *op. cit.*, p. 36.

buried in the rocks of the earth's crust, into petroleum has been investigated by several scientists in recent years, and the American Petroleum Institute sponsored a research project in this field.[100]   The radioactive elements that may be present in sediments in adequate amounts are uranium, thorium, and potassium.   Some organic shales have been found to be highly radioactive.   The radium content of limestone decreases with increasing purity, whereas that of sandstones is highly variable.   Some crude oils are radioactive.   But it remains to be seen whether these occurrences are genetically related or merely fortuitous.

It has been demonstrated by the physicists that bombardment of organic material, especially fatty acid, by alpha rays produces hydrocarbons, free hydrogen, and carbon dioxide.   From this fact the conclusion has been reached that alpha-ray bombardment of carbonaceous shales is adequate to explain the origin of petroleum.   A possible example that has been cited is the radioactive and carbonaceous Antrim shale of Michigan. According to calculations, based upon a number of assumptions and therefore "to be accepted with some caution", this shale could contribute 208 barrels of crude oil per acre-foot in 10 million years.[101]

So far, at least, the arguments opposed to petroleum genesis on a large scale by radioactivity far outweigh the arguments in favor of it.   One wonders, for example, why the Devonian Antrim shale has not yielded 200-plus barrels per acre-foot in the last 10 million years.   This shale is over 200 feet in thickness and underlies thousands of square miles.   Not only should the overlying sandstones be full of oil, but also Michigan should be exuding oil from every pore.   A similar example is the kolm of Sweden, a Cambrian deposit containing 20 per cent volatile organic matter and nearly 0.5 per cent uranium.   An extreme case is the occurrence of thucolite, a solid hydrocarbon, in strongly radioactive pegmatite at Parry Sound, Ontario.   In neither the Antrim shale, the Swedish kolm, nor the Ontario thucolite is any liquid petroleum present, although at each locality solid organic matter, or hydrocarbon, has been exposed to

[100]Clark Goodman, K. G. Bell, W. L. Whitehead, "The Radioactivity of Sedimentary Rocks and Associated Petroleum," *Econ. Geol.*, Vol. 34 (December, 1939), p. 941; K. G. Bell, Clark Goodman et al., "Radioactivity of Sedimentary Rocks and Associated Petroleum," *Bull. Am. Assoc. Petrol. Geol.*, Vol. 24 (September, 1940), pp. 1529–1547; W. J. Mead, Director, "Review of Am. Petrol. Institute, Research Project 43C (Studies of the Effect of Radioactivity on the Transformation of Marine Organic Materials into Petroleum Hydrocarbons," *Fundamental Research on Occurrence and Recovery of Petroleum, 1943*, (American Petroleum Institute, 1944); C. W. Sheppard, "Radioactivity and Petroleum Genesis," *Bull. Am. Assoc. Petrol. Geol.*, Vol. 28 (July, 1944), pp. 924–952; Roland F. Beers, "Radioactivity and Organic Content of Some Paleozoic Shales," *Bull. Am. Assoc. Petrol. Geol.*, Vol. 29 (January, 1945), pp. 1–22; C. W. Sheppard and W. L. Whitehead, "Formation of Hydrocarbons from Fatty Acids by Alpha-Particle Bombardment," *Bull. Am. Assoc. Petrol. Geol.*, Vol. 30 (January, 1946), pp. 32–51.
[101]C. W. Sheppard and W. L. Whitehead, *op.cit.*

strong radioactive bombardment for hundreds of millions of years. As a matter of fact, there is no correlation between the degree of radioactivity and the amount of oil. In addition to organic sediments exposed to radioactivity but without oil[102] are oil deposits where there is, and always has been, little or no radioactivity.

Several writers[103] have pointed out the usual absence of free hydrogen and carbon dioxide in petroleum and natural gas and associated sediments, yet radioactive bombardment should produce copious quantities of these substances, as well as hydrocarbons. Knebel wonders "why after the fatty acids are decarboxylated by radioactivity, the same radiations do not further break up the resulting hydrocarbon molecules, with production of hydrogen and unsaturated hydrocarbons."[104]

Catalysis.[105] The use of clay in the catalytic cracking of crude oil is standard petroleum refinery procedure. Brooks[106] states that carbonium-ion (a hydrocarbon group) reactions do not require the high temperatures reached in the industrial catalytic cracker. He believes that the natural evolution of petroleum is brought about through the influence of acid silicate catalysts encountered by the oil during its underground history. Natural catalysis would explain:

1. The complex character and variety of petroleum compounds.
2. The virtual absence of olefins.
3. The formation of napthenes in quantity.
4. The presence of aromatics. Their presence has not been explained in any other way since a high temperature history was disqualified by the discovery of porphyrins in crude oils.
5. The formation of light hydrocarbons including methane.

---

[102] V. C. Illing, "The Origin of Petroleum," *Science of Petroleum* (Oxford University Press, 1938), Vol. 1, pp. 32–38.

[103] Benjamin T. Brooks, "The Chemical and Geochemical Aspects of the Origin of Petroleum," *Science of Petroleum* (Oxford University Press, 1938), Vol. I, p. 47; Ben B. Cox, "Transformation of Organic Material into Petroleum under Geologic Conditions," *Bull. Am. Assoc. Petrol. Geol.*, Vol. 30 (May, 1946), p. 650; William L. Russell, *op. cit.*

[104] G. M. Knebel, "Progress Report on API Research Project 43, The Transformation of Organic Material into Petroleum," *Bull. Am. Assoc. Petrol. Geol.*, Vol. 30 (November, 1946), p. 1946.

[105] Benjamin T. Brooks, "Origin of Petroleums," *The Chemistry of Petroleum Hydrocarbons* (Reinhold, New York, 1954), Vol. I, Chapter 6, pp. 83–102; Benjamin T. Brooks, "Catalysis and Carbonium Ions in Petroleum Formation," *Science*, Vol. III (June 16, 1950), pp. 648–650; Paul D. Torry, "Origin, Migration, and Accumulation of Petroleum and Natural Gas in Pennsylvania," *Problems of Petroleum Geology* (American Association of Petroleum Geologists, Tulsa, Oklahoma, 1934), p. 447.

[106] *Op. cit.*, p. 96.

6. Differences in the character of crude oils due to differences in catalytic environments.

7. The changes in composition with geologic time.[107]

Possible arguments in opposition to evolution-by-catalysis are: (1) petroleum is rarely if ever in actual contact with the enclosing rock, because of the thin film of "wetting" water which coats the pore walls (Chapter 7); and (2) catalysts appear to be far less effective when wet than when dry. Therefore, acceptance of catalysis in the evolution of oil also involves acceptance of the concept that catalytic activity can take place even though the catalyst is water-coated. Perhaps the explanation is that activity too slow to be measureable in the laboratory may be adequate in geologic time.

A number of substances have been suggested as possible catalysts in nature's subterranean refinery.[108] Brooks[109] has published a list showing the polymerizing effect of Oklahoma sedimentary rocks, plus some minerals, on turpentine. As a general rule fuller's earth, clays, and shales ranked at the top of the list; sandstones ranged from high to intermediate, but pure quartz is inert. Among the clay minerals, montmorillonite is very active and kaolinite is not. Pure calcite and dolomite are non-catalytic.

Some *trace* impurities have been found to give very high catalytic activity.[110] Furthermore the active part of a catalyzing material is apparently on or in its surface, so a film or scattered particles of a catalytically active impurity on quartz or carbonate mineral grains could convert a non-catalytic rock into an active catalyst. This statement may explain the evolution of petroleum in some sandstone and limestone reservoirs.

Brooks[111] cites as examples of petroleums that did not have the benefit of catalysts to an adequate degree the heavy oils found in some limestone reservoirs in Mexico, Oklahoma, Kansas, and Wyoming, and the "tar," actually a very heavy oil, occurring in clean (quartz only) sandstone in the Athabaska, Alberta, area.

There are numerous examples of oils of different character occurring in the same petroliferous province, but in rocks ·of different sedimentary

[107] These points have been extracted from a letter from Benjamin T. Brooks to me dated July 28, 1955. They are repeated here with his permission.

[108] Ralph E. Grim, "Relation of Clay Mineralogy to Origin and Recovery of Petroleum," *Bull. Am. Assoc. Petrol. Geol.*, Vol. 31 (August, 1946), pp. 1495–1496; Ben B. Cox, "Transformation of Organic Material into Petroleum under Geological Conditions," *Bull. Am. Assoc. Petrol. Geol.*, Vol. 30 (May, 1946), p. 653.

[109] *Op. cit.*, Table 4, p. 94.

[110] Benjamin T. Brooks, *letter*, Aug. 12, 1958.

[111] Benjamin T. Brooks, "Catalysis and Carbonium Ions in Petroleum Formation," *Science*, Vol. III (June 16, 1950), p. 648.

facies. Varied facies are of course, significant of varied environments at time of deposition. They also imply differences in catalytic agent content. Bornhauser [112] explains the presence of lighter oils in the shale facies, or in the transition zone between shale and sand facies, in certain reservoirs of the Louisiana Gulf Coast, to a higher clay content. Russian geologists investigating the character of the hydrocarbons and the lithology of the containing rocks in the oil fields of the Baku district have noted an increase in the natural gas percentage and a decrease in the density of the petroleum with increasing amounts of clay. [113]

Geologic time as a factor in the generation of oil. It has been demonstrated by several investigators that the formation of bitumen from carbonaceous matter is a function of both temperature and time. [114] The theory that a moderate increase in temperature produces the same results in geologic time that a relatively high temperature accomplishes in a few minutes has been likened to the fact, known to all cooks, that a low oven temperature, given time enough, does just as thorough a job of cooking as a hot oven. [115] But Illing [116] questions the general conclusion that bitumen formation takes place at low temperatures regardless of the period of time involved. He found that the reaction velocities of a typical cannel coal that he investigated diminished much more rapidly than the calculated rate below 250° C and approached "virtual stagnation" at about 220° C. Brooks [117] calls attention to the fact "that aromatics cannot be formed from paraffins by heat alone at temperatures below about 500° C; below this temperature the thermal decomposition reactions are different in kind, so that aromatics cannot be formed from paraffins by

[112] Max Bornhauser, "Oil and Gas Accumulation Controlled by Sedimentary Facies in Eocene Wilcox to Cockfield Formations, Louisiana Gulf Coast," *Bull. Am. Assoc. Petrol. Geol.*, Vol. 34 (September, 1950), pp. 1887–1896.

[113] SH. F. Mekhtiev and G. P. Tamrazyan, "Distribution of Petroleum and Natural Gas in Oil Deposits of Apsheron Peninsula in Relation to Lithology of Enclosing Rocks," Review by George V. Chilingar, *Bull. Am. Assoc. Petrol. Geol.*, Vol. 39 (October, 1955), pp. 2094–2096.

[114] C. G. Maier and S. R. Zimmerly, "The Chemical Dynamics of the Transformation of the Organic Matter to Bitumen in Oil Shale," *Bull. Univ. Utah*, Vol. 14 (1924), pp. 62–81; C. David White, "Exchange of Time for Temperature in Petroleum Generation," *Bull. Am. Assoc. Petrol. Geol.*, Vol. 14 (September, 1930), pp. 1227–1228; Parker D. Trask, "Time Versus Temperature in Petroleum Generation," *Bull. Am. Assoc. Petrol. Geol.*, Vol. 15 (January, 1931), pp. 83–84.

[115] W. G. Woolnough, "Geological Extrapolation and Pseudabyssal Sediments," *Bull. Am. Assoc. Petrol. Geol.*, Vol. 26 (May, 1942), p. 769.

[116] V. C. Illing, "The Origin of Petroleum," *Science of Petroleum* (Oxford University Press, 1938), Vol. I, p. 36.

[117] Benjamin T. Brooks, "Origin of Petroleum," *The Chemistry of Petroleum Hydrocarbons*, (Reinhold, New York, 1954), Vol. I, Chapter 6, pp. 83–102.

heat alone in any long period of time at the temperatures known to prevail in oil-producing strata."

Both Illing and Brooks[118] point out that the survival of organic material in oil shales as old as Precambrian is inconsistent with the theory that such matter would be converted into bitumen at moderate temperatures if given sufficient time. Similarly, there is no significant difference between oil shales of Tertiary age and those deposited during the Paleozoic era.[119]

However, the conception of time is of considerable importance to the petroleum evolution-by-catalysis theory. The longer oil is in contact with catalytic materials the more it evolves. This fact is supported by the general axiom: "The older the oil, the better the oil." The companion axiom: "The deeper the oil, the better the oil" is probably often true because the deeper oil may be older. The higher temperatures existing at greater depths may accelerate the catalytic activity also.

The fact that our youngest known commercial petroleums are of early Pleistocene age, which is old as man measures time, is further evidence of importance of time in oil evolution.

Contamination and devolution of petroleum. Most of the chemical compounds that make up crude oil are susceptible of reaction with foreign substances. The contaminating material may be either compounds in solution in circulating waters, or oxygen and other gases in the atmosphere.

Contamination of crude oil by sulfur or oxygen would reverse natural evolution and produce heavy, unsaturated oils.[120] Exposure, or for that matter a lessening of the reservoir pressure as the surface is approached, would permit escape of dissolved volatile constituents that would leave a heavier asphaltic residue. Weathering processes and escape of volatiles may have taken place in the geological past below what are now unconformities. Bienner[122] et al state: "Some crude oils exposed to air and sunlight evolve

[118] Benjamin T. Brooks, "Origin of Petroleums, Chemical and Geochemical Aspects," *Bull. Am. Assoc. Petrol. Geol.*, Vol. 20 (March, 1936), p. 280.

[119] Alex. W. McCoy and W. Ross Keyte, "Present Interpretations of the Structural Theory for Oil and Gas Migration and Accumulation," *Problems of Petroleum Geology* (American Association of Petroleum Geologists, 1934), p. 269.

[120] Wallace E. Pratt, *Oil in the Earth* (University of Kansas Press, Lawrence, Kansas, 1942), p. 16.

[121] John Emery Adams, "Origin, Migration, and Accumulation of Petroleum in Limestone Reservoirs in the Western United States and Canada," *Problems of Petroleum Geology* (American Association of Petroleum Geologists, 1934), p. 361.

[122] F. Bienner, E. Bonnard, J. J. Burger, R. Gayral, R. Levy, M. Louis, R. Metrot, C. Salle et Ste Pechelbronn, "Contribution de la geochimie a l'etude de l'evolution des huiles brutes dans les bassins sedimentaires," *Fourth World Petrol. Congress* (1955), Section I, pp. 337–357.

gradually to the formation of asphalts . . . This evolution is accompanied by a change in its chemical nature; thus a paraffinic or mixed oil produces a napthenic asphalt." Examples include Chelif in Algeria and Pechelbronn, France.

Summary. At the time of the burial of petroleum source sediment the organic matter present is largely in solid state, but a small percentage is hydrocarbon, probably residual from life processes, in heavy liquid or semisolid condition. The manner in which any of the solid organic matter is converted to liquid is not yet known. Regardless of origin, petroleum can and does evolve while in nature's reservoirs and through geologic time. The initial petroleum is probably a semiliquid asphaltic material which evolves into a petroleum containing lighter oils, paraffins, and aromatics. This evolution cannot be brought about by temperatures above 200°C. for chlorophyl porphyrins found in crude oils would be destroyed at those heat levels. The most plausible theory so far developed for the natural evolution of petroleum in that certain minerals, such as acid silicates, acted as catalysts through geologic time.

METAMORPHIC PROCESSES AND THE DESTRUCTION OF PETROLEUM

Heat and pressure, with the aid of interstitial water, are the agents of rock metamorphism. However, there is a wide difference between the tolerance limits of oil and rock. Pressure and heat of such magnitude as to cause visible rock metamorphism are too intense to permit the existence of petroleum within the rock. Only enough temperature and pressure increase to produce a certain degree of *incipient*[123] metamorphism of rock, measured by the increased fixed-carbon content of associated coals, apparently is necessary to bring about petroleum destruction. A slightly higher degree of metamorphism results in the disappearance of natural gas from the rock.

Petroleum in place underground is ordinarily not affected by the pressures to which the enclosing rock is subjected. It occupies the interstices between the grains of the rocks, and the static weight of the overlying rock, or the lateral forces created by diastrophism, are borne by the rock itself and not by the contained fluids. However, the petroleum is subjected to hydrostatic pressures created by the weight of an overlying

[123] M. R. Campbell, "Coal as a Recorder of Incipient Rock Metamorphism," *Econ. Geol.*, Vol. 25 (November, 1930), pp. 675–696.

column of water extending to the water table.   In deeply buried rock these are considerable.

Both the oil and the enclosing rock experience the same earth temperatures.   The usual cause of heat is the rise of the isogeotherms, levels of equal earth temperature, which takes place as sediments are buried.   The average temperature increment with depth is 1° F for each 60 feet. Obviously, deeply buried oils are heated to temperatures in excess of the sea-level boiling point of water.   Other possible causes of heat include diastrophism and magmatic activity.   But diastrophism, in order to produce heat, must be of shearing intensity, and magmatic heat is so local in nature that neither of these causes can be considered to be of widespread application in the evolution of oil.

Distribution of Oil Fields and Metamorphic Rocks.   Oil has not been found in commercial amounts in belts of metamorphic rock.   It has not even been found where the coal has been altered to anthracite. Thom[124] estimates the chances of finding oil in areas of semianthracite and semibituminous coals to be of the order of 1 in 1000; the chances for gas, however, are ten times greater.

An exception to the generalization that oil does not occur in metamorphic rocks may be found where porous and permeable metamorphic rocks are in contact with younger source rocks.

Carbon ratio.   Where the metamorphism is invisible but incipient, it can be determined by the carbon ratio of coals.   The carbon ratio is the ratio between the fixed carbon and the volatiles, which can be obtained from the proximate analysis.   The percentage of fixed carbon, after recalculating on an ash- and moisture-free basis, is the carbon ratio.   The carbon ratio of lignite is below 60, bituminous coal ranges from 60 to 85, anthracite from 86 to 98, and graphite, the end point in the metamorphism of coal, has a carbon ratio of 100.

The carbon-ratio concept can be applied in regions, such as the Appalachians and eastern Oklahoma, where both oil and coal occur.   The usual procedure is to enter on a map the localities and carbon ratios of coals for which analyses are available.   Then isovols[125] or isocarbs, lines through points of equal carbon ratio, are drawn, usually with an interval of 5 per cent.   Actually, of course, the lines mark the outcrop of planes of equal carbon ratio.   These planes dip away from the area of more

[124] William Taylor Thom, Jr., "Present Status of Carbon Ratio Theory," *Problems of Petroleum Geology* (American Association of Petroleum Geologists, Tulsa, Oklahoma, 1934), pp. 69–95.

[125] C. David White, "Some Relations in Origin Between Coal and Petroleum," *Jour. Wash. Acad. Sciences*, Vol. 5 (1915), pp. 189–212.

intense deformation so that at a given point the fixed-carbon percentage increases with depth.

The isocarbs in the Appalachian province parallel the strike of the rocks and increase in value to the east in the direction of most intense diastrophism. Both here and elsewhere it has been found that there is very little oil above the 65 isocarb and not much gas above the 70 isocarb. Exceptions do occur, but they are infrequent. An outstanding exception was the discovery of a gas field in Rockingham County, Virginia, close to the 85 isocarb.

The nature of the rocks is a factor in the degree of metamorphism. If the sediments are largely unconsolidated, they can be intensely folded without much metamorphism. For example, the Tertiary sediments of California are more highly folded and less metamorphosed than the indurated Paleozoic rocks of western Pennsylvania. Also rocks which yield readily by folding or faulting are less metamorphosed than more rigid associated rocks.

The generally accepted reason as to just why there is little or no oil above the 65 per cent fixed-carbon line is that the metamorphism, even though it may be only incipient in terms of rocks, has destroyed the oil that existed there at one time. This idea is supported ". . . by evidence that oil pools once existed in the region between the present fields and the metamorphosed belt. Oil-impregnated sands crop out at the surface and have been found by drilling, but the oil is only a residue, wrapped around individual sand grains. . . ."[127]

In conclusion, it is obvious that even the early phases of metamorphism are inimical to the survival of hydrocarbons, especially oil. Where coals are available for analysis, the carbon-ratio concept has decided value, although negative in nature, in the search for new oil and gas fields. Deposits may be found in regions lying above the fixed-carbon limits stated, but the chances of such discovery are very much less than they are in areas of lower carbon ratio.

ORIGIN OF NATURAL GAS

Much less, proportionately, has been written about the origin of natural gas than about the origin of petroleum. Apparently it has been assumed

---

[127] K. C. Heald, "Essentials for Oil Pools," *Elements of the Petrol. Industry* (American Institute of Mining and Metallurgical Engineers, New York, 1940), Chapter 4, p. 42.

that petroleum and natural gas have a like origin, but this is not necessarily so.

The term "natural gas" usually implies hydrocarbons in gaseous form.
The commonest and most stable of these gaseous hydrocarbons is methane,
CH_4. Also present may be other gaseous hydrocarbons including ethane,
propane, and butane, and perhaps pentane, hexane, and heptane as
vapors. In addition, certain non-hydrocarbon gases either may be present
with the hydrocarbon gases or may occur in separate accumulations.
Among these erratic constituents are hydrogen sulfide, carbon dioxide,
nitrogen, and helium. Their origin will be discussed subsequently.

Natural gas may occur in five different environments: (1) dissolved in
petroleum, (2) with the oil but overlying it ("gas cap"), (3) in the same
trapping structure as oil but occupying different (usually higher) reservoir
beds, (4) in oil-producing districts but occupying separate traps, and (5)
in accumulations remote from known oil deposits. History has shown that
many gas fields which originally fell in the fifth category have been found,
upon further exploration, to belong in either the third or fourth. A few
large gas fields, for example those in northern Louisiana and southwestern
Wyoming, are some distance removed from the nearest known oil. Many
geologists believe that it is just a matter of time before large oil deposits
are found either beneath or in the vicinity of these gas accumulations.

There are two general theories concerning the origin of natural gas: (1)
the gas has separate genesis from oil and may never have been associated
with liquid petroleum; (2) gas is a by-product, or an end product, of the
origin and evolution of petroleum and was at one time in liquid phase.
Natural Gas with Separate Genesis from Petroleum. There
is no doubt that some gas has formed directly from putrefying organic
matter without passing through a liquid hydrocarbon phase. The so-
called "marsh gas," which is generated during the decay of vegetal matter
in bogs and swamps, is a well-known example of this. Bacteria aid in the
generation of marsh gas, which is nearly pure methane. A similar gas
may be produced by decaying animal matter. Methane from clay or mud
that contains an abundance of clams has been reported from at least two
places; in one of them, the gas is present in sufficient volume to permit
its local use in gas ranges.[128]

Another well-known occurrence of methane gas not connected in any
way with the formation of petroleum is the coal gas which is generated
from the fresh-water plant remains composing the coals and which
accumulates both in the coal seam and in any overlying porous rocks that
may be present. Such an origin is postulated for the gas found in glacial
drift immediately above the bedrock in parts of Illinois in sufficient abun-

[128] F. M. Van Tuyl and Ben H. Parker, "The Time of Origin and Accumulation of
Petroleum," *Quarterly Colo. School of Mines,* Vol. 36 (April, 1941), p. 49.

dance for farmhouse cooking and heating. An alternate source for this gas is decomposing younger organic matter such as peat beds.[129]

However, in recent years a number of geologists in northwestern Europe have come to believe that the gas fields of The Netherlands, the North Sea, and perhaps onshore England as well, had their source in underlying coal fields.

It has been suggested (1) that oil is derived from organic matter accumulating under marine conditions but that gas comes from land plants, and (2) that oil is generated from source beds which carry a "rich microflora" and that gas is derived in large part from coarse vegetal matter.[130]

Methane can also occur through inorganic processes. It has been detected in the gases given off by volcanoes. Baker[131] suggests that methane and other hydrocarbon gases may be produced by contact metamorphism through the combination of carbon from carbonaceous rocks with dissociated water vapor caused by the intrusion of molten rocks.

Natural Gas Derived from Petroleum. According to the commonly held concept of the evolution of petroleum, methane and the other hydrocarbon gases are products of that natural up-grading. These gases are probably given off throughout the evolutionary history of the petroleum, perhaps through catalysis. If this is true one would expect that, other things being equal, the older the hydrocarbon accumulation the higher the percentage of natural gas. This statement is certainly true in the Gulf Coast where the gas-oil ratio increases downward with geologic age as well as depth. A similar situation exists in many other parts of the world. The ultimate result of this natural progression would be the complete conversion of liquid petroleum to natural gas. Perhaps the best example of the application of this theory is the presence of natural gas in considerable abundance in the Appalachian province beyond (to the east of) the "extinction zone," where the carbon ratios run above 65 and the oil fields tend to disappear.[132]

The best testimony, however, for the generation of gas during the natural evolution of petroleum is the almost invariable association of the two together in nature, as outlined in the introduction to this section. Most natural gas occurs either within oil, in the gas cap of an oil deposit,

[129] Wayne F. Meents, "Tiskilwa Drift-Gas Area, Bureau and Putnam Counties, Illinois," *Illinois Geological Survey*, Circular 253 (1958).

[130] Paul D. Torrey, "Origin, Migration, and Accumulation of Petroleum and Natural Gas in Pennsylvania," *Problems of Petroleum Geology* (American Association of Petroleum Geologists, 1934), pp. 447–484.

[131] Charles Laurence Baker, "Possible Distillation of Oil from Organic Sediments by Heat and Other Processes of Igneous Intrusions; Asphalt in the Anacacho Formation of Texas," *Bull. Am. Assoc. Petrol. Geol.*, Vol. 12 (October, 1928), pp. 995–1003.

[132] C. David White, "Effects of Geophysical Factors on the Evolution of Oil and Coal," *Jour. Inst. Petrol. Technologists*, Vol. 21 (April, 1935), pp. 301–310.

or in a nearby reservoir. Natural gas has considerably greater mobility than oil, and once it has separated from the oil, it may travel through paths which the oil cannot follow and thus accumulate in separated reservoirs. In all probability, the "mother lode" of each currently isolated gas field will be discovered some day. Because of the generally wider spread of gas deposits, the discovery of gas has preceded that of oil in most areas in the past.

The direct generation of gas from decaying cellulose material is not and cannot be denied. It is doubtful, however, that the major accumulations of natural gas have had this origin.

Origin of the non-hydrocarbon natural gases.[133] Hydrogen sulfide is the only one of the four "erratic" gases considered here that commonly occurs with and has distribution approaching that of petroleum and hydrocarbon gas. At the same time, it is the only non-hydrocarbon gas which may come from the same original source materials as methane gas. Hydrogen sulfide and other sulfur compounds are produced by the decay of organisms under reducing (euxinic) conditions. This sulfur may become "fixed" as FeS_2 (marcasite or pyrite); it may escape in gaseous form or dissolved in water; or it may become trapped with the hydrocarbon-yielding material, to reappear later in natural gas or in high-sulfur crude oil.

Carbon dioxide occurs in a few instances with petroleum and hydrocarbon gas, but this association is probably fortuitous. Most of the known carbon dioxide accumulations, including all the large ones, are in localities which are not (as yet at least) productive of hydrocarbons. Carbon dioxide has been found in volcanic gases and in fumaroles. The calcination of limestone by intrusive magmas should also yield this compound. Generation by bacterial action and release from limestones undergoing ground-water solution are among other possible origins that have been suggested for this gas. Recent work by Lang [133a] with the mass spectrometer showed that the carbon dioxide gases in the samples analyzed were derived from limestone. He notes that igneous rocks are in close association with limestone in areas where carbon dioxide has been found in quantity, and he favors the theory that "the carbon dioxide gases

[133] C. E. Dobbin, "Geology of Natural Gases Rich in Helium, Nitrogen, Carbon Dioxide, and Hydrogen Sulphide," *Geology of Natural Gas* (American Association of Petroleum Geologists, 1935), pp. 1053–1072; F. A. Paneth, "Helium," *The Science of Petroleum* (Oxford University Press), Vol. 2 (1938), pp. 1511–1516; R. R. Bottoms, "Occurrence and Production of Helium in the United States," *ibid*, pp. 1517–1523; C. C. Anderson and H. H. Hinson, "Helium-bearing Natural Gases of the United States," *U. S. Bureau Mines Bull.* 486 (1951).

[133a] Walter B. Lang, "The Origin of Some Natural Carbon Dioxide Gases," *Jour. Geophysical Research*, Vol. 64 (January, 1959), pp. 127–131.

originated from the breakdown of limestone in contact with or adjacent to igneous instrusions."

Helium and nitrogen are usually, but not necessarily, together. Helium is much less abundant than nitrogen, but because of its value, it has received far more attention in the technical publications. Its presence is also easier to explain. Helium is a product of radioactive disintegration. During geologic time, a considerable volume of this gas, which became trapped in the overlying sediments, was released from radioactive minerals in the Precambrian crystalline rocks, from hydrothermal vein deposits, or from sedimentary layers containing clastic grains of these minerals. In this regard it is interesting to note that a pegmatite deposit famous for its radioactive minerals lies at the surface of the stripped Precambrian in the Central Mineral Region of Texas, a few hundred miles southeast of the Cliffside structure in the Panhandle, where Permian reservoirs overlying the buried Precambrian Amarillo Mountains contain the largest known helium deposit. Other helium accumulations have been found in Utah, where the Colorado Plateau is notable for its sedimentary uranium deposits. Again it is probably coincidental that some helium occurs with hydrocarbon natural gas; both types of gas merely happen to get caught in the same trap.

It is probable that nitrogen, like the helium, comes originally from the crystalline basement rocks. Nitrogen has been detected escaping from a number of metal mines, in some places in lethal (because of its non-respiratory characteristics) quantities.[134] Ruedemann and Oles,[135] have suggested that the nitrogen may be residual from air trapped in the strata at the time of deposition. However, there is a question whether air could be trapped in adequate volume in sediment accumulating beneath water. Fixed nitrogen has been reported as a consistent constituent of igneous rocks, both granites and basalts.[136] The quantity present is in the neighborhood of 25 parts per million.

PROBABLE SOURCE ROCKS

Among the prerequisites for the commercial occurrence of petroleum in the earth's crust (source rocks, reservoir rocks, reservoir seals, and traps) source rocks are much the most important. The other three essentials may be present

---

[134] C. E. Dobbin, *op. cit.*, *p. 1061*.

[135] Paul Ruedemann and L. M. Oles, "Helium — Its Probable Origin and Concentration in the Amarillo Field, Texas," *Bull. Am. Assoc. Petrol. Geol.*, Vol. 13 (July, 1929), pp. 799–810.

[136] Robert S. Ingols and Alfred T. Navarre, " 'Polluted' Water From the Leaching of Igneous Rock," *Science*, Vol. 116 (Nov. 28, 1952), pp. 595–596.

in great abundance, but without source rocks there will be no oil or gas. In petroleum genesis there is no such thing as immaculate conception.

Source rocks vary widely in fertility. In rare cases sediments may be virtually devoid of organic matter, as in evaporites at one extreme and deposits of boulders, gravel, and sand dumped into a graben at the other. Fortunately, in many parts of the world the paleogeologic conditions were favorable for the deposition of organic deposits, and rich source beds occur in the stratigraphic section. However, the abundance of the potential source rocks is of equal importance. This involves both thickness and areal (lateral) extent. Where these are minimal small oil or gas fields may be found, but the big prizes simply are not there.

Where both adequate source rock fertility and volume are concerned we have commercial oil and gas fields, some of them in the super-giant class. As an example of the latter, the ten top oil fields in ultimate recovery are the Persian Gulf countries, with six, the USSR with two, and Venezuela and the United States (Prudhoe Bay, Alaska), with one each. The top ten gas fields are the USSR with 6, and The Netherlands-West Germany, Iran, U.S. (Kansas-Oklahoma-Texas), and Algeria with one each.[137]

Below the top ranking fields are accumulations of oil and gas of adequate size to be commercially profitable. Here lie most of the world's petroleum fields. Unfortunately, below this group is a graveyard of frustrated hopes and considerable monetary loss due to a paucity of adequate source rocks. In these areas there may be enormous topographically naked anticlines, as in the Negev of southern Israel, or there may be oil or gas seeps at the surface, or there may be encouraging "shows" of hydrocarbons in wildcat wells drilled. In some instances this encouragement leads to the development of minuscule oil or gas fields of questionable profitability.

There are many examples of probable inadequate source rock areas. Parenthetically, I'll be very happy, if the future proves me to be wrong on any of these. First there is onshore Israel, which has one relatively small oil field with satellite on the coastal plain, and a nest of small gas fields in the hills to the west of the Dead Sea graben. The source rock for the oil deposits is a Lower Cretaceous shale; for the gas it is an Upper Jurassic shale. Both are relatively local in the areal extent of their fertility. Nearby Jordan, Lebanon, and western Syria have had numerous dry holes but no producers. Perhaps the best examples in the United States are the sedimentary basins of western Oregon western Washington.[138]

Identification of Source Rocks. This has long been a problem for petroleum geologists. The easiest identification of source rocks has been where the

[137]Halbouty, Michel T., Editor, Geology of Giant Petroleum Fields: Am. Assoc. Petroleum Geologists, Memoir 14, Tables 1 and 2, 1970.
[138]K. K. Landes, Petroleum Geology of the United States, John Wiley and Sons, New York, p. 502–505, 1970.

oil or gas occurs in lenticular sandstone bodies, as in the "shoestring" sand fields of southeastern Oklahoma and northeastern Oklahoma. In these examples, the sand lenses are completely surrounded by the highly carbonaceous Cherokee shale of Pennsylvania age.[139]

Hedberg, Sass, and Funkhouser[140] believe that the shales immediately above and below each of the productive sands in the Greater Oficina area of Venezuela were the source rocks. They point out that these shales are gray, owing to the presence of finely divided carbonaceous matter, whereas the overlying Freites shales, which are associated with barren sands, are green and lack carbonaceous matter. In 1954 Hunt, Stewart, and Dickey[141] reported on the hydrocarbons in the Eocene sediments of the Uinta Basin, Utah, which include the Green River "oil shales" (actually these rocks are bituminous siliceous dolomites). They found that the natural hydrocarbons occurring in this area are either ozocerite, albertite, gilsonite, or wurtzilite. Each one of these petroleum types is found closely associated geologically with a stratigraphic unit, and chemical examination of the latter shows that it contains the same characteristic hydrocarbon type. Therefore it is assumed that although geological connection cannot be traced, it may still be possible to connect an oil with its mother rock by chemical studies. On the other hand Brenneman and Smith have not been able as yet to develop a generally applicable method of correlating crude oils with their source rocks.[142]

In 1956 Hunt and Jamieson[143] reported as follows:

Practically all shales and carbonate rocks contain indigenous organic matter disseminated in three forms: (1) soluble hydrocarbons, which are similar in composition to the heavier fractions of crude oil found in reservoir rock, (2) soluble asphalt, which is similar to the asphaltic constituents of crude oil, and (3) insoluble organic matter (kerogen), which is pyrobituminous in nature. Non-reservoir ancient sediments have been found to contain up to five

---

[139] L. N. Neumann, N. W. Bass, R. L. Ginter, S. F. Mauney, Charles Ryniker, and H. M. Smith, "Relationship of Crude Oils and Stratigraphy in Parts of Oklahoma and Kansas," *Bull. Am. Assoc. Petrol. Geol.,* Vol. 25 (September, 1944), pp. 1801–1809.

[140] H. D. Hedberg, L. C. Sass, and H. J. Funkhouser, "Oil Fields of Greater Oficina Area, Central Anzoategui, Venezuela," *Bull. Am. Assoc. Petrol. Geol.* Vol. 31 (December, 1947), p. 2137.

[141] John M. Hunt, Francis Stewart, and Parke A. Kickey, "Origin of Hydrocarbons of Uinta Basin, Utah," *Bull. Am. Assoc. Petrol. Geol.,* Vol. 38 (August, 1954), pp. 1761–1798.

[142] M. C. Brenneman and P. V. Smith, Jr., "The Chemical Relationships Between Crude Oils and Their Source Rocks," *Habitat of Oil* (American Association of Petroleum Geologists, Special Vol., 1958), pp. 818–849.

[143] John M. Hunt and George W. Jamieson, "Oil and Organic Matter in Source Rocks of Petroleum," *Bull. Am. Assoc. Petrol. Geol.,* Vol. 40 (March, 1956), pp. 477–488; Reprinted in *Habitat of Oil* (American Association of Petroleum Geologists Special Vol., 1958), pp. 735–756; James P. Forsman and John M. Hunt, "Insoluble Organic Matter (Kerogen) in Sedimentary Rocks of Marine Origin," *ibid*, pp. 747–778.

times as much oil as that reported from recent unconsolidated sediments off the Gulf and California coasts. A typical ancient petroleum source rock such as the Frontier shale in the Powder River Basin of Wyoming, which has yielded millions of barrels of oil to reservoirs in the past, still contains 6 barrels of oil, 20 barrels of asphalt, and about 250 barrels of kerogen per acre-foot.

The distribution of this oil, asphalt, and kerogen within the non-reservoir rocks of a sedimentary basin varies between different facies of the same formation.

In 1969[144] and 1973[145] two papers were published that cover relatively simple and practical chemical procedures for identifying probable source rocks. We now know that source rocks do contain non-expelled oil and gas, in some instances perhaps more than was squeezed out in the first place.

There is no argument but that shales are the number-one source rocks, since (1) shale is by far the most abundant sedimentary rock, and (2) the environmental conditions governing shale despoition also favor the deposition and entombment of organic matter. Second in importance to shale as source rock is limestone. However, contrary to considerable popular opinion on that subject reef limestones are very unlikely source rocks. The reef-making organisms are never present in great abundance at any one time. Only the penthouse apartments of the reef are occupied, and bacteria and scavengers have removed all traces of possible hydrocarbon producing organic matter from the lower, abandoned apartments.

The limestones that are source rocks are those that were originally either a calcareous ooze or a fine clastic sediment derived from a carbonate rock terrain. The latter may have considerable intermixed clay, and the lithified rock is classified as an argillaceous limestone.

The evidence is unmistakable in several areas that limestone has been the source rock. A richly organic limestone, once a calcareous ooze, but containing some organic matter as well, unconformably overlies the principal oil reservoir in Michigan, also a limestone. Storage is in the top few feet of this limestone, due to leaching prior to the deposition of the overlying ooze. Below the leached zone the limestone is tight.

Possible Fresh-Water Source Rocks. The traditional view that source materials must be deposited in marine environment has been losing ground in recent years. Noble[146] classified this belief as one of the "prejudices" restricting the discovery of new oil fields. Certainly as far

[144]R. Shoresh, Geochemical Project Source Rocks Investigation, The Institute for Petroleum Research and Geophysics, Holon, Israel, Report No. 1039, August, 1969.
[145]L. R. Snowdon and R. G. McCrossan, Identification of Petroleum Source Rocks Using Hydrocarbon Gas and Organic Carbon Content, Geol. Survey of Canada, Paper 72-36, Ottawa, 1973.
[146]Earl B. Noble, "Geological Masks and Prejudices," *Bull. Am. Assoc. Petrol. Geol.*, Vol. 31 (July, 1947), pp. 1109-1117.

as potential source materials in recent deposits are concerned, lake sediment can be just as promising as sea-floor sediment, as discussed earlier in this chapter. In fact, presumably indigenous oil is present in sufficient quantity and concentration in some modern lacustrine deposits as to be possibly commercial.[147]

The tradition of a marine source was so strong up until a few years ago that all accumulations of oil or gas in continental reservoirs were attributed to migration from marine source rocks. Recently, however, oil has been discovered in occurrences difficult to explain by migration from marine source rocks. In the Rocky Mountain province, oil and gas have been found in non-faulted, non-marine, Tertiary strata in both the Powder Wash[155] and the East Hiawatha[156] fields. A field in Shensi province in China has been described[157] in which both the reservoirs and the probable source rocks occur in a thick series of early Mesozoic continental rocks. Several recently discovered California fields are also producing from continental rocks, and the probable source rocks have similar origin.

A symposium was held in 1954 on "Oil and Gas in Continental Beds." Papers were given describing the occurrence of oil and gas in non-marine sediments in Colombia,[158] West Pakistan,[159] Utah,[160] and other parts of the intermontane west[161] and England.[162] In every instance the problem of bringing the oil from a marine source rock is difficult if not impossible. In the Green River and Uinta basins of the Wyoming-Utah-Colorado corner, for example, the oil is in Tertiary lake beds. If not indigenous but of marine origin, this oil would have had to pass through

[147] F. M. Swain, "Relationship of Study of Modern Lake Sediments to Recognition of Non-Marine Source Beds," *Am. Assoc. Petrol. Geol.* (Abstract), Annual Meeting Program (1956), pp. 21-23.

[155] W. T. Nightingale, "Petroleum and Natural Gas in Non-Marine Sediments of Powder Wash Field in Northwest Colorado," *Bull. Am. Assoc. Petrol. Geol.*, Vol. 22 (August, 1938), pp. 1020-1047.

[156] C. E. Dobbin, "Exceptional Oil Fields in Rocky Mountain Region of United States," *Bull. Am. Assoc. Petrol. Geol.*, Vol. 31 (May, 1947), pp. 797-823.

[157] C. H. Pan, "Non-marine Origin of Petroleum in North Shensi, and the Cretaceous of Szechuen, China," *Bull. Am. Assoc. Petrol. Geol.*, Vol. 25 (November, 1941), pp. 2058-2068.

[158] W. S. Olson, "Source-Bed Problem in Velasquez Field, Colombia," *Bull. Am. Assoc. Petrol. Geol.*, Vol. 38 (August, 1954), pp. 1645-1652.

[159] E. S. Pinfold, "Oil Production from Upper Tertiary Fresh-Water Deposits of West Pakistan," *Bull. Am. Assoc. Petrol. Geol.*, Vol. 38 (August, 1954), pp. 1653-1660.

[160] John M. Hunt, Francis Stewart, and Parke A. Dickey, "Origin of Hydrocarbons of Uinta Basin, Utah," *Bull. Am. Assoc. Petrol. Geol.*, Vol. 38 (August, 1954), pp. 1671-1698.

[161] Wayne M. Felts, "Occurrence of Oil and Gas and Its Relation to Possible Source Beds in Continental Tertiary of Intermountain Region," *Bull. Am. Assoc. Petrol. Geol.*, Vol. 38 (August, 1954), pp. 1661-1670.

[162] P. E. Kent, "Coal Measures of England," *Bull. Am. Assoc. Petrol. Geol.*, Vol. 38 (August, 1954), pp. 1699-1713.

a thick series of relatively impervious materials in order to end in sand-stone lenses.

In addition to distinguishing continental deposits from marine sediments by means of fossils (which aren't always present), such strata also can be distinguished by definite geochemical differences, according to Degens, Williams, and Keith.[163]

Geologic Age of Alleged Source Rocks. No doubt potential source rocks were deposited somewhere during every geological period, including the Precambrian.[164] As would be expected, the periods of greatest marine submergence contain the largest number of possible source rocks. Periods of benign climates with most luxuriant plant growths also have an unusual number of possible source rocks.

The better known of the alleged source rocks are organic shales of wide areal distribution. Among the most famous of these are the Utica and other Ordovician shales, the widespread Chattanooga shale, which is of Devonian-Mississippian age, the Cherokee shale of the Mid-Continent Pennsylvanian, the Cretaceous Eagle Ford shale of Texas, and the Miocene Monterey shale of the California oil fields.

Relationship between Presence of Possible Source Beds and Distribution of Oil Fields. The importance to be attached to the presence or absence of possible source beds in an area undergoing exploration depends to a considerable extent upon the distance through which one concedes that oil can migrate, a subject to be discussed in Chapter 9. Those who believe that oil can migrate no great distance consider the nature and abundance of source material of utmost importance.[165]

The erratic distribution of oil can be most easily explained by variations in the amount of source material locally available. Heald[166] cites a number of examples of empty or near empty traps in the immediate vicinity of traps filled with oil. In the Conroe, Texas field the trap is full of oil and gas, but nearby are two equally good traps that contain but little oil in comparison to Conroe. The north flank of the regional anticline that

---

[163] E. T. Degens, E. G. Williams, and M. L. Keith, "Environmental Studies of Carboniferous Sediments—Part I: Geochemical Criteria for Differentiating Marine From Fresh-Water Shales," *Bull. Am. Assoc. Petrol. Geol.*, Vol. 41 (November, 1957), pp. 2427-2455; Part II: "Geochemical Criteria," Vol. 42 (May, 1958), pp. 981-997.

[164] F. M. Swain, A. Blumentals, and N. Prokopovich, "Bituminous and Other Organic Substances in Precambrian of Minnesota," *Bull. Am. Assoc. Petrol. Geol.*, Vol. 42 (January, 1958), pp. 173-189; S. A. Tyler, E. S. Barghoorn, and L. P. Barrett, "Anthracitic Coal from Precambrian Upper Huronian Black Shale of the Iron River District, Northern Michigan," *Bull. Geol. Soc. Am.*, Vol. 68 (October, 1957), pp. 1293-1304.

[165] Frank R. Clark, "Origin and Accumulation of Oil," *Problems of Petroleum Geology* (American Association of Petroleum Geologists, Tulsa, Oklahoma, 1934), p. 334.

[166] K. C. Heald, "Essentials for Oil Pools," *Elements of the Petroleum Industry* (American Institute of Mining and Metallurgical Engineering, 1940), Chapter IV, pp. 27-28.

overlies the Amarillo Mountains and crosses the Texas Panhandle from east to west is full of oil and gas. Very little oil or gas has been found on the south flank (so far, at least), yet the geological conditions are the same on both sides. Similar relationships between full traps and empty traps exist in the Great Valley of California. Quoting Heald: "The absence of source must be responsible for the failure to discover oil in extensive areas where all other conditions exist." Also, "Recognition that source conditions are variable may justify the search for oil pools in areas where the other requisites are believed to be mediocre or poor, for if a great deal of oil has been formed it will accumulate if given any encouragement."

6

RESERVOIR ROCKS

The accumulation of oil or gas into a commercial deposit requires a combination of reservoir rock, seal rock, and trap. The reservoir rock is the container; it is usually much more extensive than the hydrocarbon deposit that has been localized by a trap. Beyond the confines of the oil or gas pool, the reservoir rock is almost always filled with water.

General Qualifications for Reservoir Rocks.[1] The qualifications of a reservoir rock are simple: it must have room enough to store a worthwhile volume of hydrocarbons, and the storage facilities must be such that the contained oil or gas discharges readily when the reservoir is penetrated by a well. Any buried rock, whether it is igneous, sedimentary, or metamorphic, that meets these specifications may be utilized by migrating hydrocarbons as a reservoir. Actually, however, most of the world's oil and gas occurs in sandstones or carbonate rocks simply because these are by far the most common rocks that qualify as reservoirs in those segments of the earth's crust containing generating or migrating hydrocarbons. The reservoir character of a rock may be an original feature of the rock (intergranular porosity of sandstones) or a secondary character resulting from chemical changes (solution porosity of limestones) or physical changes (fracturing of any brittle type of rock). The secondary changes may merely add to the storage capacity of an original reservoir or they make a reservoir out of an originally inhospitable rock.

In order to contain enough oil or gas to make extraction profitable, a reservoir rock must exceed a minimum porosity and a minimum thickness.

---

[1] David Donaghue, "Fundamental Data on Subsurface Reservoirs," *Bull. Am. Assoc. Petrol. Geol.*, Vol. 28 (December, 1944), pp. 1754–1755; P. G. Nutting, "Some Physical and Chemical Properties of Reservoir Rocks Bearing on the Accumulation and Discharge of Oil," *Problems of Petroleum Geology* (American Association of Petroleum Geologists, 1934), pp. 825–832; G. E. Archie, "Introduction to Petrophysics of Reservoir Rocks," *Bull. Am. Assoc. Petrol. Geol.*, Vol. 34 (May, 1950), pp. 943–961.

Porosity is the percentage of total reservoir rock volume occupied by voids (interstices). A common method of determining porosity is to "take a sample of rock, extract the fluids present, and obtain the bulk volume of the sample either by direct measurement or by observing the volume of liquid displaced when the sample is totally submerged in the liquid. A dry weight of the sample is taken, and the interstices of the rock are filled with a liquid of known density. The weight of the rock, plus the liquid, is thus obtained. The pore volume can then be calculated by subtracting the dry weight of the rock from the weight of the rock plus liquid and dividing this difference in weight by the density of the liquid. The pore volume divided by the bulk volume is the fractional porosity, which when multiplied by 100 expresses the porosity in per cent." [2]

The value of the porosity and thickness minima depend upon local conditions. Most producing reservoir rocks have porosities above 10 per cent and thickness greater than 10 feet. However, a rock with lesser porosity may be exploitable if the thickness is great, or a thinner rock may be developed successfully if the porosity is unusually large. The value of the oil and the production cost also enter into the question of whether a reservoir may be exploitable.

In addition to adequate porosity and thickness, a reservoir rock must have a certain degree of lateral continuity, or the volume of oil stored will not be adequate. In some areas lateral persistence of porosity cannot be taken for granted. Many wildcat wells have failed to become discovery wells because the reservoir rock was locally "tight." The first well drilled on the Eldorado dome in Butler County, Kansas, was a dry hole, but it was later completely surrounded by producing wells. On the other hand, other wildcatters have discovered oil only to find subsequently that they had a one-well pool with dry offsets owing to a decrease in porosity in the reservoir rock north, south, east, and west of the discovery well.

Some sheet sandstones and some porous carbonate rocks are true regional reservoir rocks, containing water or hydrocarbons everywhere in the subsurface. At the other extreme are porous zones of such limited lateral extent that they are, in themselves, traps for oil or gas accumulation. Many carbonate rock formations are notorious for their erratic porosities, and in some areas sandstones are equally unreliable. In districts too numerous to mention, not only must the oil-seeker locate a trap, but also the trap must be baited with a porous zone in which hydrocarbons can accumulate.

The ability of a rock to discharge its hydrocarbon content is dependent

[2] Charles D. Russell and Parke A. Dickey, "Porosity, Permeability, and Capillary Properties of Petroleum Reservoirs," *Applied Sedimentation* (John Wiley & Sons, New York, 1950), Chapter 32, pp. 586–587.

upon its *permeability*. There are three requisites to permeability: (1) porosity, (2) interconnecting pores, and (3) pores of supercapillary size. Although a permeable rock also must be porous, a porous rock does not have to be permeable. Pumice is porous, but not permeable, because the voids are not interconnecting. Shale may be quite porous but impermeable because the pores are capillary or subcapillary in size, thereby preventing free movement of the contained fluids.

Permeability is defined as

the ability of the reservoir rock to transmit fluid. It is normally expressed in millidarcys. A medium has a permeability of one millidarcy if it will transmit one milliliter per second of a fluid of 1 centipose vicosity through a cross section of 1 square centimeter of rock, under a pressure gradient of 1 atmosphere per centimeter. The usual method of obtaining permeability is to measure the rate of flow of air across a sample of the reservoir rock. The sample of rock should be of uniform cross section and large enough so that small errors in measurement of its length and area do not significantly affect the permeability calculation.[3]

Rock possessing both porosity and permeability is referred to as having "effective porosity." The amount of effective porosity is the percentage of the total rock volume occupied by interconnecting voids of supercapillary size. It is the effective porosity that determines the amount of liquid or gas that can move out of a reservoir rock into a well.

Porosity[4] is both created and destroyed by natural geological processes. *Primary porosity* in sedimentary rocks is that resulting from the accumulation of detrital or organic material in such a manner that openings or voids are left between grains of sand or fragments of shells. As a matter of fact, it is impossible to pack such material, especially spheroidal grains, without leaving considerable interconnecting void space. Primary porosity is of greatest importance in sandstone reservoirs. *Secondary porosity* is the result of some type of geological activity after the sediment has been converted into rock. It is of great importance in carbonate rock reservoirs. The most common types are solution cavities, which range in size from that of a pinhead to the Carlsbad Caverns, and fissures or fractures produced mainly by rock jointing. The fracture type of rock opening is rarely visible in well cuttings but is more readily discernible in cores. Porosity determinations made on rock samples are always *minimum* figures because of the difficulty of evaluating the void space present in fractures.

[3] John W. Tynan, "Everyday Reservoir Engineering," *Mines Mag.* Vol. 42 (November, 1952), pp. 50–52.

[4] S. E. Coomber, "The Porosity of Reservoir Rocks," *Science of Petroleum* (Oxford University Press, 1938), Vol. 1, pp. 220–223.

Heald emphasizes that bedding planes may interrupt the effective porosity:

Most of the oil and gas fields of the world are in sedimentary rocks, and most sedimentary rocks contain bedding planes or partings on which commonly, although not invariably, there is a film or selvage of clay that may be an effective barrier to the free movement of oil or gas from the rock on one side of the bedding plane to the rock on the other. Even where there is no clay film there may be induration or cementation at the bedding plane, which will reduce the permeability. These bedding planes are believed to be responsible for the observed phenomenon of very low vertical permeability as manifested by the behavior of wells even though the measurements in the laboratory on short cylinders of "plugs" of the reservoir rock may have shown high permeability values. The samples tested in the laboratory rarely include bedding planes. In the process of coring to secure samples, the rock tends to break along clay partings and bedding planes so that the fragments that are available for measurements do not include the barriers to transverse movement of fluids.[5]

The geological activities that diminish and may even destroy porosity are compaction, cementation, recrystallization, and granulation. Compaction is a squeezing effect brought about by the weight of overlying rock; some rocks are subjected to additional squeezing by diastrophic forces. In clays compaction may be as much as 50 per cent, but the shrinkage is much less with sands. Squeezing of any sediment produces a closer packing of the grains, while at the same time excess fluids are driven out.

Loose sand grains become sandstones by compaction and cementation.[6] If the cementation is carried to completion the porosity is destroyed. Fortunately, this rarely occurs and most sandstones are left with enough primary porosity to store large quantities of water, oil, or gas. Maxwell and Verrall[7] have attempted to determine experimentally what happens to sandstone porosity when the sandstone becomes deeply buried. They find that both compaction and cementation are aided by high temperatures and the presence of alkaline solutions. "Experiments indicate that a pure, well-sorted quartz sand, saturated with seawater or fluid of similar composition and pH, probably would not retain appreciable porosity ($\pm$ 10 per cent) if buried to depths of 25,000 feet or more, at temperatures above

[5] K. C. Heald, "The Petroleum Reservoir," *Conservation* (American Institute of Mining and Metallurgical Engineers, New York, 1951), Chapter III, pp. 55 and 57.

[6] Charles M. Gilbert, "Cementation of Some California Tertiary Reservoir Sands," *Jour. Geol.*, Vol. 57 (January, 1949), pp. 1–17; C. A. Fothergill, "The Cementation of Oil Reservoir Sands and Its Origin," *Proc. Fourth World Petrol. Congress* (1955), Sec. I, pp. 301–314.

[7] John C. Maxwell and Peter Verrall, "Low Porosity May Limit Oil in Deep Sands," *World Oil*, Part 1 (April, 1954), pp. 106 et seq; Part 2 (May, 1954), pp. 102–104.

270° C." Deeper drilling may prove this conclusion, based upon experimental data, to be on the pessimistic side.

Lowry[8] found that the early Paleozoic quartz sandstones of Virginia had suffered porosity loss due to "welding," the solution of silica at grain contacts and concurrent reprecipitation of that silica in the pore spaces as quartz outgrowths. Not only have these sandstones been buried deeply in the past, but they have also been squeezed by diastrophism. Hughes and Cooke[9] determined that the original pore volume of dry sandstone samples subjected to hydrostatic pressures up to 14,500 pounds per square inch was diminished from 3 to $7\frac{1}{2}$ per cent. As would be expected, the greater part of the contraction took place in the lower pressure ranges due to compaction packing. Another approach has been to determine the effects on permeability resulting from compression of sandstone reservoir rocks. In this instance the amount of clay in the sandstone proved to be the most important factor. "For sandstones containing large amounts of clay, preliminary tests indicated very large reductions in permeability with increasing effective overburden pressures."[10] Fatt[11] found by experiment that sandstone compressibilities are functions of pressure, mineral composition, and texture.

Recrystallization may destroy any pre-existing porosity by changing the rock into a dense interlocking aggregate of crystals. It is a common feature of metamorphism, and that is one of the reasons why metamorphic rocks are characteristically impermeable.

Granulation or crushing may lower the porosity and destroy the permeability by squeezing the rock. The rocks that have been deeply buried in the geologic past, and those that have been strongly compressed by lateral diastrophic forces, have suffered granulation to varying degrees. A study of five sandstones ranging in depth from 2885 to 8343 feet, in two Wyoming wells, shows a progressive change with depth from the original loose random packing to a tighter packing resulting from pressure.[12] Pressure effects included crushing and yielding of the mineral grains.

[8] W. D. Lowry, "Factors in Loss of Porosity by Quartzose Sandstones of Virginia," *Bull. Am. Assoc. Petrol. Geol.*, Vol. 40 (March, 1956), pp. 489–500.

[9] D. S. Hughes and C. E. Cooke, Jr., "The Effect of Pressure on the Reduction of Pore Volume of Consolidated Sandstones," *Geophysics*, Vol. XVIII, No. 2 (April, 1953), pp. 298–309.

[10] A. S. McLatchie, R. A. Hemstock, J. W. Young, "The Effective Compressibility of Reservoir Rock and Its Effects on Permeability," *Jour. Petrol. Technol.* Vol. 10 (June, 1958), pp. 49–51.

[11] I. Fatt, "Compressibility of Sandstones at Low to Moderate Pressures," *Bull. Am. Assoc. Petrol. Geol.*, Vol. 42 (August, 1958), pp. 1924–1957.

[12] Jane M. Taylor, "Pore Space Reduction in Sandstones," *Bull. Am. Assoc. Petrol. Geol.*, Vol. 34 (April, 1949), pp. 701–716.

With the exploration for oil extending to depths below 20,000 feet, the persistence of porosity with depth becomes a subject of utmost importance.[13] It is possible that sufficient intergrain void space in sandstone reservoirs carries downward as far as sedimentary basins extend into the earth's crust. It is less likely that limestone-solution cavities have comparable persistence with depth, and it is highly improbable that fractures and fissures retain much permeability at even those levels now being explored. Some relatively deep mines in crystalline rocks which have a water problem in the upper levels are dry in the lower levels because of the drawing together of the fissure walls with depth. It can be concluded that in general reservoir porosities tend to diminish with depth of burial.

Oil-field reservoir rocks are studied best by means of cores, obtained during the drilling of both exploration and exploitation wells. These cores permit the engineering department, or a custom-core laboratory, to obtain quantitative data regarding the porosity, permeability, and oil saturation (percentage of void space occupied by oil). The laboratory evaluation of cores is known as core analysis.[14] The usual procedure is to drill out small cores ("plugs") of standard size from the well core. These plugs are taken both parallel with and normal to the bedding planes. The porosity and other data are determined by testing them.

[13] James S. Cloninger, "How Deep Oil or Gas May Be Expected." *World Oil*, Vol. 130 (May, 1950), pp. 57–62.

[14] J. A. Klotz, "Principles of Core Analysis," *Jour. Petrol. Technol.*, Vol. 4 (August, 1952), pp. 28 et seq.; J. G. Crawford, "Interpretation of Core Analysis," *Mines Mag.*, Vol. 43 (October, 1953), pp. 124–126; Ben A. Elmdahl, "How to Use Core Analysis . . . to Find Oil," *Oil and Gas Jour.* (Feb. 27, 1956), pp. 104 et seq.; L. C. Locke and Jack E. Bliss, "Core Analysis Technique For Limestone and Dolomite," *World Oil* (September, 1950), pp. 204 et seq; E. DeGolyer and Harold Vance, "Bibliography of the Petroleum Industry," *Bull. Agr. and Mech. Coll. Texas*, 83 (1944), pp. 355–360 (6 pages of references); John R. Suman, "Drilling, Testing, and Completion," *Elements of the Petroleum Industry* (American Institute of Mining and Metallurgical Engineers, New York, 1940), pp. 187–196; M. D. Taylor, "Determination of the Porosity, Permeability, and Saturation of Core Samples," *Oil and Gas Jour.*, Vol. 40 (Nov. 20, 1941), pp. 40 et seq.; G. E. Archie, "Electrical Resistivity an Aid in Core Analysis Interpretation," *Bull. Am. Assoc. Petrol. Geol.*, Vol. 31 (February, 1947), pp. 350–366; John C. Calhoun, Jr., "Methods of Obtaining Porosity," *Oil and Gas Jour.*, Vol. 47 (Nov. 18, 1948), p. 121; F. B. Plummer and P. F. Tapp, "Technique of Testing Large Cores of Oil Sands," *Bull. Am. Assoc. Petrol. Geol.*, Vol. 27 (January, 1943), pp. 64–84; James A. Lewis, "Core Analysis, an Aid to Increasing the Recovery of Oil," *Am. Inst. Min. Met. Engr. Tech. Publ.* 1487 (1942), pp. 8; R. A. Morse, P. L. Terwilliger, and S. T. Yuster, "Relative Permeability Measurements on Small Core Samples," *Producers Monthly*, Vol. 11 (August, 1947), pp. 19–25; "Salt Water Yardstick," *The Link*, Vol. 12 (June, 1947), pp. 8–11; Donuil Hillis, "Colorimetric Method of Determining Percentage of Oil in Cores," *Bull. Am. Assoc. Petrol. Geol.*, Vol. 21 (November, 1937), pp. 1477–1485; John G. Caran, "Core Analysis," *Subsurface Geologic Methods* (Colorado School of Mines, 1949), pp. 238–264.

Sandstone Reservoirs.[15] Throughout the world, sandstone is a most important reservoir rock. Locally it may be exceeded in the volume of oil produced by carbonate rock, but in many great oil districts limestones and dolomites are entirely absent from the stratigraphic section. Sandstones possess the properties of porosity and permeability to a greater and more consistent extent than any other abundant rock. Furthermore, they may be thick, in some places several hundred feet thick. Sandstones may or may not have great lateral continuity.

Sandstone porosity is of two types, intergranular and fracture. The intergranular porosity is the net void space remaining between the constituent grains of the rock. Initial intergranular porosity depends mainly upon the extent to which the sand is graded (sorted).[16] Moderately rounded sand grains, such as commonly compose sandstones, which are all approximately the same size, settle in water into an aggregate having a porosity of 35 to 40 per cent. With a mixture of sizes the porosity becomes less, for the smaller grains partially fill the interstices between the larger grains, which otherwise would be left open. Ill-sorted sands have porosities of 30 per cent and less. If the sediment is a mixture of sand grains and clay, the rock may have no effective porosity whatever.

As a general rule, cementation lowers the porosity percentage from an initial 30 to 40 down to 10 to 20. Not even all this reduced space is available, however, for hydrocarbon storage, as most oil-bearing sandstones contain some interstitial water. Bartle[17] has calculated the effective porosity of a gas sandstone reservoir in northwestern Missouri at 7 per cent.

Casts made of sandstone pores have shown that, although some sandstones have suffered no postdepositional changes other than cementation, others have been leached so that the pores may be even larger than the largest grains. Some of the pores in the Bradford "sand," Bradford, Pennsylvania, could hold from 10 to 100 of the surrounding sand grains.[18] Although solution cavities are commonly thought of for carbonate reservoirs, they obviously also can be important in sandstone reservoirs.

[15] S. E. Coomber, "The Porosity of Reservoir Rocks," *Science of Petroleum* (Oxford University Press, 1938), Vol. 1, pp. 220–223; P. G. Nutting, "Some Physical and Chemical Properties of Reservoir Rocks Bearing on the Accumulation and Discharge of Oil," *Problems of Petroleum Geology* (American Association of Petroleum Geol., 1934), pp. 825–832; Harry M. Ryder, "Character of Pores in Oil Sand," *World Oil*, Vol. 127 (April, 1948), pp. 129–134; S. M. Paine, "Petrophysical Analysis of Some Wilcox Wells," *Petrol. Technol.*, Vol. 8 (October, 1956), pp. 25–31.

[16] Wilbur F. Cloud, "Effects of Sand Grain Size Distribution upon Porosity and Permeability," *Oil Weekly*, Vol. 103 (Oct. 27, 1941), pp. 26–32.

[17] Glenn G. Bartle, "Effective Porosity of Gas Fields in Jackson County, Missouri," *Bull. Am. Assoc. Petrol. Geol.*, Vol. 25 (July, 1941), pp. 1405–1409.

[18] P. G. Nutting, "Some Physical and Chemical Properties of Reservoir Rocks Bearing on the Accumulation and Discharge of Oil," *Problems of Petroleum Geology* (American Association of Petroleum Geologists, 1934), p. 827.

"Tight" sandstones, or tight zones within a sandstone, may be due to nearly complete cementation, but more often they are the result of inadequate sorting of the detrital material at time of deposition. Intermixed clay or silt, or flakes of mica, make a sandstone virtually impermeable. Some apparently tight sandstones carry water or oil in fractures, in which event the actual porosity is many times the measurable porosity of a core sample. Sandstone is a competent and brittle rock, and it is just as subject to fissuring as any other rock of comparable competence. If the sandstone is not tight but has a normal porosity of 10 to 20 per cent, it too may have a greater actual porosity due to fractures, but their existence is less obvious. Finn[19] believes that the permeability of the Oriskany sandstone in Pennsylvania and New York has been "augmented in many producing areas by the presence of small open fractures, some of which have the character of open joint planes which are partly sealed by projecting quartz crystals. The presence of these fractures or slightly open joint planes has been the chief reason for the very large open flows in some Oriskany sand wells, and has caused the Oriskany to have a generally higher productive capacity than the average producing sand in the Appalachian area."

The storage capacity of a sandstone reservoir for oil as against water is a function of relative pore size, according to Krynine.[20] When the sand grains are very fine (diameter 0.125–0.0625 millimeter) water is stored preferentially over oil.

The original source of most sand grains is granite or granite gneiss. Heald[21] points out that the first sand to result from the wasting of granitic rock is not clean quartz sand but a mixture of quartz grains, clay particles, and accessory minerals in various stages of decomposition. A much better reservoir rock is produced after the sand has been reworked one or more times. Probably most sandstones, except those overlapping the crystalline basement rocks, consist of sand grains derived from older sandstones.

Environments of deposition. The dimensions of a sandstone body depend upon the conditions of its sedimentation. The most extensive sheet sandstones are deposited by a transgressing sea. They are continuous bodies throughout the area of overlap even though the sand deposited at the end of the transgression is younger than the sand deposited

[19] Fenton H. Finn, "Geology and Occurrence of Natural Gas in Oriskany Sandstone in Pennsylvania and New York," *Bull. Am. Assoc. Petrol. Geol.*, Vol. 33 (March, 1949), pp. 303–335.

[20] Paul D. Krynine, "Geology of the Arctic Slope of Alaska," *U. S. Geol. Survey*, Oil and Gas Investigations Map OM 126, Sheet 2.

[21] K. C. Heald, "Essentials for Oil Pools," *Elements of the Petroleum Industry* (American Institute of Mining and Metallurgical Engineers, 1940), p. 30.

when transgression started.   Some of the major reservoirs of the Gulf
Coast are of this type.   Malkin and Jung[22] point out that, as the sand
was being deposited along the strand line, organic muds were being deposited to the seaward.   Oil generated in the organic sediments could
migrate into the laterally adjacent sand or sandstone with the greatest ease.

Most sandstones are not sheet sands but are lenticular.   At one extreme
are lenses many miles across, and at the other ex..·eme are the "shoestring"
sands, which may measure but a few feet in width.   The latter are so
small as to constitute traps as well as reservoirs.   Lenticular sands are
deposited in regressing seas, along stagnant strand lines,[23] in offshore
bars or shallow "banks,"[24] in deltas[25] and river flood plains,[26] and on
lake floors.

Some of these lenticular-sand occurrences are fresh or brackish water
deposits, and we now have numerous examples of oil in sandstones of
continental origin.[27]   Some of the oil-bearing reservoir rocks both on the
Gulf Coast[28] and in California change up-dip from marine to continental
facies.

Conglomerate reservoirs.   Conglomerates can be looked upon as
coarse-grained sandstones.   They have all the qualifications for a reservoir rock, but this variety of sedimentary rock is rare, especially out in

[22] Doris S. Malkin and Dorothy A. Jung, "Marine Sedimentation and Oil Accumulation
on Gulf Coast, I. Progressive Marine Overlap," *Bull. Am. Assoc. Petrol. Geol.*, Vol. 25
(November, 1941), pp. 2010–2020.

[23] George V. Cohee, "Lateral Variation in Chester Sandstones Producing Oil and Gas in
Lower Wabash River Area, with Special Reference to New Harmony Field, Illinois and
Indiana," *Bull. Am. Assoc. Petrol. Geol.*, Vol. 26 (October, 1942), pp. 1594–1607; R. H.
Nanz, "Grain Orientation in Beach Sands: A Possible Means for Predicting Reservoir
Trend," *Am. Assoc. Petrol. Geol. Convention Program* (1955), pp. 107–108.

[24] John L. Rich, "Submarine Sedimentary Features on Bahama Banks and Their Bearing
on Distribution Patterns of Lenticular Oil Sands," *Bull. Am. Assoc. Petrol. Geol.*, Vol. 32
(May, 1948), pp. 767–779.

[25] H. N. Fisk, "Sand Facies of Recent Mississippi Delta Deposits," *Proc. Fourth World
Petrol. Congress* (1955), Sec. I, pp. 377–398; Daniel A. Busch, "Deltas Significant to Subsurface Exploration," *World Oil*, Part I (December, 1954), pp. 95 et seq.; Part II (January,
1955), pp. 82 et seq; Daniel A. Busch, "The Significance of Deltas in Subsurface Exploration," *Tulsa Geological Society Digest*, Vol. 21 (1953), pp. 71–80; Robert H. Nanz, Jr.,
"Genesis of Oligocene Sandstone Reservoir, Seeligson Field, Jim Wells and Kleberg Counties, Texas," *Bull. Am. Assoc. Petrol. Geol.*, Vol. 38 (January, 1954), pp. 96–117.

[26] Melville R. Mudge, "Sandstones and Channels in Upper Pennsylvanian and Lower
Permian in Kansas," *Bull. Am. Assoc. Petrol. Geol.*, Vol. 40 (April, 1956), pp. 654–678.

[27] "Symposium on Oil and Gas in Continental Beds," *Am. Assoc. Petrol. Geol.*, Vol.
38 (August, 1954), pp. 1654–1713.

[28] Max Bornhauser, "Oil and Gas Accumulation Controlled by Sedimentary Facies in
Eocene Wilcox to Cockfield Formations, Louisiana Gulf Coast," *Bull. Am. Assoc. Petrol.
Geol.*, Vol. 34 (September, 1950), pp. 1887–1896.

the great sedimentary basins. However, some basal sandstones may be coarse enough to be termed conglomerates; the oil-bearing Sooey conglomerate of central Kansas is of this type. A conglomerate consisting of schist detritus from the underlying crystalline basement rock is a reservoir rock in two fields in the Los Angeles Basin.[29]

Carbonate Rock Reservoirs. By carbonate rock is meant limestone, dolomite, and rocks intermediate between these two. In a few areas, notably the Lima-Indiana district and the Michigan basin, carbonate rocks are practically the sole reservoirs. In some other regions, as in the Middle East, Mexico, and the U. S. Mid-Continent, both limestones and sandstones contain prolific quantities of oil and gas. It has been estimated that 50 per cent of our known oil reserves[30] are in carbonate rock, and 60 per cent of our annual production[31] comes from carbonate reservoirs.

Carbonate reservoirs differ in several respects from sandstone reservoirs. Porosity is more likely to be localized, both laterally and vertically, within the rock layer. Although sheet porosity is possessed by a few carbonate rocks, this condition is exceptional. Within a given carbonate rock formation, even if several hundred feet in thickness, the porosity is in many places confined to the uppermost 25 to 50 feet. On the other hand, the pores may be much larger than in sandstone reservoirs, giving the rock an unusual permeability. For this reason, wells drilled into carbonate reservoirs hold the records for high initial yields, and limestone pools tend to be shorter-lived than sandstone pools.

Origin and character of carbonate rock porosity.[32] The porosity of carbonate rocks is the net result after pore-producing and pore-

[29] Harold W. Hoots, "Origin, Migration, and Accumulation of Oil in California," *Calif. Div. of Mines, Bull.* 118 (August, 1941), p. 267.

[30] John Emery Adams, "Oil in the Limestone Cycle," *Tulsa Geological Society Digest*, Vol. 22, (1954), pp. 34–35.

[31] R. C. Craze, "Performance of Limestone Reservoirs," *Jour. Petrol. Technol.* Vol. 2 (October, 1950), pp. 287–294.

[32] W. V. Howard and Max W. David, "Development of Porosity in Limestones," *Bull. Am. Assoc. Petrol. Geol.*, Vol. 20 (November, 1936), pp. 1389–1412; W. V. Howard, "Accumulation of Oil and Gas in Limestone," *Problems of Petroleum Geology* (American Association of Petroleum Geologists, 1934), pp. 365–376; W. V. Howard, "A Classification of Limestone Reservoirs," *Bull. Am. Assoc. Petrol. Geol.*, Vol. 12 (December, 1928), pp. 1153–1161; W. A. Waldschmidt, P. E. Fitzgerald, and C. L. Lunsford, "Classification of Porosity and Fractures in Reservoir Rocks," *Bull. Am. Assoc. Petrol. Geol.*, Vol. 40 (May, 1956), pp. 953–974; W. C. Imbt, "Carbonate Porosity and Permeability," *Applied Sedimentation*, P. D. Trask, editor, (John Wiley & Sons, New York, 1950), pp. 616–632; W. J. Burgess, "Limestone Reservoirs," *Oil and Gas Jour.* (April 15, 1957), pp. 198 et seq; Alfredo Sotomayor Costaneda, "Distrobución causas de la porosidad en las calizas del Cretácico Medio en la región de Tampico," *Bol. Asoc. Mex. Geol. Petrol.*, Vol. 6 (May-June, 1954),

reducing processes have completed their work.   Positive porosity is either primary or secondary.   The negative, or pore-reducing, processes include cementation (or other precipitation) and recrystallization.

Primary porosity is that resulting from the original deposition of carbonate rock.   In all probability, much limestone is clastic, and consists of either shell detritus or calcite "sand" grains derived from older carbonate rocks.   Theoretically such rock should have the same interstitial voids as those possessed by sandstone, but usually the original intergranular porosity has been reduced by infilling through the precipitation of calcite or dolomite by circulating solutions.   There are four other types of primary porosity in limestones.   These are the initial voids in oölitic limestones, coquinas, reef limestones, and reef-flank detrital zones.   Oölites[33] ordinarily range from 0.1 to 0.5 millimeter and are ellipsoidal.   They are cemented together by calcium carbonate, and since, as a general rule, the degree of cementation is greater than it is for sandstones, the porosity and permeability are lower.   In many places the oölites occur in discontinuous zones or lenses completely surrounded by relatively dense limestone. Coquinas are clastic shell deposits; their initial porosities (and effective porosities) may be very large, due to the jackstraw arrangement of the various shaped shells.

A limestone reef[34] is a deposit, mainly of organic origin, that "has been built upward at a more rapid rate than the contemporaneous sediments deposited about its margins."[35]   Although modern reefs are mostly built by algae or corals, many other types of organisms, such as crinoids,[36] contributed to reef development in the geologic past.   Exploration for hydrocarbons has led to the discovery of many ancient buried reefs, some containing phenomenal quantities of oil.   Many of the oil fields in Alberta, including Leduc (Fig. 14-9), are of this type.

Reefs are characteristically porous, but it is probable that the initial

pp. 157-206; R. A. Bramkamp and R. W. Powers, "Classification of Arabian Carbonate Rocks," *Bull. Geol. Soc. Am.*, Vol. 69 (October, 1958), pp. 1305-1318; Samuel P. Ellison, Jr., "Origin of Porosity and Permeability," *Improving Oil Recovery*, University of Texas, Chapter 7, pp. 83-91.

[33] F. B. Plummer, "Pore Systems in Reservoir Rocks," *Oil and Gas Jour.*, Vol. 43 (Nov. 18, 1944), p. 245; D. L. Graf and J. E. Lamar, "Petrology of Fredonia Oölite in Southern Illinois," *Bull. Am. Assoc. Petrol. Geol.*, Vol. 34 (December, 1950), pp. 2318-2336.

[34] W. E. Pugh, *Bibliography of Organic Reefs, Bioherms, and Biostromes*, Seismograph Service Corporation (Tulsa, Oklahoma, March, 1950).

[35] W. H. Twenhofel, "Characteristics and Geologic Distribution of Coral and Other Organic Reefs," *World Oil*, Vol. 129 (July 1, 1949), pp. 61-64.

[36] John W. Harbaugh, "Mississippian Bioherms in Northeast Oklahoma," *Bull. Am. Assoc. Petrol. Geol.*, Vol. 41 (November, 1957), pp. 2530-2544.

porosity, consisting both of intra-shell space, where the animal itself lived, and inter-shell voids resulting from the loose intergrowth of shell-producing marine organisms, has been augmented by secondary processes, especially solution of calcite.[37] The principal reef producer at Leduc has been entirely converted to dolomite.[38] Of growing importance as an oil reservoir is detrital reef material[39] which may accumulate on the flanks and about the reef perimeter. In the Marine pool (Fig. 11–16) of Illinois the principal reservoir is not the reef rock itself but the "coquina-like detrital limestone which forms the mantling deposit of a Niagaran reef."[40] Perhaps the greatest accumulation of clastic shell debris is the Horseshoe atoll[41] in west Texas. Here there is a horseshoe-shaped mass of calcareous detritus with maximum measurements of 90 miles by 70 miles by 3000 feet. The porosity, however, is mainly secondary; the percentage of voids is noticeably lower in argillaceous parts of the limestone body.

The most striking feature about some of the reefs is the great thickness of the porous zones. The discovery well of the Golden Spike, Alberta, pool was brought into production in April, 1949, with a rated initial production of 10,000 barrels per day after having cored 544 feet of porous Devonian reef material. Later, a gas well near the Berland River in the Alberta foothills penetrated 551 feet of the same reef rock. The maximum thickness of porous reef rock reported for the Scurry County, west Texas, fields first discovered in mid-1948 is "approximately" 600 feet.[42]

Older reef fields include the Hendrick pool of Winkler County, Texas,[43] the Southern Field of Mexico,[44] Kirkuk, in Iraq, and others.

With the exception of reef porosities, which are at least, in part, pri-

[37] K. C. Heald, "Essentials for Oil Pools," *Elements of the Petroleum Industry* (American Institute of Mining and Metallurgical Engineers, 1940), p. 31.

[38] D. B. Layer, et al., "Leduc Oil Field, Alberta, a Devonian Coral Reef Discovery," *Bull. Am. Assoc. Petrol. Geol.*, Vol. 33 (April, 1949), pp. 572–602.

[39] John W. Harbaugh, *op. cit.*; F. Stearns MacNeil, "Organic Reefs and Banks and Associated Detrital Sediments," *Am. Jour. Science*, Vol. 252 (July, 1954), pp. 385–401.

[40] Heinz A. Lowenstam, "Marine Pool, Madison County, Illinois, Silurian Reef Producer," *Ill. Geol. Survey, Rept. Investigations* 131 (1948).

[41] Donald A. Myers, Philip T. Stafford, and Robert J. Burnside, "Geology of the Late Paleozoic Horseshoe Atoll in West Texas," *Univ. of Texas Publ.* 5607 (April 1, 1956), pp. 1–113.

[42] D. H. Stormont, "Scurry County, West Texas, Limestone Reef Development," *Oil and Gas Jour.*, Vol. 48 (July 7, 1949), pp. 54 et. seq.

[43] K. C. Heald, "Essentials for Oil Pools," *Elements of the Petroleum Industry* (American Institute of Mining and Metallurgical Engineers, 1940), p. 31.

[44] John M. Muir, "Limestone Reservoir Rocks in the Mexican Oil Fields," *Problems of Petroleum Geology* (American Association of Petroleum Geologists, 1934), pp. 377–398.

mary, the porosity in most carbonate rock reservoirs is largely secondary. Processes which produce subsequent porosity include recrystallization, solution, dolomitization, and fracturing. Recrystallization, although it tends to obliterate primary porosity, produces new voids between crystals and between cleavage faces; this state is known as intercrystalline porosity.[45] Ohle[46] believes that the permeability of carbonate rocks may be increased by recrystallization of a fine grained limestone: "Coarsening of the grain greatly reduces the number of pores but the newly created interstices are larger and in addition they are much straighter and smoother, and longer from offset to offset. The routes presented to transmitted fluids by such recrystallized rocks are far less devious than those available in the original fine grained limestone. Since the coarse grains tend to have plane boundaries, the flow is essentially between parallel walls." (p. 673).

The greatest carbonate rock porosity-producing (or expanding) agent is solution. Calcite and dolomite, especially when above the water table, are leached by percolating waters which utilize whatever pre-existing porosity may be present. The resultant solution cavities range in size from minute pores to gigantic caverns. Regardless of size, the openings are interconnecting and extremely irregular. A reversal from dissolving to precipitating produces dripstone deposits in caves and comparable deposits in the smaller pores, all of which tends to reduce the pore space available. Infiltration of the overlying sediment may also lessen or even obliterate carbonate rock porosity. The red shale of the Molas (Pennsylvanian) formation has penetrated the solution cavities of the underlying Ouray (Mississippian) limestone in the Rattlesnake (Fig. 11–13), New Mexico, field to such an extent that only the lowest 5 feet of a weathered zone 35 to 55 feet thick has adequate porosity and permeability to function as reservoir rock.[47]

In most carbonate reservoirs the solution cavities are of modest size, little larger than the interstices between grains of sand, but they range from this size upward to actual caverns. The Dollarhide field in Andrews County, west Texas, is an example of cavern accumulation. In drilling the Fusselman pay zone, the drill dropped as much as 16 feet into oil-

[45] Samuel P. Ellison, Jr., "Origin of Porosity in Carbonate and Chert Reservoirs," A Symposium on Carbonate Reservoirs (Agricultural and Mechanical College of Texas, Bull. 11, 1951), p. 41.

[46] Ernest L. Ohle, "The Influence of Permeability on Ore Distribution in Limestone and Dolomite," Econ. Geol., Vol. 46, Part 1 (November, 1951), pp. 667–706; Part 2 (December, 1951), pp. 871–908.

[47] H. H. Hinson, "Reservoir Characteristics of Rattlesnake Oil and Gas Field, San Juan County, New Mexico," Bull. Am. Assoc. Petrol. Geol., Vol. 31 (April, 1947), pp. 731–771.

filled openings in the limestone.[48] Other west Texas[49] fields have also been developed in truly cavernous zones, as well as some fields in Kentucky. Such accumulations come close to the usual layman's concept of an underground "lake" of oil.

When the openings are large, it is obviously impossible to determine the porosity percentage by the usual core-analysis methods. When it is possible to measure the porosity of actual reservoir limestones and dolomites, it ranges from 5 to 20 per cent.[50]

Solution porosity is caused by circulating ground waters. These waters take advantage of any primary porosity by enlarging the already existing pores.[51] They also follow and enlarge joints and bedding planes.[52] It can be observed readily at the outcrop that fossil shells may be exceptionally vulnerable to solution so that fossiliferous carbonate rocks develop a pitted appearance.

Because most solution porosity is developed above the water table,[53] a carbonate rock must be not only emergent but also exposed, or nearly so, to subareal erosion. Therefore an unconformity should overlie every limestone with solution porosity, which explains why carbonate rock porous zones tend to lie near the top of the formation. They do not, however, have to lie at the very top; in the present cycle of erosion, solution leaching by circulating ground waters has taken place in exceptionally vulnerable layers several hundred feet below the surface.[54]

A list of fields producing from solution cavities in carbonate rocks would include most of the limestone and dolomite fields of the world.

Because dolomites are comparable in solubility to limestones, they are subject to the same leaching by percolating meteoric waters. However, there are some places in which limestones have been locally dolomitized, and the dolomite zones are porous and permeable, whereas the non-dolomitized limestone is dense and impervious. Such porous zones make

---

[48] D. H. Stormont, "Huge Caverns Encountered in Dollarhide Field Make for Unusual Drilling Conditions," *Oil and Gas Jour.*, Vol. 47 (April 7, 1949), pp. 66 et seq.

[49] H. P. Bybee, "Possible Nature of Limestone Reservoirs in the Permian Basin," *Bull. Am. Assoc. Petrol. Geol.*, Vol. 22 (August, 1938), pp. 915–924.

[50] S. E. Coomber, "The Porosity of Reservoir Rocks," *Science of Petroleum* (Oxford University Press, 1938), Vol. 1, pp. 220–223.

[51] John Emery Adams, "Origin, Migration, and Accumulation of Petroleum in Limestone Reservoirs in the Western United States and Canada," *Problems of Petroleum Geology* (American Association of Petroleum Geologists, 1934), pp. 347–363.

[52] Jean M. Berdan, "Hydrology of Limestone Terrane in Schoharie County, New York," *Trans. Am. Geophys. Union*, Vol. 29 (April, 1948), pp. 251–253.

[53] Jean M. Berdan, *op. cit.*

[54] B. C. Moneymaker, "Some Broad Aspects of Limestone Solution in the Tennessee Valley," *Trans. Am. Geophys. Union*, Vol. 29 (February, 1948), pp. 93–96.

both oil reservoirs and traps (Chapter 14). Obviously the development of this particular porosity is tied up with the process of dolomitization. The traditional theory that the porosity is due to volume shrinkage accompanying replacement of calcite by dolomite, molecule by molecule, is untenable for at least four reasons:[55] (1) replacement is always volume for volume and not molecule for molecule; (2) many dolomitized zones are not porous; (3) porosity, when present, varies widely from the calculated 12.3 per cent; and (4) the cavities are not like any other shrinkage openings in shape but are much more like solution cavities. They may be lined with euhedral crystals of dolomite. It has been suggested that in this instance, at least, the dolomitization was brought about by circulating ground waters, and the porosity was the result of an excess of solution over precipitation during the replacement process.[56] Hohlt,[57] and Chilingar and Terry,[58] state that in primary carbonate rocks the crystals have like orientations (with the *c* or vertical axes parallel), but replacing or secondary calcite or dolomite crystals have random orientations. Since the vulnerability of the random crystals to solution is greater than the oriented crystals, they tend to leach first.

The Madison limestone reservoir rock in the Beaver Lodge field of North Dakota has had its primary porosity increased from 3 to 5 times by dolomitization, according to Towse.[59] On the other hand Bybee[60] cites an instance where dolomitization of an initially porous and permeable oölitic limestone has completely destroyed the porosity.

Carbonate rocks are brittle and in many places are extensively fractured. In some fields the fractures augment the cavity porosity of the carbonate reservoir rock, but in others the limestone or dolomite is internally quite impervious so that all the hydrocarbon deposit is stored in joint cracks and other types of crevices. The latter type of reservoir is described subsequently in a separate section. One example of fractures adding to the overall porosity of a limestone reservoir is in the Marine pool of Illinois.

[55] A. N. Murray, "Limestone Oil Reservoirs of the Northeastern United States and of Ontario, Canada," *Econ. Geol.*, Vol. 25 (August, 1930), pp. 452–469.

[56] Kenneth K. Landes, "Porosity through Dolomitization," *Bull. Am. Assoc. Petrol. Geol.*, Vol. 30 (March, 1946), pp. 305–318.

[57] Richard B. Hohlt, "The Nature and Origin of Limestone Porosity," *Quarterly Colo. School of Mines*, Vol. 43 (October, 1948), pp. 1–51.

[58] George V. Chilingar and Richard D. Terry, "Relationship Between Porosity and Chemical Composition of Carbonate Rocks," *Petrol. Eng.*, Vol. 26 (September, 1954), pp. B 53–54.

[59] Donald Towse, "Petrology of Beaver Lodge Madison Limestone Reservoir, North Dakota," *Bull. Am. Assoc. Petrol. Geol.*, Vol. 41 (November, 1957), pp. 2493–2507.

[60] H. H. Bybee, "Hitesville Consolidated Field, Union County, Kentucky," *Bull. Am. Assoc. Petrol. Geol.*, Vol. 32 (November, 1948), pp. 2063–2082.

"The average daily production and the cumulative production . . . demonstrate that none of the discontinuous porous streaks appears thick enough to have the storage capacity correlative with the amount of fluid produced. It is the writer's opinion that the network of secondary porosity zones lining the fissure system and a crevice system connect the discontinuous producing streaks with each other and with the main reef core underneath to form one common reservoir."[61] Boyd[62] has noted a gas reserve in excess of the calculated capacity of the Silurian dolomite in the Howell gas field, Michigan, and ascribes this discrepancy to the presence of fissures in the reservoir rock which add to its storage capacity. Towse[63] notes abundant fracturing in the denser layers of a limestone reservoir and believes that these fractures permitted movement of hydrocarbons between the porous lentils. The importance of fracture porosity in carbonate reservoirs is emphasized by Rose.[64]

Nuss and Whiting[65] succeeded in obtaining a plastic model of fracture porosity enlarged by solution, in the Devonian limestone reservoir rock of the South Fullerton field, Texas.

Environments of deposition. The ecology of reef development has received more attention than that of non-reef carbonate rocks, probably because of their prominence in modern oceans. Counselman[66] sums up the optimum conditions for reef development as follows:

1. A relatively stable platform on which to build.
2. Shoal water (less than 100 fathoms).
3. Absence of contamination by physical or chemical impurities.
4. Warm (about 70 degrees Fahrenheit).
5. Mild surf or swell.
6. Absence of strong currents.

The non-reef limestones constitute about 95 per cent of the carbonate rock in the earth's crust, according to Adams.[67] Without doubt, most of these were formed under marine conditions on continental shelves. Some

[61] Heinz A. Lowenstam, "Marine Pool, Madison County, Illinois, Silurian Reef Producer," *Ill. Geol. Survey, Rept. Investigations* 131 (1948).

[62] Harold E. Boyd, *informal communication.*

[63] Donald Towse, *op. cit.*

[64] Walter D. Rose, "Study of Porosity in Carbonate Rock," *A Symposium on Carbonate Reservoirs* (Agricultural and Mechanical Coll. of Texas, Bull. 11, 1951), pp. 51–87.

[65] W. F. Nuss and R. L. Whiting, "Technique for Reproducing Rock Pore Space," *Bull. Am. Assoc. Petrol. Geol.*, Vol. 31 (November, 1947), pp. 2044–2049.

[66] Frank B. Counselman, "Origin and Geology of Carbonate Reservoirs," *Bull. Agr. and Mech. Coll. of Texas*, 11 (1951), pp. 27–39.

[67] John Emery Adams, "Non-Reef Limestone Reservoirs," *Bull. Am. Assoc. Petrol. Geol.*, Vol. 37 (November, 1953), pp. 2566–2576.

were clastic, some evaporitic, some organic, but most were probably genetic mixtures. Shallow warm waters accelerated organic growth and chemical precipitation, but cold and deep water limestones are also possible. By far the greatest volume of carbonate rocks has been deposited relatively close to land, for the emergent areas supplied not only most of the original food consumed by limestone-making organisms, but the calcium carbonate itself, both in clastic and dissolved form.

Non-marine limestones were also deposited in the past, but they are quantitatively insignificant compared with marine carbonate rocks.

Igneous Rock Reservoirs. Although the total of known occurrences of hydrocarbons in igneous rocks is large,[68] the number of commercial occurrences is much smaller. Sellards[69] lists two in Cuba, one in Mexico, and fourteen in Texas. Production from the Cuban fields is obtained from fractures in serpentine. The igneous rock field of Mexico is the Furbero in Vera Cruz. A sill of gabbro has been intruded into shale, metamorphosing the shale both above and below the sill. Oil occurs in porous zones in both the gabbro and the metamorphosed shale.

Five Texas counties in the Coastal Plain, Williamson, Travis, Bastrop, Caldwell, and Medina, contain oil reservoir rocks that were originally igneous (Fig. 6–1). In every instance, the volcanic activity that produced the igneous material took place during the Cretaceous period, and the volcanic rocks are embedded in Cretaceous sediments. "In some instances apparently the lava was erupted in the Cretaceous sea and formed a submarine volcanic cone. Some of the volcanic cones projected above sea-level or were subsequently so elevated as to be exposed and subjected to erosion. Some possibly were entirely submarine. Many of the igneous masses in this region, originally embedded in the Cretaceous strata, are now exposed. None of the exposed igneous rock produces oil, and of the embedded igneous masses many are likewise non-productive.[70] The porosity is apparently in part primary, due to the vesicular character of the flow rocks, and in part secondary due to the alteration of the volcanic material and its intensive fracturing.[71] Over 25 million barrels of oil have been produced from these igneous rock reservoirs in south-central Texas. The first discovery was made in 1913, the latest one in 1955.

[68] Sidney Powers et al., "Symposium on Occurrence of Petroleum in Igneous and Metamorphic Rocks," *Bull. Am. Assoc. Petrol. Geol.*, Vol. 16 (August, 1932), pp. 717–858.

[69] E. H. Sellards, "Oil Accumulation in Igneous Rocks," *Science of Petroleum* (Oxford University Press, 1938), Vol. 1, pp. 261–265.

[70] E. H. Sellards, *op. cit.*, p. 261.

[71] Marcus A. Hanna, "Fracture Porosity in Gulf Coast," *Bull. Am. Assoc. Petrol. Geol.*, Vol. 37 (February, 1953), pp. 274–275.

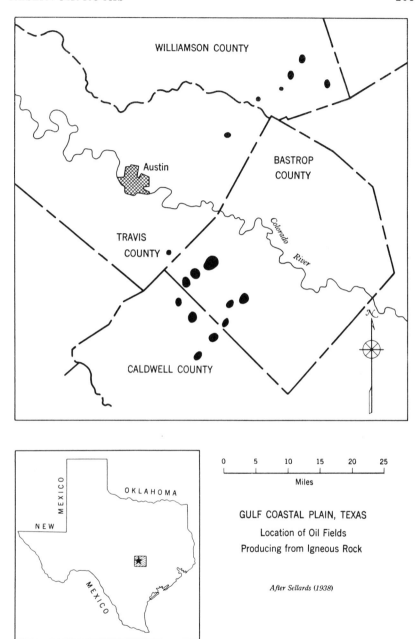

Fig. 6–1. Group of oil fields of Texas with igneous rock reservoirs. *Courtesy Texas Bureau of Mines and Geology.*

Exclusively Fracture Reservoirs.[72]   We have already noted that fractures may increase the storage capacities of sandstones, carbonate rocks, and serpentines.   We now will consider those reservoirs in which virtually all of the accumulation is within fractures.   Obviously in sandstones and limestones, there can be every gradation between exclusively non-fracture porosity and exclusively fracture porosity.

Fracture porosity is the void space between the walls of a crack or fissure.   This space has very finite thickness, but the other two dimensions are indefinite.   Except where cracks have been widened by solution the crack widths (void thicknesses) are probably in the one-tenth to one-fiftieth of an inch (0.25 to 0.05 centimeter) range.

All rocks are brittle, to varying degrees, above the zone of flowage, so any movement of the earth's crust, from intense folding to gentle settling, tends to fracture them.   Topography exerts some control on jointing near the surface, according to Chapman[73]   There is also the fracturing produced by volume shrinkage, which in turn is brought about by cooling (igneous rocks) and dessication (sedimentary rocks).

Fracturing tends to follow a geometric pattern.   Although the shape of the pattern depends upon the nature of the forces producing the fractures, all patterns have one feature in common: the fractures are interconnecting.   The result is a network or system of voids with extreme permeability, providing that the crack widths are supercapillary.

The fracture porosity of a given volume of rock is the product of the average crack width and the area of the fracture planes.   The latter is dependent upon the pattern and its periodicity in space.   Although fracture porosity, even at its best, falls far below the maxima reached by sandstone and carbonate rock reservoirs, it may still be adequate for the accumulation of hydrocarbons in commercial quantities, as will be shown subsequently by examples.   There is nothing unusual about this phenomenon.   Millions of people, scattered over the world, are dependent upon fracture porosity for their water supply.

The discovery of fracture porosity accumulations of oil and gas has lagged because of the improbability of intersecting a vertical or nearly vertical fracture plane with a vertical borehole.   The chances are many times greater that the drill will penetrate the massive rock between cracks than that it will hit one.   When an oil or gas accumulation of the fracture

[72] Edward A. Koester and Herschel L. Driver, chairmen, "Symposium on Fractured Reservoirs," *Bull. Am. Assoc. Petrol. Geol.*, Vol. 37 (February, 1953), pp. 201–330; M. King Hubbert and David G. Willis, "Important Fractured Reservoirs in the United States," *Proc. Fourth World Petrol. Congress* (1955), Section I, pp. 57–82.

[73] Carleton A. Chapman, "Control of Jointing by Topography," *Jour. Geol.*, Vol. 66 (September, 1958), pp. 552–558.

type has been discovered, it has been because luck, or the law of probability, led to the penetration of a fracture by the drill. Who knows how many oil or gas fields were *not discovered* by the unlucky wildcats that drilled between the cracks of a fracture type reservoir? The number must be many times the number of those discovered.

It was pointed out in Chapter 2 that hydraulic fracturing has provided a means of breaking through the tight rock lying between a borehole and the nearest fracture. Widespread use of this technique in wildcat wells should diminish the percentage of oil and gas fields penetrated by the drill without being discovered.

Oil or gas in paying quantities has been found in exclusively fracture reservoirs in shale, chert, siltstone, sandstone, carbonate rock, and various types of basement rocks.

Fractured shale. Shale can have no effective porosity other than by fracturing. Shale gas in Paleozoic formations is of widespread occurrence in the eastern half of the United States; it attains considerable economic importance in eastern Kentucky where over 3800 wells have been drilled to tap gas accumulations in fractured Devonian shale.[74] Production here has been stimulated by shooting the wells with nitroglycerine, which breaks through to fractures not penetrated by the borehole. Oil occurs in fractured Cretaceous and Eocene shales in various Rocky Mountain states including the Florence and Cañon City fields of south central Colorado, the Rangely (Fig. 11-11) shallow field[75] of northwestern Colorado, and in the Roosevelt and Duchesne fields of eastern Utah.

Fractured chert. Miocene fractured cherts and siliceous shales in California have produced about 300 million barrels of oil (with substantial reserves remaining) in two areas, Santa Maria on the coast, and the southwestern part of San Joaquin Valley.[76] These reservoirs are "characterized by low porosity and high permeability." Initial well productions commonly range from 200 to 1000 barrels per day, with a maximum of about 2500 barrels.

Fractured siltstone. Siltstone is too tight, because of the subcapillary size of the interstices between the fine silt particles, to give up any appreciable volume of the oil that may be contained within, although

---

[74] Coleman D. Hunter and David M. Young, "Relationship of Natural Gas Occurrence and Production in Eastern Kentucky (Big Sandy Gas Field) to Joints and Fractures in Devonian Bituminous Shale," *Bull. Am. Assoc. Petrol. Geol.*, Vol. 37 (February, 1953), pp. 282–299.

[75] Victor E. Peterson, "Fracture Production from Mancos Shale, Rangely Field, Rio Blanco County, Colo.," *Bull. Am. Assoc. Petrol. Geol.*, Vol. 39 (April, 1955), p. 532.

[76] Louis J. Regan, Jr., "Fractured Shale Reservoirs of California," *Bull. Am. Assoc. Petrol. Geol.*, Vol. 37 (February, 1953), pp. 201–216.

FRACTURED SPRABERRY CORES

Fig. 6–2. Near-vertical fracture planes are responsible for the splintering of cores cut through some parts of the Spraberry siltstone of west Texas. *Courtesy M. King Hubbert and Shell Development Company.*

such rock may be a gas reservoir. Therefore oil production from siltstone formations must come from fractures. The most notable such accumulation is in the Middle Permian Spraberry formation[77] of west Texas. Actually the Spraberry is a thousand foot section of alternating layers of black shales and silty shales (86 per cent), siltstones (13 per cent), and limestone or dolomite (less than 1 per cent). Parts of the formation are sliced by steeply dipping fracture planes (Fig. 6–2) with crack widths ranging from less than $\frac{1}{1000}$ inch (0.0025 centimeter) up to one-fourth inch (0.6 centimeter). The aggregate thickness of the fractured Spraberry is about 300 feet; the proven area exceeds 750 square miles and may be up to ten times that area. For some years the various Spraberry fields have been producing over 2,000,000 barrels of oil each month.

[77] Walter M. Wilkinson, "Fracturing in Spraberry Reservoir, West Texas," *Bull. Am. Assoc. Petrol. Geol.,* Vol. 37 (February, 1953), pp. 250–265; George T. Schmitt, "Genesis and Depositional History of Spraberry Formation, Midland Basin, Texas," *Bull. Am. Assoc. Petrol. Geol.,* Vol. 38 (September, 1954), pp. 1957–1978; Joseph W. Marshall, "Spraberry Reservoir of West Texas," *Bull. Am. Assoc. Petrol. Geol.,* Vol. 36 (November, 1952), pp. 2189–2191; G. Frederick Warn, "Spraberry Structural Conditions," *World Oil* (April, 1953), pp. 100 et seq.; George R. Gibson, "Relation of Fractures to the Accumulation of Oil," *Oil and Gas Jour.* (Nov. 29, 1951), pp. 107 et seq.; Roy F. Carlson, "Spraberry Fractures," *Oil and Gas Jour.* (Oct. 25, 1951), pp. 76 et seq.

The fractures through the Spraberry siltstone are both oil-storage reservoirs and conduits for the oil which "bleeds" out of the adjacent rock walls. Production can be and is being stimulated by artificial fracturing on a grand scale.

Fractured sandstone. Sandstone reservoirs with production exclusively from fractures are unusual, because sands cemented to the degree that they lose all interpore permeability are scarce. However, Pinfold[78] has described one oil field in West Pakistan in which tightly cemented fresh-water sands of late Tertiary age produce commercially from fractures.

Fractured carbonate rocks. Good examples of accumulation exclusively in fissures cutting carbonate rocks are some of the foreign fields. Muir states the following regarding the Tamaulipas limestone in the northern fields of Mexico: "Due to the dense nature of the limestone, the oil (12.5° A.P.I.) does not penetrate it, but is found in joint planes or other openings of induced character. . . . According to the lack, or presence, of 'induced porosity,' wells drilled into the Tamaulipas limestone vary in size from a mere showing of oil to gushers of 30,000 barrels per day, or larger. This variation occurs between wells which may be only 200 feet apart, horizontally."[79]

Coomber[80] has noted that the Asmari limestone, the reservoir rock in the Iranian oil fields, is similarly lacking in obvious porosity but produces great quantities of oil from cracks and fissures. A somewhat different impression of the reservoir properties of the Asmari limestone is given by Lane:

The intense folding and flexing of this rock mass has cracked and fissured it extensively and this fissure system is responsible for the very free fluid connection throughout the reservoir which is the feature of these fields. The late Lord Cadman in one of his addresses referred to it as 'the transport organization of the underground reservoir system.' These fissures are, however, small, being normally less than 0.1 inch in width and they do not contribute a large proportion of the storage space in spite of the fact that when a well is drilled no appreciable production is obtained until a fissure is penetrated, even though the drill may pass through bands of highly porous rock. Though opinions vary, it is generally estimated that at least 80 per cent of the recoverable oil is stored in the porous limestone while only 20 per cent exists in the fissures.[81]

[78] E. S. Pinfold, "Oil Production From Upper Tertiary Fresh-Water Deposits of West Pakistan," *Bull. Am. Assoc. Petrol. Geol.*, Vol. 38 (August, 1954), pp. 1653–1660.

[79] John M. Muir, "Limestone Reservoir Rocks in the Mexican Oil Fields," *Problems of Petroleum Geology* (American Association of Petroleum Geologists, 1934), p. 382.

[80] S. E. Coomber, "The Porosity of Reservoir Rocks," *Science of Petroleum* (Oxford University Press, 1938), Vol. 1, p. 221.

[81] H. W. Lane, "Oil Production in Iran," *Oil and Gas Jour.*, Vol. 48 (Aug. 18, 1949), p. 128.

Perhaps the best example of an exclusively fracture carbonate rock reservoir is the "First Pay" in the Ain Zalah field of northern Iraq. This limestone is Upper Cretaceous and the reservoir itself is the top 250 to 300 feet of the formation. According to Daniel:[82] "Ain Zalah is a field where the reservoir rock is extremely tight and, though it has a little porosity, has practically no permeability; fractures are all-important and without them there would be no production; they both hold and yield the oil, and large productions can be drawn from individual wells for limited periods."

The Selma (Upper Cretaceous) chalk produces from a fractured section on the downthrown side of the Gilbertown fault in Alabama.[83]

Fractured basement rocks. Basement rock is the geologically old and crystalline rock on which younger sedimentary formations have been deposited. Up until a short time ago it was assumed that when the drill bit hit the basement all chances for oil production had been used up. However, the basement rocks, which are also brittle, are fractured too. If original topography or subsequent structure movements have elevated these rocks above flanking, oil-bearing sedimentary formations, the oil may leak into, and become trapped in, the fractured crystallines. Examples of this are known in Kansas, California, Venezuela, Morocco, and elsewhere.

Over a million barrels of oil have been produced from the fractured top of the basement Precambrian quartzite by 16 wells in the Orth field of Rice County, Kansas.[84] Elsewhere on the Central Kansas Uplift are 20 or more other wells, likewise producing from quartzite and four wells in which the fractured reservoir rock is pink biotite granite. It is now customary to give wells penetrating the Precambrian a fracture treatment if they yield one or more gallons of oil per hour. Some wells have thereby been made into commercial producers which formerly would have been plugged and abandoned as dry holes.

Five California fields produce from fractured pre-Cretaceous basement schist.[85] Two of these, Edison and Mountain View, are in the San

[82] E. J. Daniel, "Fractured Reservoirs of Middle East," *Bull. Am. Assoc. Petrol. Geol.*, Vol. 38 (May, 1954), p. 774.

[83] Jules Braunstein, "Fracture-Controlled Production in Gilbertown Field, Alabama," *Bull. Am. Assoc. Petrol. Geol.*, Vol. 37 (February, 1953), pp. 245–249.

[84] Robert F. Walters, "Oil Production From Fractured Precambrian Basement Rocks in Central Kansas," *Bull. Am. Assoc. Petrol. Geol.*, Vol. 37 (February, 1953), pp. 300–313.

[85] Richard G. Reese, "El Segundo Oil Field," *Calif. Div. of Mines, Bull.* 118 (March, 1943), pp. 295–296; J. C. May and R. L. Hewitt, "The Nature of the Basement Complex Oil Reservoir, Edison Oil Field, California," *Bull. Am. Assoc. Petrol. Geol.*, Vol. 31 (December, 1947), pp. 2239–2240; J. H. Beach, "Geology of Edison Oil Field, Kern County, California," *Structure of Typical American Oil Fields* (American Association of Petroleum Geologists, 1948), Vol. 3, pp. 58–85.

Joaquin Valley. The other three, Wilmington, El Segundo, and Playa del Rey, are in the Los Angeles Basin. ⟩ The "schist" discovery well in the Edison field of California, completed in 1945, was drilled 83 feet into the basement crystallines for an initial potential production of 528 barrels per day. Within the next two years 106 producers were completed in the fractured metamorphics in this field, with reservoir thicknesses ranging up to 1350 feet. The basement rocks which are productive are in a prominent topographic dome surrounded by younger sediments. They have produced about 30 million barrels of oil. It was estimated in 1948 by Eggleston[86] that 15,000 barrels per day, nearly 2 per cent of the total for California, was coming from basement reservoirs, a source of oil formerly considered to be entirely impossible.

The champion of basement rock producers is the La Paz-Mara district in western Venezuela.[87] After several dry holes had been drilled previously into the basement complex of metamorphic and intrusive igneous rocks, a well drilled in 1953 came in for 3900 barrels per day from a depth of 8889 feet, 1089 feet below the basement rock surface. Two years later this field was producing 80,000 barrels per day from 29 wells in the basement reservoir. The cumulative production was already over 31 million barrels. The storage in the basement rock is exclusively in fractures.

Oil also occurs in fractured basement Paleozoic schist in French Morocco.[88]

So far almost all of the fractured basement rock discoveries have been by accident. One wonders how many fields have been missed because of inadequate exploration of the barely scratched basement by unsuccessful wildcats. Certainly hydraulic fracturing is indicated wherever a borehole finds the basement rock elevation relatively high, regardless of the inhospitable appearance of that rock.

Age of Reservoir Rocks.[89] The geologic age of reservoir rocks is but one of several ages or times that the petroleum geologist must consider. Of comparable importance is the age of the source materials and the time of hydrocarbon generation. Of greater import is the time at

---

[86] W. S. Eggleston, "Summary of Oil Production from Fractured Rock Reservoirs in California," *Bull. Am. Assoc. Petrol. Geol.*, Vol. 2 (July, 1948), p. 1353.

[87] J. E. Smith, "Basement Reservoir of La Paz-Mara Oil Fields, Western Venezuela," *Bull. Am. Assoc. Petrol. Geol.*, Vol. 40 (February, 1956), pp. 380–385.

[88] E. J. Brill, "Un Gisement dans les schistes fractures du socle l'oued beth (Maroc Francais)", *Proc. Third World Petrol. Cong.* (1951), Sec. I, pp. 315–328.

[89] W. A. J. M. van Waterschoot van der Gracht, "The Stratigraphical Distribution of Petroleum," *Science of Petroleum* (Oxford University Press, 1938), Vol. 1, pp. 58–62; "Oil Zones of the United States," in various issues of *Oil and Gas Jour.* (March 8, 1943, Cambrian and Lower Ordovician, to Oct. 28, 1943, Miocene and Pliocene); L. B. Kellum, "Petroleum Stratigraphy" (privately published, 1944).

which accumulation takes place (Chapters 10 to 14). But reservoir rock age has one outstanding advantage: it is the only age that can be determined consistently with some degree of exactness.

Actually the age of an oil- or gas-filled reservoir rock is more or less accidental. The presence of the hydrocarbons is due to a combination of reservoir rock and trap which afforded sanctuary to the oil or gas. Inasmuch as one or both of these conditions may be the result of geological activities taking place long after deposition, the age of the host rock is not too significant. The source organisms may or may not have been deposited during the same geological period.

In the past, the age of the reservoir rock was considered to be much more important than it is today.. As a matter of fact, too great attention to geologic age of producing formations led to the development of a series of prejudices that impeded the search for new deposits. It was categorically stated in almost every oil district that there was "no oil below the blank formation," the blank being the name of whatever the stratigraphically deepest reservoir for that area happened to be at the moment. The discovery of oil in commercial quantities in Precambrian rocks should bring an end to the game of limiting reservoir rocks to certain favored periods.

Reservoirs range in age from Precambrian to Pleistocene. Some periods are, however, much more important than others as a time of reservoir-rock deposition. The greatest production comes from rocks deposited during periods of thick and widespread sedimentation, with a considerable volume of porous and permeable rock present in the section. Benign climates are not essential to the deposition of reservoir rock. The greatest periods of reservoir-rock deposition were also times of widespread flourishing life, but the time involved in a geologic period is great and the two were not necessarily contemporaneous.

Over half (53 per cent) of the oil reserves in the major fields of the free world is in Mesozoic reservoirs. Miocene and Oligocene reservoirs are second with 29 per cent, and the Paleozoic third with 9 per cent. The balance of the reserve is in Tertiary reservoirs older and younger than the Miocene–Oligocene.[90]

The following review of the age of reservoir rocks has been summarized from van der Gracht,[91] with some newer information added.

Cenozoic reservoirs. Although the Pliocene is usually listed as the age of the youngest productive reservoirs, there are a few minor exceptions. Both in south-eastern Europe and in California some of the reservoirs transgress the Pliocene/Pleistocene boundary. Recent sands in

---

[90] G. M. Knebel and Guillermo Rodriguez-Eraso, "Habitat of Some Oil," *Bull. Am. Assoc. Petrol. Geol.*, Vol. 40 (April, 1956), pp. 547–561.

[91] W. A. J. M. van Waterschoot van der Gracht, *op. cit.*, pp. 61–62.

California and Texas contain oil in a few places, and small productions have been reported from two fields in Pecos County in the latter state.[92] The Tertiary, mainly Miocene and Pliocene formations, is almost the sole source of oil in California. It is the only oil-producing period along the Gulf Coast margin in Texas and Louisiana. There, as a general rule, the reservoir-containing epochs progress from Pliocene to Eocene from the coast inland. Non-marine strata of Paleocene (or Eocene) age produce oil or gas in 19 fields in the Uinta, Piceance Creek, and Sand Wash basins of Utah and Colorado.[93]

To list the foreign fields that produce from Tertiary reservoirs is to call the roll of almost all the major oil districts outside the United States. In South America major oil fields of Trinidad, Venezuela, Colombia, and Peru produce from Tertiary rocks. In Europe the fields of Romania and southern Russia, and some of the Middle Eastern production, including prolific fields in Iraq and Iran as well as the oil in India, Burma, and the Netherlands East Indies comes from the Tertiary.

Mesozoic reservoirs. The Cretaceous is a prolific source of oil in the United States in the Gulf Coast interior and in the Rocky Mountains. It is also of great importance in Mexico, western Venezuela, and in the Persian Gulf area including Bahrain Island, Safaniya field of northern Saudi Arabia, the spectacular Burgan and other Kuwait fields, and the fields of southern Iraq. Europe and Africa have some Cretaceous production.

Jurassic and Triassic reservoirs are relatively unimportant in North America, but some production is obtained in the Rocky Mountain province, especially from sandstones of Jurassic Age. Some flank production in the north German salt domes comes from Jurassic strata, and the Emba district in Russia northeast of the Caspian produces mainly from Jurassic rocks. The Jurassic reaches its greatest oil productivity in the great Arab limestone fields of Saudi Arabia, including especially Abqaiq and Ghawar.

Paleozoic reservoirs. Most of the Paleozoic oil produced to date comes from fields in the United States and Canada. However, there is also extensive Paleozoic production in Russia, and the recent discoveries in the Sahara are in rocks of the Paleozoic Age. The Permian produces in the Texas Panhandle and in the salt basin of western Texas and southeastern New Mexico. Reservoir rocks of Pennsylvanian age are of rela-

[92] Frank J. Gardner, "Texas Oil Flows From Rocks of Every Geological Age," *Oil and Gas Jour.* (Feb. 21, 1955), pp. 107–114.

[93] M. Dane Picard, "Tertiary Oil and Gas Fields in Utah and Colorado," *Bull. Geol. Soc. Am.*, Vol. 67 (December, 1956), p. 1800; W. T. Nightingale, "Petroleum and Natural Gas in Non-marine Sediments of Powder Wash Field in Northwest Colorado," *Bull. Am. Assoc. Petrol. Geol.*, Vol. 22 (August, 1938), pp. 1020–1047.

tively minor importance in the state of Pennsylvania, but they have yielded enormous quantities of oil and gas in the Mid-Continent, especially in Oklahoma and Kansas. The Rocky Mountain province produces some oil from Pennsylvanian rocks, and southeastern New Mexico and western Texas have both had discoveries in the rocks of that period. Relatively insignificant quantities of oil are obtained in Europe from Carboniferous and Permian reservoirs.

The Mississippian is outstanding in Illinois. Rocks of this period are also productive in other parts of the eastern United States, in the northern Mid-Continent, and in the northern Rocky Mountain province, including especially Wyoming, Montana, and Alberta. Devonian reservoirs produce most of the oil in the Appalachian fields and in Michigan. One Siluro-Devonian formation in the Mid-Continent is an important reservoir. The prolific fields of Alberta produce from Devonian reefs. Silurian reservoirs containing oil or gas occur on the west side of the Appalachian district and across southwestern Ontario into Michigan. Texas became a Silurian producer in 1940 when a successful wildcat in Ward County was completed in limestone of that age.[94]

Ordovician reservoirs, several in number, have superseded Pennsylvanian as the leading source of oil in the Mid-Continent. The once great Lima-Indiana district of Ohio and Indiana produces from Ordovician dolomites. Other reservoirs of this age occur in Kentucky and Tennessee. Gas has been obtained from Cambrian reservoir rocks in New York for many years. However, it did not become an important oil source until mid-1948, when oil was discovered in considerable volume in a Cambrian reservoir rock in the Lost Soldier field, Wyoming.[95] In August, 1953, a 1300 barrel well (initial daily production) was completed in Cambrian sandstone in west Texas. Subsequently there has been an active and successful search for Cambrian reservoir fields in both west Texas and the northern Rockies.[96] There is also, and has been for some years, production in the Mid-Continent from basal Paleozoic sandstone which may be of Cambrian age but also may be younger.

Oil is produced commercially from the Precambrian in Rice, Barton, and Russell counties, Kansas.

[94]C. D. Cordry and M. E. Upson, "Silurian Production, Shipley Field, Ward County, Texas," Bull. Am. Assoc. Petrol. Geol., Vol. 25 (March, 1941), pp. 425–427.

[95]E. W. Krampert, "Commercial Oil from Cambrian Beds in Lost Soldier Field, Wyoming," (abstract), Am. Assoc. Petrol. Geol., Program Annual Meeting (St. Louis, 1949), pp. 12–13; Gilbert M. Wilson, "Cambrian Production at Lost Soldier is Significant Wyoming Discovery," World Oil, Vol. 128 (November, 1948), pp. 76–77.

[96]Norman S. Morrisey, "Cambrian Has the Stage in West Texas," Oil and Gas Jour. (July 5, 1954), pp. 116–119; Aaron W. Cook, "Cambrian Production Opens New Trend in

RESERVOIR FLUIDS

The voids in all rocks are filled with water, water and oil, or water and gas. The only exception is where the rock lies above the water table and its capillary fringe. Even then the voids are water-filled following a rain, and after the downward-percolating water has passed by, the pore walls remain coated with adhering or wetting water which may last until the next rain, or may evaporate or be sucked back to the surface by plant life.

The fluids are not producible by boreholes except from reservoir rocks for reasons covered in the preceding chapter. Unless a reservoir rock is still undergoing some compaction, the reservoir fluids are merely occupants of rooms within a competent structure.

We shall first explore the pressures upon these room-dwelling fluids, and then discuss the fluids themselves.

Reservoir Pressures.[1] Reservoir pressure is the pressure exerted by the fluids and gases contained in the reservoir rock. This pressure can be measured during flow, or by shutting in the well and lowering a gauge to the reservoir face. Values range from but little above atmospheric pressure (14.7 pounds per square inch) to a maximum *recorded* reading of 12,635 pounds per square inch.[2] The *estimated* reservoir pressure at

[1] Stanley C. Herold, *Oil Well Drainage* (Stanford University Press, 1941); C. V. Millikan, "Production Practice," *Elements of the Petroleum Industry* (American Institute of Mining and Metallurgical Engineers, New York, 1940), pp. 255–257; V. C. Illing, "The Origin of Pressure in Oil Pools," *Science of Petroleum* (Oxford University Press, London, 1938), Vol. 1, pp. 224–229; Committees on Reservoir Development and Operation (Standard Oil Co. of New Jersey, Affiliated companies, and Humble Oil and Refining Company), *Joint Progress Report on Reservoir Efficiency and Well Spacing* (Standard Oil Development Co., 1943); George Dickinson, "Geological Aspects of Abnormal Reservoir Pressures in Gulf Coast Louisiana," *Bull. Am. Assoc. Petrol. Geol.*, Vol. 37 (February, 1953), pp. 410–432 (36 references).

[2] Lee S. McCaslin, Jr., "Tide Water's Record Breaker — Bottom Hole Pressure 12,635 Psi," *Oil and Gas Jour.* (Sept. 8, 1949), pp. 58 et se~

20,741 feet in the Richardson and Bass test in Plaquemines Parish, Louisiana, was 17,500 pounds. As a general but not invariable rule, the reservoir pressure declines during the lifetime of an oil or gas field from a maximum at time of discovery to a minimum at time of abandonment. The degree of the decline depends upon several factors. One is the amount of gas present. The most rapid loss of pressure takes place when very little gas is present and recovery is largely by expansion of the liquid phase only. Other factors are the presence of and the pressure of the edgewater and the percentage of the total volume of the reservoir occupied by the hydrocarbons. If the edgewater pressures are high and the aquifer thick and of wide extent, the initial reservoir pressure may be maintained throughout the history of the oil field, even after all the oil wells have become water wells. If the aquifer is limited in size and the hydrocarbons occupy the greater part of that reservoir, their removal will obviously be accompanied by a pressure drop because no new supplies of water are available to occupy the relinquished space.

Reservoir pressure is of great importance to the oil producer. The initial pressures may be of such magnitude as to cause the well to flow, thereby saving pumping expense. Even if the well ceases to flow, or fails to flow initially, the difference in pressure between the reservoir and the well causes the hydrocarbons to flow into the well and to rise in the tubing toward the surface so that the pumping depth may be considerably less than the reservoir depth. Furthermore, the maintenance of high reservoir pressure during the lifetime of a pool frequently results in a greater yield than is possible if the pressure is allowed to drop appreciably. The principal reason for the higher eventual recovery with maintenance of reservoir pressure lies in the relationship between the gas and oil within the reservoir. The greater the pressure, the more gas the oil can carry in solution. The more gas the oil has in solution, the lower its viscosity. The escape of gas to the surface in disproportionate amounts to the oil produced results in more and more sluggish oil being left behind in the reservoir. Also, the dropping of the pressure below the point at which the gas begins to come out of solution causes the formation of minute gas bubbles, which in turn retard the flow of oil. The proportion of gas to oil is known as the gas-oil ratio,[3] and it is modern production practice to attempt to maintain it as near the initial ratio possible. If the gas in the reservoir initially overlies the oil ("gas cap") or if during exploitation it escapes from the oil upward to the top of the reservoir instead of to the surface, it still maintains a useful function. The expansive force of the gas in the gas cap tends to drive the oil downward into wells that have

[3] R. V. Higgins, "Productivity of Oil Wells and Inherent Influence of Gas: Oil Ratios and Water Saturation," *U. S. Bur. Mines, Rept. Investigations* 3657 (September, 1942).

been completed low in the oil zone. If, however, the wells are completed in the gas cap, only gas is produced, and the greater part of the reservoir energy becomes dissipated.

Normally the greater the depth of the reservoir fluids the greater the pressure and the higher the temperature. Under these conditions natural gas increases in density, because of compression, but oil decreases in density because of the increasing quantity of gas forced into solution. As a result these fluids merge in deep reservoirs. The oil in phase with the gas is called "condensate" because it returns to liquid form when the pressure is lowered to below the dew point. With more and more hydrocarbons being produced from deep reservoirs, the number of condensate fields is rising, and continued increase in both number and percentage of total production is to be expected.

Possible causes of reservoir pressure. Ordinarily the pressure on the water in an aquifer (water reservoir) is the "hydrostatic head" or weight of the column of overlying water. Likewise in oil and gas reservoirs the basic pressure is the hydrostatic pressure of the water which almost invariably underlies an oil or gas accumulation. If the reservoir rock is continuously permeable to the outcrop, the pressure exerted by the edgewater is approximately equivalent to that of a vertical column of water extending from the oil-water interface upward to the level of the water table below the outcrop of the reservoir rock. This pressure amounts to 44 to 47 pounds per square inch for each 100 feet of water column. The variation is due to differences in dissolved solids content. In some oil fields lateral permeability of the reservoir rock apparently does extend to the outcrop, for the reservoir pressure is very close to the calculated pressure. On the other hand, abnormal pressures indicate a lack of (or at least an inadequate) connection between the reservoir fluids and the water table. Probably in most fields there is at least an imperfect hydraulic connection with the water table, either by way of the reservoir rock or a transstratification fracture system, so that hydrostatic head is one of several forces involved in the reservoir pressure.

When the aquifer which is doubling as a hydrocarbon reservoir has an outlet at a lower elevation than the intake, then the fluid pressures are hydrodynamic (Chapter 13) rather than hydrostatic. The fluid pressure within the aquifer at any point depends upon the depth of that point below the piezometric surface which slopes from the water level at the intake to the outlet. Many reservoir pressures are classified as abnormal because they do not coincide with the pressure calculated for the well depth below the local water table. Actually they are not really abnormal unless they fail to be in accord with the differences in elevation between the reservoir and the water level at the intake, in the case of hydrostatic pressure, or between the reservoir and the piezometric surface where pressures are hydrodynamic.

Although pure water is only slightly compressible, reservoir brines have higher compressibilities because of the presence of salts and gases in solution. It has been suggested that water expansion, which has accompanied the lowering of reservoir pressure due to the exploitation of the East Texas field, has been the mechanism by which the edgewater has followed behind the oil as the oil has been removed.[4] The artesian movement of water from the outcrop of the Woodbine sand to the zone of oil removal in East Texas is not considered adequate to fill immediately the pore spaces vacated by the oil drawn to the surface. Bugbee[5] emphasizes that only where the volume of edgewater is large (as at East Texas) can water expansion be of great importance. He also points out that in a lenslike reservoir hydrostatic pressures cannot be present, but expansion of compressed water can and no doubt does occur. A decrease in the confining pressure also causes undersaturated petroleum to expand until the bubble point is reached, as was pointed out in Chapter 2.

Because the hydrocarbons occur in the pore spaces between rocks walls, they do not carry any of the weight of the overlying rock, providing that the reservoir rock is completely compacted. It is quite possible that some of the younger sandstone reservoirs are still undergoing some compaction, in which event the reservoir fluids are holding up a part of the weight of the overlying rock. Where this situation exists, the load pressure may be as much as one pound per square inch per vertical foot—over twice the pressure for water alone. An outstanding example of such abnormally high pressures is the Gulf Coast[6] province in the United States. In addition to the compaction of the reservoir rock, if the rock overlying the reservoir has not compacted to its maximum extent, the reservoirs receive a pressure increment as the result of the squeezing out of the liquid contained in the compacting clay, mud, or calcareous ooze. Another suggested explanation for the abnormally high pressures in the Gulf Coast is that they are due to compaction of fluids in reservoirs which are completely sealed from the surface.[7]

In a few places, as at Long Beach, California, and at Goose Creek, Texas, the production of oil has been accompanied by a settling or a sub-

[4] R. J. Schilthuis and William Hurst, "Variations in Reservoir Pressure in the East Texas Field," *Trans. Am. Inst. Met. Engrs.* Vol. 114 (Petroleum Development and Technology, 1935), pp. 164–173; John S. Bell and J. M. Shepherd, "Pressure Behavior in the Woodbine Sand," *Jour. Petrol. Tech.*, Vol. 3 (January, 1951), pp. 19–28.

[5] J. M. Bugbee, "Reservoir Analyses and Geologic Structure," *Trans. Am. Inst. Min. Met. Engrs.* Vol. 151 (Petroleum Development and Technology, 1943), pp. 99–111.

[6] George Dickinson, *op. cit.*

[7] Donald L. Katz, *memorandum* dated Nov. 17, 1958.

sidence of the surface, perhaps because of accelerated compaction made possible by the lowering of pressure in the reservoir.   In many cases of surface subsidence of oil fields, however, the removal of large quantities of reservoir sand along with the hydrocarbons has been responsible for the settling.

Various miscellaneous causes of reservoir pressure have been suggested. According to Waldschmidt,[8] "much of the initial high pressure in some fields may directly result from precipitation of crystalline minerals within the voids of the reservoir rock."   He points out that the volume of water with anhydrite in solution is less than the total volume of water with the calcium sulfate precipitated.   The natural cementation of clastic reservoir rock would therefore result in an increase in the pressures existing within the reservoir rock, provided that it was a closed system at the time the precipitation of the natural cement took place.   The hydration of minerals in the surrounding rocks and the adsorption of water by clay would withdraw water from the reservoir and result in a decrease in pressure.   The gradual denudation of an oil field would also result in a decrease in pressure due to the lessening of hydrostatic head and the lowering of temperature.

It can be concluded from this discussion of reservoir pressures that the extent of the connection between the fluids in the reservoir and the water table is of utmost importance.   In a completely open system, where a sandstone is laterally continuous and uniformly permeable from trap to outcrop, the pressure is exclusively due to either the hydrostatic or to the hydrodynamic head.   In a closed system, with no connection between reservoir and ground-water circulation, any or all of the other pressure causes enumerated could be responsible.   Probably of great importance is the pressure created by the expansive force of water trapped with the hydrocarbons in a closed reservoir.   Timing controls the importance of such possible pressure causes as cementation and petroleum generation. If cementation can take place in volume *after* the hydrocarbons have accumulated in the reservoir, then cementation becomes a potent pressure cause.   Similarly, if lithifaction is complete *before* oil is formed, the generation of oil would produce a volume change that would result in an increase in pressure.

In most reservoirs the hydraulic system is probably neither wide open nor tightly closed.   Instead an imperfect, tortuous, high-friction connection with the water table may exist so that the reservoir pressure is a combination of hydrostatic and other forces.

[8] W. A. Waldschmidt, "Cementing Materials in Sandstones and Probable Influence of Migration and Accumulation of Oil and Gas," *Bull. Am. Assoc. Petrol. Geol.*, Vol. 25 (October, 1941), pp. 1839–1879.

WATER

Water, occurring both by itself in all porous rocks and with oil and gas in hydrocarbon accumulations, is by far the most abundant of the reservoir fluids. Almost from the beginning of the petroleum industry it has been the custom to collect and analyze oil-field waters, and many reports and scientific papers have resulted from this activity.[9]

In earlier days of water analysis, a number of misconceptions arose concerning the relationship between the occurrence of oil and the salinity of the associated water. It is now known that there is no relationship between the two substances. The associated water in some of the Michigan oil fields is highly concentrated brine, whereas in part of the Rocky Mountain district it is almost drinkable.

Importance and Distribution. Formation waters are associated with all reservoir rocks and complicate many of the problems of oil production. A knowledge of their composition often helps in the solution of these problems. Water analysis is, therefore, an important tool which provides chemical and physical data for the interpretation of electric logs, for reservoir studies and, by correlation with analyses of known waters, may determine the source of water entering the borehole.

Petroleum reservoirs contain water and oil in the capillary pore spaces. The exact nature of the relationship, however, is complex and depends upon such factors as the geometry of the rock pores and the physical and chemical properties of the fluids. In general oil field terms, however, waters are classified according to their position relative to the producing zones, top water above them, intermediate water between two producing zones, and bottom water below. The latter is referred to with two terms, bottom water or edgewater. Bottom water, in the restricted sense, underlies an accumulation over its entire extent, which is the case when the reservoir bed is thicker than the petroleum accumulation. Edge water occurs in the producing layer around the border of the accumulation in layers that are thinner than the vertical interval of the accumulation

[9] L. C. Case et al., "Selected Annotated Bibliography on Oil Field Waters," *Bull. Am. Assoc. Petrol. Geol.*, Vol. 26 (May, 1942), pp. 865–881; C. W. Washburne et al., "Oil Field Waters," *Problems of Petroleum Geology* (American Association of Petroleum Geologists, 1934), Part VI, pp. 833–985; C. E. Reistle, Jr., "Identification of Oil Field Waters by Chemical Analysis," *U. S. Bur. Mines Tech. Paper*, 404 (1927), pp. 1–25; A. R. Bowen, "Interpretation of Oil Field Water Analyses," *Science of Petroleum* (Oxford University Press, 1938), Vol. 1, pp. 653–656; C. E. Wood, "Methods of Analysis of Oil Field Waters," *ibid.*, pp. 645–652; H. A. Stiff, Jr., "The Interpretation of Chemical Water Analysis by Means of Patterns," *Jour. Petrol. Technol.*, Vol. 3 (October, 1951); J. F. Sage, "Water Analysis," *Subsurface Geology in Petroleum Exploration* (Colorado School of Mines, 1958), pp. 251–263; J. C. McKinnell, "Identification of Mixtures of Waters from Chemical Water Analyses," *Jour. Petrol. Technol.*, Vol. 10 (September, 1958), pp. 79–85.

(oil and gas column). In complex oil zones made up of several porous units, intermediate water may be found within the productive section as bottom or edgewater in the individual members or as an entirely water-bearing stratigraphic layer.

Oil-field waters are usually saline except at shallow depth. However, alkaline waters occur in some areas, for example in the younger Tertiary fields of the Bolivar Coast in Venezuela. The dissolved salts present are primarily chlorides, sulphates, carbonates, and bicarbonates of sodium, magnesium, and calcium, but minor amounts of other elements are usually also present. Normally, few of these other elements are determined by chemical analysis, although Emery[10] has shown that by spectrographic analysis they may be used to differentiate between waters. Waters range from essentially fresh with less than 1000 milligrams per liter of dissolved solids to brines with more than 250,000 milligrams per liter.

The characteristics of water encountered in the various porous zones of an oil field may differ sharply in chemical constituents, ion concentration, or both. These variations may occur laterally; that is, they may occur between different reservoirs in the same stratigraphic interval, or vertically, between different producing zones lying one over the other.

Practical Uses of Water Analysis. In order to make best use of this variation in water composition, the production geologist must have available a file or library of analyses of waters from a wide variety of sources. The more complete the file, the more useful it becomes. The water analyses should be obtained from company wells, from competitors, and from such sources as the U.S. Bureau of Mines.

The applications are many and varied. Water from an unknown source, such as an unidentified formation in a wildcat well, or water produced from a well in which no known water-bearing section is open to production, can be compared with analyses of the known waters in the file and identified. In the wildcat, the new reservoir may be correlated with one in a known field and, in the producing well, the rise of water in the producing zone may be distinguished from the production of water due to a casing leak or cement failure. In the latter instance, the depth of leak or failure may be estimated by recognizing the source of the extraneous water. In one area, where production is obtained from very lenticular sands occurring within a continuous shale interval, well control is not sufficient to determine the lateral limits of each lens. However, because the composition of formation water varies from one lens to another, it is possible to define the reservoirs by the use of water analyses. In the absence of more concrete evidence, a comparison of formation waters

[10] F. H. Emery, "Spectrographic Analyses of Oil Well Brines," *Oil and Gas Jour.* (March 4, 1937), pp. 53–54.

sometimes can reveal a fault or other barrier to fluid movement between apparently connected reservoirs. The source of water appearing after a fracture treatment or acidizing of a well may be doubtful but can often be determined by use of water analysis, and, if the water is from two zones, the percentage from each may be calculated. The logging engineer, lacking a measured formation-water resistivity with which to make electric log calculations, can compute a resistivity from an analysis of the waters. The reservoir engineer is also interested in the composition of formation water, particularly from a secondary recovery standpoint. The character of the water, if it is to be reinjected, determines the amount and nature of surface treatment it needs to avoid contamination or plugging of the formation.

Still another very important need for water analyses arises in connection with studies of the flow of underground waters. It has been shown by Hubbert[11] that oil accumulations occur with a tilted oil-water interface when the formation waters are flowing rather than static. Furthermore, from direct field studies it has been found that many formation waters, hitherto regarded as being static, are actually in a state of motion.

Water analyses in this connection are important principally because the density increases with the concentration of dissolved solids. In a sedimentary basin containing waters of increasing salinity with depth, if the waters are static, they are found in a horizontally stratified arrangement with the densest waters at the bottom and the surfaces of equal density, or equal salinity, horizontal. In other words, if the waters of a basin are static, the lines of equal salinity should coincide approximately with the structure contour lines. If, on the contrary, the waters are in motion, the lines of equal salinity are likely to be much lower on the side from which the water is running than on the side toward which it is flowing.

Purely qualitatively, low-salinity waters at depths of several thousand feet constitute strong presumptive evidence that the waters are flowing. For otherwise, saline waters would be expected at those depths. By low-salinity waters is meant waters which are dilute as compared with sea water. Sea water contains about 35,000 milligrams per liter, whereas, many deep formations contain waters with 15,000 milligrams per liter, or less.

Quantitatively, the water in a given formation will flow if a potential gradient exists, and the potential of water at a given point is defined by:

[11]M. King Hubbert, "Entrapment of Petroleum Under Hydrodynamic Conditions," *Bull. Am. Assoc. Petrol. Geol.*, Vol. 37 (August, 1953), pp. 1954–2026.

$$\phi = gz + \frac{p}{\rho w}, \tag{1}$$

or

$$\phi = gh, \tag{2}$$

where

$$h = z + \frac{p}{\rho wg} \tag{3}$$

Here, p is the pressure and ρw the density of the water at an underground point of elevation z, g the acceleration due to gravity, ϕ the potential of the water at the point, and h its total head, or the height above sea level that the water would rise statically in an open tube terminated at the point.

Since the density of water is involved in all these calculations, and the density is derivable from the salinity, it is quite important that such data be available.

Water-Analysis Reports. Water analyses are reported in several forms and use a number of different units. The reports may give actual ion concentrations, or the ions as combined into hypothetical salts. The most common units are:

1. Milligrams per liter (mg/1). This unit is milligrams of solute per liter of solution.

2. Parts per million (ppm.). This unit is milligrams of solute per kilogram of solution. (Equals milligrams per liter divided by the specific gravity of the solution.)

3. Percent by weight (numerically equal to ppm. divided by 10,000).

4. Grains per gallon (to convert to mg/1, multiply by 17.12 if in U.S. gallons or by 14.26 if in Imperial gallons).

5. Reaction value. (Expressed either in milliequivalents per liter or milliequivalents per kilogram.)

For low concentrations, there is little difference between milligrams per liter and parts per million and the two terms are often used interchangeably. For high concentrations, however, there is an appreciable difference. For example, 200,000 ppm. sodium chloride in solution equals 230,000 milligrams per liter, a difference of 15 per cent. The reaction value of an ion is the equivalent amount of a hydrogen ion with which it reacts or which it replaces, and is expressed in milliequivalents per liter or per kilogram. It is determined by multiplying the ion concentration in milligrams by the appropriate reaction coefficient.[12] Reaction coefficients for the more common ions are given in Table 1.

[12] Reaction coefficient equals reciprocal of equivalent weight. Equivalent weight equals molecular weight divided by effective valence. An equivalent equals one gram equivalent weight. A milliequivalent equals one thousandth of an equivalent.

TABLE 1.   REACTION COEFFICIENTS OF ACTIVE RADICALS

Positive Radicals	Reaction Coefficients	Negative Radicals	Reaction Coefficients
Sodium (Na)	0.04348	Sulfate (SO_4)	0.02082
Potassium (K)	0.02558	Chloride (Cl)	0.02820
Calcium (Ca)	0.04990	Nitrate (NO_3)	0.01613
Magnesium (Mg)	0.08224	Carbonate (CO_3)	0.03333
		Bicarbonate (HCO_3)	0.01639
		Sulfide (S)	0.06238
		Iodine (I)	0.00788

The comparison of water analyses using the numerical values for the constituents is an extremely tedious process.   Hence, a number of graphical methods of representing the composition of water have been devised. The Stiff diagram, which is becoming the most popular of these methods, and which is included sometimes as part of the production laboratory report form greatly facilitates the use of the analysis and the identification of the water.

Graphical representation of analyses.   A number of methods of identifying waters by means of their analyses are in general use.   Most of these methods employ a graphic presentation to facilitate comparison of one analysis with another.   The oldest is the Palmer system which was designed for studying the reacting characteristics, salinity, alkalinity, and hardness, rather than for correlating and identifying waters, although it does have some application as a means of correlation.   It classifies the ions as follows:

Positive ions	**Negative ions**
A. Alkalies: Na and K	A. Strong Acids: Cl and SO_4
B. Alkaline Earths: Ca and Mg	B. Weak Acids: CO_3, HCO_3, and S

The ion concentrations are converted to percentage reaction values, and these values are combined according to the basic principles shown in Fig. 7–1, whereby a combination of strong acids with the alkalies gives primary salinity and with alkaline earths gives secondary salinity.   Similarly, the weak acids combined with the alkalies and alkaline earths give primary and secondary alkalinity respectively.   It is obvious from the diagrams that it is not possible for a water to have both secondary salinity and primary alkalinity and, if the strong acids equal the alkalies, the water has only primary salinity and secondary alkalinity.

One of the simplest and best methods of plotting oil-field water analyses for identification and correlation is the one proposed by Stiff, in which the reaction values of the ions are plotted on a system of rectangular

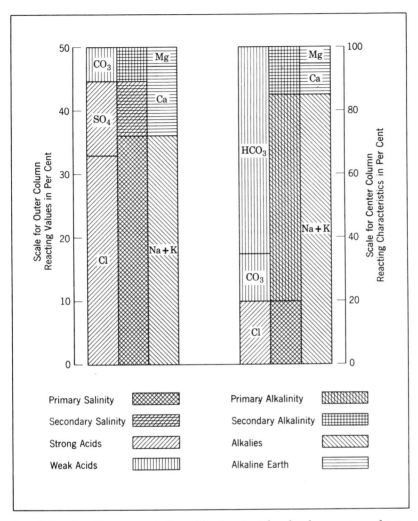

Fig. 7-1. Graphic representation of basic principle of Palmer system of interpreting water analyses. *Courtesy Shell Oil Company.*

coordinates, as shown in Fig. 7–2. The positive ions are plotted to the left and the negative ions to the right of a vertical zero line, and the points are connected by straight lines to form a closed diagram. The significance of the diagrams shown has already been discussed. As has been mentioned, this type of diagram may be used to show the effects of the mixing of waters, such as contamination of formation water by drilling-mud filtrate. This contamination may be important when studying electric logs as well as for determining the source of water.

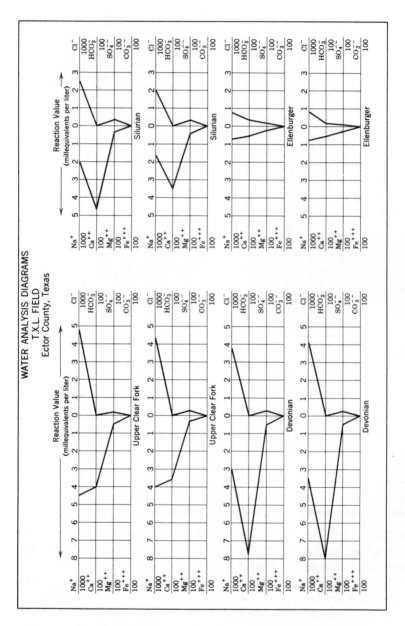

Fig. 7-2. Stiff method of water analysis interpretation. *Courtesy Shell Oil Company.*

OIL

Unfortunately for both the petroleum explorer and the petroleum consumer oil is infinitely less abundant in underground reservoirs than water. However, it is stored in the same manner, and water is an almost invariable associate.

The geologist, in his search for new deposits of petroleum and natural gas, need not be an organic chemist. Nevertheless, some of the physical and chemical properties of the natural hydrocarbons do concern him. An elementary knowledge of the chemistry of petroleum is necessary for proper recognition and evaluation of seeps, for an adequate understanding and interpretation of the natural history of oil, and for an intelligent study of the underground movement of oil once it has been formed. Furthermore, the market value of the oil or gas discovered is dependent upon its chemical constitution, and the economic geologist can never afford to lose sight of the marketability of his wares.

Chemical Nature of Petroleum.[13] Petroleum, or "rock oil" is a naturally occurring liquid consisting dominantly and essentially of hydrocarbon compounds. It is a *mixture* of widely varying proportions of these compounds. Some of the hydrocarbons are gases and some are solids; both types are in solution in the liquid hydrocarbons which predominate. Because petroleum is a mixture it does not have a definite chemical composition, and neither does it have fixed physical properties. However, in spite of an indefinite chemical composition, crude oils are remarkable in their consistency of elementary composition. No matter how they may differ in appearance, properties, and hydrocarbon compounds present, nearly all natural petroleums range from 83 to 87 per cent carbon and from 11 to 14 per cent hydrogen.

Hydrocarbon constituents. The percentage of hydrocarbons in a crude oil is also the measure of its purity. In most crude oils all but 0.3 per cent to 3.0 per cent by weight of the compounds present are hydrocarbons. Practically all of these hydrocarbons are composed of paraffin, napthene, or aromatic groups, or mixtures or combinations thereof. As

[13] Benjamin T. Brooks, editor, *The Chemistry of Petroleum Hydrocarbons* (Reinhold, New York, 1954), Vol. 1; Frederick D. Rossini and Beveridge J. Mair, "Composition of Petroleum," *Progress in Petroleum Technology*, Advances in Chemistry Series No. 5 (American Chemical Society, 1951), pp. 334–352; Frederick D. Rossini, "Hydrocarbons from Petroleum," *Jour. Inst. Petrol.* Vol. 44 (April, 1958), pp. 97–107; Benjamin T. Brooks and A. E. Dunstan, editors, "Crude Oils, Chemical and Physical Properties," *Science of Petroleum* (Oxford University Press, 1950), Vol. 5, Part 1; Stewart P. Coleman, "Physical and Chemical Properties of Petroleum and Its Products," *Elements of the Petroleum Industry* (American Institute of Mining and Metallurgical Engineers, 1940), pp. 4–20; E. DeGolyer and Harold Vance, "Bibliography of the Petroleum Industry," *Bull. Agr. and Mech. Coll. Texas*, 83 (1944), pp. 137–144.

the aromatics are the least abundant, most oils are either paraffin base, napthene base, or mixed (paraffin-napthene) base. A paraffinic oil has a relatively high hydrogen content in relation to the carbon; the reverse is true of a naphthenic oil. The paraffin oils are generally of low density, and residues of paraffin wax, or petrolatum, are obtained in refining such oils. The naphthenic oils are heavier and contain a large percentage of viscous, but volatile, lubricating oils. The decomposition of the liquid constituents of the naphthenic oils during distillation produces a semi-solid or solid asphaltic residue during the refining process. For this reason napthenic oils may be referred to as having an asphalt base.

Typical of paraffinic-base crude oils are those found in Pennsylvania and elsewhere in the Appalachian region. They also occur in Michigan and in parts of northern Louisiana and southern Arkansas. Elsewhere, crude oils of this type have been found in Peru and the Argentine, as well as in the Near East. The Dutch East Indies has also produced some paraffin-base crude, as well as great quantities of naphthenic petroleums. The latter is the principal type of crude oil produced in Venezuela and Colombia. California and the Gulf Coast are the major sources of napthenic-base oil in the United States. Mixed-base crude oils are called the Mid-Continent type because of their prevalence in Illinois, Kansas, Oklahoma, and in Texas outside of the Gulf Coast. Mexico produces both naphthenic- and mixed-base petroleums, and the same is true in Romania. Since deeper drilling into older formations in all parts of the world has resulted in the discovery of crude oils that are lighter in weight and more paraffinic in base than those found in shallower formations, it is becoming much less practicable to classify crude oils geographically.

The number of individual hydrocarbon compounds that may occur in petroleum is not yet known. The petroleum industry, through the American Petroleum Institute, has been subsidizing research on the composition of crude oil since 1927. One project (Research Project No. 6) has been to determine the hydrocarbon compounds present in a single representative petroleum and then to learn to what extent the major constituents vary in different petroleums. The crude oil chosen to be the guinea pig is a petroleum classified as intermediate in composition from a well in the Ponca City field of Oklahoma. So far, from the gas and gasoline, kerosene, light gas-oil, and heavy gas-oil and light lubricating distillate fractions alone, 159 separate hydrocarbon compounds have been isolated. These range from CH_4 (methane) with a boiling point of $-164°$ C to $C_{24}H_{50}$ which boils at $389°$ [14] The identification of the individual hydrocarbons in the lubricant fraction (14 per cent of the crude oil) has not yet been accomplished because of the difficulties in isolating these readily decomposable high boiling point compounds.

[14]Frederick D. Rossini, *op. cit.*, pp. 104–106.

In extending this study to the gasoline fractions of seven other representative petroleums it was found that "while the relative amounts of the main classes of hydrocarbons in the gasoline fraction varied from petroleum to petroleum, the relative amounts of the individual compounds within a given class were of the same magnitude for the different petroleums".[15] Therefore the major differences in crude oils are due to variations in the relative amounts of the main classes of hydrocarbons. Some petroleums consist mostly of the more volatile fractions, and others are preponderantly made up of the less volatile lubricant fractions.

Sulfur, oxygen, and nitrogen. Sulfur is the most common impurity in crude oil. About 7 per cent of the crude oil being produced in the United States circa 1950 contained over 2 per cent sulfur by weight; nearly 40 per cent, however, ran less than 0.25 per cent.[16] Sulfur occurs in crude oils in various compounds, such as hydrogen sulfide, and even (rarely) as elemental sulfur. Because sulfur compounds are heavy, only the heavier crude oils contain them, and, conversely, most of the heavy crudes have a relatively high sulfur content. They present a serious problem to the refiner.

Oxygen is generally lower, but in a few cases assays have run as high as 2 per cent by weight. Crude oils containing oxygen compounds in measurable quantities are also high in specific gravity. Nitrogen is even more insignificant, ranging from less than 0.05 per cent to a maximum of 0.8 per cent in analyses of crude oils from 150 fields in the United States.[17] The highest percentages were in California petroleums.

Metallic constituents. The complete combustion of crude petroleum results in an ash residuum ranging from 0.01 to 0.05 per cent.[18] As much as one third of this ash may be chlorides, especially sodium chloride. Many elements have been detected in the ash. Some of these were no doubt present in the original crude oil in compounds dissolved in water droplets dispersed through the oil, but others were in solution in the oil itself.[19] Among the metals reported as consistently present in petroleum ash are vanadium, nickel, copper, cobalt, molybdenum, lead, chromium, manganese, and arsenic. Uranium is more erratic in its occurrence.[20]

The metals present in largest percentage are vanadium and nickel. It has been suggested that the uranium and other minor metals present in the crude oil are impurities picked up from the host rocks, whereas the

[15] Frederick D. Rossini and Beveridge J. Mair, *op. cit.*, p. 349.
[16] H. M. Smith, cited by Rossini and Mair, *op. cit.*, p. 335.
[17] H. M. Smith, *op. cit.*
[18] W. A. Gruse and D. R. Stevens, cited by Rossini and Mair, *op. cit.*
[19] Frederick D. Rossini and Beveridge J. Mair, *op. cit.*, p. 337.
[20] R. L. Erickson, A. T. Myers, and C. A. Horr, "Association of Uranium and Other Metals With Crude Oil, Asphalt, and Petroliferous Rock," *Bull. Am. Assoc. Petrol. Geol.*, Vol. 38 (October, 1954), pp. 2200–2218.

vanadium and nickel were present in the original petroleum source material.[21] Some, and perhaps all, of the vanadium and nickel in crude oil are present as metal-organic porphyrin complexes.[22]

Physical Properties of Petroleum. Because crude petroleum is a mixture, its physical properties vary widely depending upon the type and relative abundance of hydrocarbons and impurities present. The property most often referred to is the density or "gravity".

Density. Density may be calculated as specific gravity with unity represented by water. With most liquids, however, it is common to describe the density in terms of degrees Baumé or API. By this method of calculation, water has a density of 10. Numbers lower than 10 are used for the density of substances heavier than water and numbers above 10 for lighter substances. The common scale in the petroleum industry is the API (American Petroleum Institute) scale, in which the specific gravity is equal to the fraction 141.5 ÷ (degrees API + 131.5). This scale differs from the Baumé scale, which is used in determining the density of many liquids only in the value of the two constants. In the Baumé scale the numerator is 140 instead of 141.5, and the denominator constant is 130 instead of 131.5.

Density is determined very simply by merely dropping a hydrometer with the API scale calibrated upon the stem into the crude oil. However, because the volume and hence the density of oil varies considerably with temperature, it is necessary to take the temperature of the liquid at the same time so that the density reading can be converted to the temperature standard. In the United States the standard temperature for recording densities of petroleum is 60° F. Conversion tables are published by the American Society for Testing Materials.

The density of crude oil from different fields ranges from less than 10° to as high as 60° and even higher. An oil below 10 in the API scale is, however, still lighter than the water with which it is associated. Only pure water has a density of 10°; the highly mineralized water with which most oils are associated is considerably heavier. As a general rule, the lighter oils contain a higher than average percentage of the profitable gasoline hydrocarbons and a lower percentage of deleterious compounds which increase refining expense. For this reason, much crude oil is priced on a density basis. The density of the oil may even determine whether a wildcat well of low capacity has discovered a new field or will have to be

[21]Harold J. Hyden, A. T. Myers, C. A. Horr, et al, "Uranium and Other Trace Elements in Crude Oils From the Western States," *Am. Assoc. Petrol. Geol.*, Annual Meeting (April, 1957).

[22]Gordon W. Hodgson and Bruce L. Baker, "Vanadium, Nickel, and Porphyrins in Thermal Geochemistry of Petroleum," *Bull. Am. Assoc. Petrol. Geol.*, Vol. 41 (November, 1957), pp. 2413–2426.

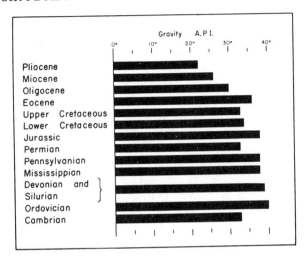

Fig. 7-3. U. S. Crude Oil Gravities and Geologic Time (*after Hopkins*).

plugged and abandoned because of the unprofitable character of the oil discovered.

The evolution of oil, discussed in Chapter 5, involves a decrease in napthenic hydrocarbons and a corresponding increase in paraffinic hydrocarbons—in other words, a gradual change from napthene to paraffin base. This change also produces a marked decrease in density (or increase in degrees API). Rather striking evidence that such an evolution has taken place can be shown by plotting the average API densities for over 7600 oil fields in the United States by the geologic age of the reservoir rock (Fig. 7-3). The non-conformity of the Cambrian density is probably due to the inadequacy of the sample (only 0.7 of one per cent of the total number of oil pools). The anomalous average density of the Eocene oils is probably due to the fact that most of the Eocene oil in the United States comes from South Texas reservoirs which are more deeply buried than the average domestic Cretaceous reservoir.[23] The other interruptions to a steadily increasing quality of oil with geologic time may be due to different climates (such as the Permian), which resulted in different source materials with different maturation rates, or to varying environmental controls on the deposition of potential catalytic materials.

Other physical properties. Petroleums vary in *color* from colorless to black. The valuable Pennsylvania oil noted for its high yield of premium lubricating fractions is amber in color. Medium-density oils such

[23]George R. Hopkins, "A Projection of Oil Discovery," *Jour. Petrol. Technol.*, Vol. 2 (June, 1950), p. 7.

as those of the Mid-Continent are predominantly green, and heavy oils are commonly black. The color of crude oil appears to depend upon the presence of unsaturated hydrocarbons and of hydrocarbons that contain nitrogen, oxygen, or sulfur in addition to the hydrogen and carbon. As a general rule, the lighter the oil in color, the higher its degrees API and the greater its value.

The *odor* of crude oil depends upon the character and the volatility of the hydrocarbons within the mixture. Some crude oil has a decidedly aromatic odor, owing to predominance of the aromatic group of hydrocarbons. High-sulfur crudes may have the odor of hydrogen sulfide. On the other hand, light crude oils may have a very noticeable gasoline odor due to the escape from the mixture of this highly volatile material.

Viscosity, or resistance to flow, is measured by the time taken for a given quantity of oil to flow through a small opening. Some oils are so viscous at ordinary temperatures that they must be heated before they can be pumped. On the other hand, some crude oils that are so light at ordinary temperatures that they are only slightly more viscous than water may become highly viscous during the winter because of the congealing of dissolved wax.

The *volatility* of crude oil is dependent upon the boiling points of the various compounds present in the mixture. One common constituent, methane, has a boiling point of $-164°$ C, whereas the boiling points for some of the paraffin compounds are too high to be measured by usual procedures. Inasmuch as petroleum is always at a temperature above the boiling points of some of its hydrocarbons, the more volatile compounds are continuously exerting pressure and tend to escape unless the oil is in a completely sealed unit. Under such circumstances, it is obviously impossible to refer to a single boiling point for any crude oil. Nor is it possible to refer to a freezing point for petroleum. The hydrocarbons present tend to solidify separately at different temperatures. The viscosity increases with lower temperatures until it can no longer be measured and the oil is in a "plastic solid" state, but if the cooling is continued, a truly solid state is finally reached.

The geologist is interested in the *surface tension and capillary force* of petroleum, because these physical properties play a role both in the migration of oil through the rocks of the earth's crust and in the permanent retention of some oil in considerable volume in the underground reservoir in spite of most efficient processes of exploitation. The surface tensions of petroleums containing dissolved gas in deep reservoirs are extremely low. "Low surface tensions tend to minimize the effects of capillary forces in the displacing of crude oil from porous media by high pressure gases."[24] Water has greater capillary force than oil; consequently this fluid may be expected to draw itself into the finest pore spaces of the rock and to force the oil into the larger pores.

Oil in the Reservoir. There are great physical and environmental differences between crude oil in a test tube and crude oil in a reservoir rock. Now that we have discussed oil as seen in the chemical laboratory, let us attempt to visualize it in its precaptive state and setting.

Physical nature of oil underground. Crude oil underground is under a pressure varying from several hundred to several thousand pounds per square inch, depending mainly upon depth. Under such pressures, the oil can and does contain in solution a considerable quantity of gas. As the oil is produced and rises from reservoir to surface, and while it is in storage at the surface, much of this gas escapes from solution with the result that the oil has much less mobility at the surface than it had when it was under high pressure in the reservoir rock. The dissolved gas not only makes the crude oil lighter in density but also considerably increases its fluidity.

Second in importance to the mobility of oil underground is the effect of higher earth temperature. Although there is considerable variation from one place to another, the general average increment in temperature is 1° F for every 60 feet of additional depth. In other words, if the reservoir is at a depth of 6000 feet, the oil is at a temperature approximately 100° above the average temperature at the surface. It is both a commonly observed fact and an oft-repeated experiment that heating an oil decreases its viscosity. As a general rule, the heavier the oil the greater the degree of the decrease. Therefore, oil in the reservoir must have far greater fluidity than oil examined under room temperatures at the surface. Since the fluidity of water also increases with higher temperatures, the "flow-ability" of both these liquids is higher the deeper the reservoir rock.

Fluid relationships of oil underground. The reservoir assembly where oil has accumulated includes rock, voids, water, oil, and gas. The voids are full of water before the accumulation takes place. The oil, which is always both lighter than the water and immiscible with it, makes room for itself by pushing the water downward in the reservoir rock. However, *it does not displace all of the original water.* A film of adsorbed or "wetting" water remains lining the pore walls. This water has been called the "irreducible interstitial water."[25] Where any of the voids

---

[24]D. L. Katz, R. R. Monroe, R. P. Trainer, "Surface Tension of Crude Oils Containing Dissolved Gases," *Am. Inst. Mining Met. Engrs.* Tech. Publ. 1624 (1943).

[25]W. V. Engelhardt, "Interstitial Water of Oil Bearing Sands and Sandstones," *Proc. Fourth World Petrol. Congress* (1955), Section I, pp. 2–18.

are of capillary size, the original water remains there also. As a result there is not only water in the reservoir *below* the oil accumulation, but there is also interstitial water within the zone of accumulation itself. Some petroleum engineers refer to this water as connate water, which is an unfortunate misapplication of a geological term referring to the origin of ground water and not to its ecology.

Because of the usual presence of wetting water in oil reservoirs, the oil is not in contact with the rock itself. It occupies the interstices between the water envelopes, and the storage space occupied by oil is less than the total porosity of the reservoir rock. If, however, air can get to the oil accumulation (as when denudation takes place), then the wetting water evaporates and the oil is let down onto the rock walls of the voids, resulting in oil-wet reservoirs.[25a]

The relationship between oil and gas in the reservoir depends upon the degree to which the oil is saturated with gas. If the oil is undersaturated, all gas entering the reservoir eventually is taken into solution in the oil. The diffusion of the newly dissolved gas through the oil takes time. If the oil is saturated, the gas accumulates above the oil, pushes the oil (and water beneath the oil) downward in the same manner as the oil displaced the water, and creates thereby a gas "cap".

The saturation point of oil for gas depends upon the reservoir pressure and temperature and upon the composition of the oil. Therefore changes in one or more of these factors after accumulation has taken place may result in the formation or increase of a gas cap, or in its diminution or complete disappearance. Because a gas cap by its presence signifies that the underlying oil is saturated with gas, one can expect the oil to have minimum density and maximum fluidity for that temperature and pressure.

As noted in Chapter 2, the changes in properties of oil and gas phases in crude oils are considerable with increasing depth. "With increasing pressure, more natural gas dissolves in the oil phase and more normally liquid hydrocarbons dissolve in the gas phase."[26]

Conclusion. Oil underground has a fluidity and mobility approaching that of water at the surface. Actually some light oils have less viscosity in the reservoir than boiling water has at the surface, but it is also true that some heavy oils are considerably more viscous in the reservoir than surface water. Oil occupies the interstices between water-wet void walls in the reservoir rock. It is lighter than water and immiscible with

[25a]R. O. Leach, O. R. Wagner, H. W. Wood, and C. F. Harpke, "A Laboratory and Field Study of Wettability Adjustment in Water Flooding," Jour. Petrol. Technology, Feb., 1962, p. 212.

[26]Donald L. Katz and Brymer Williams, "Reservoir Fluids and Their Behavior, *Bull. Am. Assoc. Petrol. Geol.*, Vol. 36 (February, 1952), p. 342. Further discussion in A. E. Hoffmann, J. S. Crump, and C. R. Hocott, "Equilibrium Constants for a Gas-Condensate System," *Jour. Petrol. Technol.*, Vol. 5 (January, 1953), pp. 1–10.

it. It is heavier than gas and miscible with it up to the point of saturation. Much of the earlier experimentation in migration of oil that was carried on in various laboratories did not take into account the vastly greater mobility due to higher temperature and dissolved gas. Thus, the results obtained are of little or no value in studying the probable movement of oil underground. Only by simulating as nearly as possible all the environmental conditions (including the wetting effect of interstitial water) can experimental studies of oil migration be of other than questionable value.

Geological Applications of Oil Analyses. Ever since Professor Silliman of Yale made an analysis of western Pennsylvania seep oil for the promoters of the Drake well, it has been the practice, where oil is discovered, to obtain samples for examination and analysis.[27] Although the primary purpose of making these analyses is to determine the value of the oil and to supply essential data to the refiner, geologists have found the character of oil to be of value in solving problems of correlation and migration. "Reservoir petroleum engineers measure the pressure-volume, viscosity, and phase behavior of crude oil and gas mixtures underground as an aid to predicting the future reservoir performance, or the volumes and rates of various stream constituents which will be producible through a specific number of wells."[28]

For geological study it is now customary to transform the analytical data into a "correlation index;"[29] these data may be applied directly[30] or may be used in grouping the crude oils into various types.[31] When only a small sample of oil is available in core or cuttings, the index of refraction[32] may be used as a substitute for gravity determination and even chemical analysis.

As a general but not invariable rule, crude oils from the same stratigraphic levels in a single oil province are similar, and those from different ages of rock are dissimilar.[33] For this reason, the analyses of petroleums have been used for purposes of stratigraphic correlation. They also can

[27]C. H. McKinney and O. C. Blade, "Analyses of Crude Oils from 283 Important Oil Fields in the United States," *U. S. Bur. Mines Rept. Investigations* 4289 (May, 1948).

[28]E. K. Schluntz, *memorandum* dated October 28, 1958.

[29]H. M. Smith, "Correlation Index to Aid in Interpreting Crude Oil Analyses," *U. S. Bur. Mines Tech. Paper*, 610 (1940).

[30]L. M. Neuman et al., "Relationship of Crude Oils and Stratigraphy in Parts of Oklahoma and Kansas," *Bull. Am. Assoc. Petrol. Geol.*, Vol. 25 (September, 1941), pp. 1801–1809.

[31]J. G. Crawford and R. M. Larsen, "Occurrences and Types of Crude Oils in Rocky Mountain Region," *Bull. Am. Assoc. Petrol. Geol.*, Vol. 27 (October, 1943), pp. 1305–1334.

[32]Hollis D. Hedberg, "Evaluation of Petroleum in Oil Sands by Its Index of Refraction," *Bull. Am. Assoc. Petrol. Geol.*, Vol. 21 (November, 1937), pp. 1464–1476.

[33]Donald C. Barton, "Correlation of Crude Oils with Special Reference to Crude Oil of Gulf Coast," *Bull. Am. Assoc. Petrol. Geol.*, Vol. 25 (April, 1941), pp. 561–592; L. M. Neumann et al., "Relationship of Crude Oils and Stratigraphy in Parts of Oklahoma and

be used to prove the absence of vertical (transformational) migration, as at Oklahoma City, where the oils in the Pennsylvanian formations are different from those in the underlying Ordovician rocks. Exceptions to this general rule may signify vertical migration, as on Trinidad[34] and at Garber, Oklahoma, where similar oils are found in fourteen superimposed reservoirs.

Some thought has been given as to whether or not the presence and relative abundance of trace metals in petroleum has any geological significance. So far, at least, it has not been possible to use trace-metal content as a correlation tool. However, in the Seminole area of Oklahoma it has been found that the areal distribution of concentrations of vanadium and nickel might have paleogeographic significance. These metals are most abundant near the old shorelines, decreasing basinward.[35] Haeberle[36] has found an apparent connection between density and facies in Gulf Coast oil fields. The heavier petroleums tend to occur in continental and shallow water sediments, with the lighter oils more abundant in deeper water marine facies.

Estimation of Oil Reserves.[37] It is impossible to calculate accurately the reserves of oil not yet produced from a petroleum deposit, until enough wells have been drilled to outline the area of the accumulation. With the data obtained during the drilling, it is possible to make a fairly accurate estimate of the amount of oil that can be produced subsequently. A distinction should be made between such calculations of petroleum re-

---

Kansas," *Bull. Am. Assoc. Petrol. Geol.*, Vol. 30 (May, 1946), pp. 747–748; Charles Bohdanowicz, "Stratigraphic Comparison of Polish Crude Oils," *Bull. Am. Assoc. Petrol. Geol.*, Vol. 21 (September, 1937), pp. 1182–1192.

[34]K. W. Barr, F. Morton, and A. R. Richards, "Application of Chemical Analysis of Crude Oils to Problems of Petroleum Geology," *Bull. Am. Assoc. Petrol. Geol.*, Vol. 27 (December, 1943), pp. 1595–1617.

[35]L. C. Bonham, "Geochemical Investigation of Crude Oils," *Bull. Am. Assoc. Petrol. Geol.*, Vol. 40 (May, 1956), pp. 897–908.

[36]Fred R. Haeberle, "Relationship of Hydrocarbon Gravities to Facies in Gulf Coast," *Bull. Am. Assoc. Petrol. Geol.*, Vol. 35 (October, 1951), pp. 2238–2248.

[37]F. S. Shaw, "Review of Ultimate Recovery Factors and Methods of Estimation," *Oil Weekly*, Vol. 106 (July 13, 1942), pp. 16–26; V. Bilibin, "Methods of Estimating Oil Reserves," *International Geological Congress, 17th Session, U.S.S.R.*, 1937 (Preprint, 1936); K. Marshall Fagin, "Notes on Estimating Crude Oil Reserves," *Petrol. Engineer*, Vol. 17 (September, 1946), pp. 92–94; W. A. Bruce and H. J. Welge, "Restored–State Method for Determination of Oil in Place and Connate Water," *Tech. Publs.* (Standard Oil Co., N. J., 1947), pp. 13–33; E. R. Brownscombe and F. Collins, "Estimation of Reserves and Water Drive from Pressure and Production History," *Jour. Petrol. Geol.*, Vol. 4 (April, 1949), pp. 92–99; J. J. Arps, "Estimation of Primary Oil Reserves," *Jour. Petrol. Technol.*, Vol. 8 (August, 1956), pp. 182–191; Donald L. Katz, "Development of Techniques in Reserve Estimation," *Jour. Petrol. Technol.*, Vol. 5 (September, 1953), pp. 81–83.

serves in "proved" areas and calculations made on the basis of an insufficient number of wells to obtain all the reservoir data necessary. If a field has been outlined on three sides by dry holes, but not on the fourth side, it is possible to obtain from what is known of the structural picture of the accumulation a "probable" area estimate which can be used in making a reserve calculation. If, however, there are only one or two wells drilled in a newly discovered field, the ultimate recovery estimate is even less accurate. The estimate is based upon the available geophysical and subsurface information; subsequent exploration makes possible a proved reserve figure and the temporary estimate is dropped. However, the so-called "proved" oil-reserve figures, which are published periodically, include not only the actual proved reserves plus the probable reserves but also the best guesses as to the reserves beneath newly discovered fields. This practice is not as bad as it sounds, for the number of new and undeveloped discoveries at any one time is relatively insignificant compared with the total number of "drilled out" fields.[38]

Even in proved areas there are factors of uncertainty. An oil field is never abandoned because of complete exhaustion. It is abandoned when it is no longer profitable to pump oil from the reservoir, separate it from the accompanying water, and market it. Calculations of the recoverable oil are of necessity based on the market situation at the time the calculations are made. Fluctuations in the price of oil or in the production cost may cause the abandonment point to be reached earlier or later than originally calculated.

Methods of calculating the amount of oil that can be produced by an oil field fall into three general classes: (1) the production history of older, geologically similar fields; (2) the production history of the pool itself during its early months, and (3) volumetric methods, in which the actual bulk of the oil in the underground reservoir is calculated and the recoverable percentage estimated.

The first method cannot be used with any great degree of accuracy. Every natural reservoir has its own specifications and its own behavior pattern, and close parallels are highly improbable. This method is applied by taking the per acre-foot recovery figure for a field which had the same type of reservoir rock and a similar structural setting and multiplying it by the area and thickness of the reservoir of the newer field.

The history written by an oil well or a group of wells during their first few months of production can be used in at least three different ways to estimate future production. One of these is to take the initial daily production, or the cumulative production for the first thirty days, or the

[38]F. H. Lahee, "Our Oil and Gas Reserves: Their Meaning and Limitations," *Bull. Am. Assoc. Petrol. Geol.*, Vol. 34 (June, 1950), pp. 1283–1292.

first year's production and multiply this figure by a factor in order to obtain the probable eventual yield. Factors that have been used for the initial daily production range from 644 for wells which produce less than 100 barrels a day initially, down to 240 for wells producing over 600 barrels a day.[39] In some areas, such as California, these factors are much too conservative. According to Moore,[40] a common rule of thumb has been to multiply the wide-open production for the first thirty days by 300, in order to obtain the ultimate yield. Experience has shown that in some of the older California fields the ultimate production was four to six times the first year's production.

Inasmuch as most states now prohibit the unrestricted flow of oil from a well, methods based on wide-open flow over a period of time no longer have much applicability. Before proration the *decline-curve* method of estimating underground oil reserves was widely used.[41] In this method the daily production of the well is plotted and a curve drawn which shows the rapid tapering off from the high flush production to the more slowly diminishing "settled" production. By means of the many available curves that chart production from time of discovery until time of abandonment, it is possible to project into the future the curve from a well that is only a few months old and to make a fairly reliable estimate of its total yield. The sum of the estimates for the wells in a field gives the total eventual production of that field. A somewhat less exact procedure is to plot the production for an entire field and project it in the same way.

A third method of using history to calculate reserves is to plot the cumulative production against the decline in reservoir pressure and then project this "bottom-hole pressure-production curve" into the future. As discussed in a subsequent paragraph, the pressure-decline method is an important means of estimating reserves of natural gas. It is not nearly as suitable, however, when applied to oil deposits, but it does have the advantage over the two other methods based on yield to date, in that it is applicable to wells of restricted production as well as to those with wide-open flow. The lack of accuracy of the pressure-decline method applied to oil accumulation is based on the fact that the bottom-hole pressure in an oil reservoir may be only to a minor extent due to the volume and compression of the hydrocarbons in the reservoir. It has already been shown that with a complete natural-water drive there is little or no diminution in pressure between the time of first production from the reservoir and its complete exhaustion of oil.

[39]S. F. Shaw, "Review of Ultimate Recovery Factors and Methods of Estimation," *Oil Weekly*, Vol. 106 (July 13, 1942), p. 24.

[40]Fred H. Moore, *informal communication.*

[41]Willard W. Cutler, Jr., "Estimation of Underground Reserves by Oil-Well Production Curves," *U. S. Bur. Mines Bull.* 228 (1924).

Katz[42] has developed a method of reserve calculation for oil-gas mixtures in a closed system in which data on the oil and gas production to date and the bottom-hole pressure history are combined with information regarding the properties of the oil-gas mixtures.

The volumetric[43] approach to reserve calculations is as old as the exploitation of oil fields. All such methods are fundamentally the same, based on calculating the amount of oil in the underground reservoir from the degree of saturation of the porous volume of a reservoir rock of finite dimensions. The classical formula for determining the oil in a reservoir is $X = FAtps$, in which X is the oil in place in barrels, F the factor 7758, A the area in acres, and *tps*, respectively, average thickness in feet, percentage of porosity, and percentage of oil saturation (which is the complement of the percentage of water present). Multiplying the three factors —area, thickness, and porosity—gives the number of acre-feet of void space in the reservoir rock in which fluids can be stored. The factor 7758 transforms this volume figure from acre-feet to barrels. Multiplying by the percentage of oil saturation diminishes the barrel volume of potential oil space to the extent that water is present with the oil in the rock pores.

From electric logs, drill records, and core analyses, it is possible to calculate rather closely the area, the reservoir thickness, and porosity percentage. The porosity percentage of the massive rock can be determined accurately by core analysis, but it is not possible to obtain the percentage of *additional* porosity that may be present in fracture openings. If fracture porosity is suspected, it is necessary to estimate the total porosity.

In earlier times it was assumed that oil filled the pores so that the porosity volume was also the oil volume. This belief led to excessively high estimates of both the amount of oil originally present in the reservoir and of the amount of the unrecovered oil left behind when the field was abandoned. We now know that oil sands have an interstitial water[44] content equal to 10 to 50 per cent or more of the volume of pore space. As a general rule, the lower the permeability of the reservoir, the higher the percentage of interstitial water. Since the average figure ranges between 30 and 35 per cent, only about two-thirds of the available pore space actually contains oil. This interstitial water is commonly referred to as "connate" water, although, as mentioned earlier, that terminology is not in exact accord with the original definition. Modern core analysis permits

[42]D. L. Katz, "A Method of Estimating Oil and Gas Reserves," *Trans. Am. Inst. Mining Met. Engrs.*, Vol. 118 (1936), pp. 18–32.

[43]John F. Dodge, Howard C. Pyle, and Everett G. Trostel, "Estimations by Volumetric Method of Recoverable Oil and Gas from Sand," *Bull. Am. Assoc. Petrol. Geol.*, Vol. 25 (July, 1941), pp. 1302–1326; George R. Elliott and William L. Morris, "Oil Recovery Prediction," *Oil and Gas Jour.*, Vol. 48 (June 16, 1949), pp. 84 et seq.

[44]R. J. Schilthuis, "Connate Water in Oil and Gas Sands," *Trans. Am. Inst. Min. Met. Engrs.*, Vol. 127 (1938), pp. 199–214.

a fairly accurate determination of the volume of water present, from which the percentage of saturation (filling) of the pore space by oil can be calculated. Quantitative interpretation of electric logs provides an independent method of computing both porosity and percentage of hydrocarbon saturation. Use of both core analyses and electrical logs provides a close and substantiated control on the amount of oil in the reservoir.

The amount of oil in place is not, of course, the amount of oil that will be produced during the lifetime of the field. In order to convert oil in place to stock-tank oil, it is necessary to apply both a shrinkage factor and a recovery factor. The former compensates for the diminution in volume, due to the escape of dissolved gas, that takes place as the oil is brought to the surface. The shrinkage differs with different crude oils. The oil in place is multiplied by a factor usually ranging from 63 to 87 per cent. The recovery factor is the percentage of oil that can be extracted profitably from the reservoir. In older fields the recovery was as low as 20 per cent, but today with advanced techniques of secondary and tertiary oil recovery, the ultimate goal is as at least 80 per cent. It is not possible to forecast the economic situation 10, 15, or 20 years hence. On the other hand, it is possible to determine the type of reservoir energy that is propelling the oil from the reservoir into the well and to determine through the experience of older fields what percentage of oil can be expected with modern production methods from reservoirs with that particular type of fluid drive. Since it is inevitable, however, that this factor is built in part upon personal opinion, divergences of as much as 20 per cent in final expected recovery figures may occur.

Semisolid and Solid Hydrocarbons. In addition to the crude oils that flow into a well (some with the help of induced heat), there are oils that are too viscous to flow and solid hydrocarbons that almost certainly were once liquid. Therefore these hydrocarbons either are, or once were, reservoir fluids and so will be discussed at this point. In terms of known abundance in the earth's crust, liquid petroleum ranks first, natural gas second, and the semisolid and solid hydrocarbons third. The relationship between natural gas and petroleum, the highly viscous to solid hydrocarbons, and kerogen and coal is shown by the accompanying table.[45]

The bitumen groups belonging in the semisolid and solid hydrocarbon classification are the natural asphalts, ozocerites, asphaltites, and asphaltic pyrobitumens. Examples of natural asphalts include both highly viscous liquids, such as the brea deposits in California which ensnared many types of animals (some of them now extinct), and natural asphalt, which occurs on the island of Trinidad and elsewhere in ponds and lakes. The

[45]E. Eisma, C. F. Jackson, M. Louis, G. R. Schulze, and H. M. Smith, "Report by the Committee Dealing with Resolution 4 of the Third World Petroleum Congress," *Proc. Fourth World Petrol. Congress* (1955), Section I, pp. 1–4.

so-called rock asphalts are limestone and sandstones impregnated by asphaltic material which no doubt was originally liquid petroleum. Ozocerite is a mineral wax (natural paraffin); there are other but less common members of this group.

TABLE 2

Natural organic mineral (bitumen)	Petrobitumen (soluble in carbon disulfide)	Natural gas of petroleum Crude oil Natural asphalt Ozocerite Asphaltite
		Asphaltic pyrobitumen
	Kerobitumen (insoluble)	Organic matter of source beds Kerogen Fossil microalgae
		Cannel coal, boghead Fossil waxes and resins
	Carbonaceous matter (carbobitumen) (insoluble)	Peat Lignite Coal Anthracite coal

Asphaltites are vein asphalts, such as gilsonite, which instead of occurring in surface accumulations are found in tabular bodies filling preexisting cracks in the rocks at or near the surface. The pyrobitumens are similar in physical and chemical properties to the petroleum bitumens, but their probable derivation from petroleum cannot be as readily demonstrated. Some of the pyrobitumens, such as albertite, wurtzilite, and elaterite, occur in veins similar to gilsonite and other asphaltites. Other pyrobitumens are disseminated in rocks like bituminous schists and marbles. X-ray studies tend to confirm the belief that petroleum asphalt, squirted into cracks in the rocks, converts to asphaltites and pyrobitumens through heat and pressure.[46]

Kerogen, the organic matter in so-called oil shale, is disseminated bitumen. It yields a petroleum-like substance upon distillation. The trans-

[46]Robert S. Young, "Preliminary X-Ray Investigation of Solid Hydrocarbons," *Bull. Am. Assoc. Petrol. Geol.*, Vol. 38 (September, 1954), pp. 2017–2020.

formation from solid to liquid takes place only under high temperatures; pressures of the order of magnitude existing in the earth's crust do not produce oil from kerogen.

GAS

Physical Properties and Chemical Character. The physical properties of a natural gas include color, odor, and inflammability. The principal ingredient of most natural gases is methane, CH_4, which is colorless, odorless, and highly inflammable. The variety of substances present in natural gas is much less than in crude oil. However, in addition to methane, ethane, propane, and other hydrocarbon gases, nitrogen, hydrogen sulfide, and carbon dioxide are present in widely varying amounts, in many natural gases. A less common constituent is helium. Hydrogen sulfide has a distinct and penetrating odor, and a few parts per million are sufficient to give the natural gas a decided odor. Nitrogen and helium are not inflammable, and if one or both of these elements are present in abundance the natural gas does not burn. Gasoline in vapor form is present in most natural gas ("wet gas") and is extracted and sold as "natural gasoline".

Physically and genetically natural gas is a gaseous phase of petroleum, and its occurrence in the reservoir is very similar to that of petroleum. Its solubility in petroleum has already been discussed. Natural gas is also soluble in water; under the temperature and pressure conditions existing in reservoirs, oil-field waters have a capacity for dissolved gas amounting to about one-twentieth that of petroleum. Therefore, when a petroleum, caught in a trap, is saturated with natural gas, gas occupies not only the voids in the reservoir above the oil, but is also in solution in the interstitial water throughout the hydrocarbon-filled reservoir. Furthermore, natural gas (mainly methane) may occur in solution in water-filled reservoirs in quantities up to 14 cubic feet per barrel of water beyond the confines of oil and gas fields.

It has been found that at depths below a few thousand feet the water or brine that fills the pore space in the subsurface formations in the region bordering the Gulf of Mexico from southern Texas eastward to Alabama generally contains a detectable amount of dissolved hydrocarbon gas. The chief constituent of the dissolved gas is usually methane, but measurable proportions of ethane, propane, and butane are often present. The total quantity of gas dissolved in the water of the subsurface formations in this area probably exceeds the known proved gas reserves heretofore discovered in commercial accumulations in the area.[47]

[47]Stewart E. Buckley, C. R. Hocott, and M. S. Taggart, Jr., "Distribution of Dissolved Hydrocarbons in Subsurface Waters," *Habitat of Oil* (American Association of Petroleum Geologists, Special Volume, 1958), pp. 881–882.

A major difference between oil and gas is in mobility. Gas can move through finer openings and up lesser grades. Therefore it can travel farther from its point of origin than oil, which explains why gas may occur in traps overlying water, with no intervening petroleum.

There is considerable variation in natural gases from different reservoirs and from different parts of the same reservoir, in terms of percentages of methane, gasoline, other hydrocarbons, hydrogen sulfide, nitrogen, helium, and carbon dioxide. However, geologists have made but little use of analytical data of natural gases. It has been suggested that an effective method of search for new oil deposits might be developed from gas analyses.[48] The lightest gases tend to migrate the farthest from the source oil. Therefore, the plotting of "isoethane" lines, drawn through points of equal ethane content in a gas field, might show the direction in which to prospect for the source oil, and the spacing of the lines might be indicative of the distance.

Estimation of Gas Reserves.[49] Because gas is a highly compressible substance, the extraction of a quantity of gas squeezed into a reservoir results in a measurable drop in the reservoir pressure (except in the presence of an active water drive), and owing to the approximate applicability of Boyle's law, there is a definite ratio between the amount of gas initially in the reservoir, the drop in pressure, and the volume of gas removed. Therefore, unlike oil, it is possible to make an estimate of the gas in the reservoir after the drilling of only one well, providing that that well has produced enough gas to make a measurable decline in the reservoir pressure. The formula used is

$$R = Q \frac{(P_1 d_1 - P_3 d_3)}{(P_1 d_1 - P_2 d_2)}$$

In this formula R is the available gas reserve, Q is the amount of production between discovery and date of appraisal, P_1 is the reservoir pressure at time of discovery, P_2 is the reservoir pressure at time of appraisal, P_3

[48]H. C. Allen, "Chemistry Reveals Important Facts," *Oil and Gas Jour.*, Vol. 28 (Oct. 24, 1929), pp. 42 et seq.; Paul H. Price and A. J. W. Headlee, "Geochemistry of Natural Gas in Appalachian Province," *Bull. Am. Assoc. Petrol. Geol.*, Vol. 26 (January, 1942), pp. 19–35.

[49]P. McDonald Biddison, "Estimation of Natural Gas Reserves," *Geology of Natural Gas* (American Association of Petroleum Geologists, 1935), pp. 1035–1052; E. B. Elfrink, C. R. Sandberg, and T. A. Pollard, "Application of Compressibility Factors in the Estimation of Gas Reserves," *Oil and Gas Jour.*, Vol. 47 (March 3, 1949), pp. 89–91; John C. Calhoun, Jr., "Estimating Gas Reserves," *Oil and Gas Jour.*, Vol. 47 (Dec. 9, 1948), p. 115; Henry J. Gruy and Jack A. Crichton, "A Critical Review of Methods Used in Estimation of Gas Reserves," *Petroleum Development and Technology* (American Institute of Mining and Metallurgical Engineers, 1949), Vol. 179, pp. 249–263.

is the pressure at which abandonment is scheduled, and d_1, d_2, and d_3 are deviation factors at pressures P_1, P_2, and P_3. Since natural gases do not behave exactly according to Boyle's law, it is necessary, for very accurate estimates, to apply deviation factors. These are obtained from published formulas or tables.[50]

Natural gas reserves can also be calculated by decline-curve and by volumetric methods. A common volumetric formula obtains the cubic feet of gas per acre foot at base temperature and pressure by multiplying together six factors: (1) 43,560, the number of cubic feet per acre foot; (2) porosity percentage; (3) percentage of available void space, obtained by subtracting the interstitial water percentage from 100; (4) reservoir pressure in pounds per square inch, divided by the base pressure used, which is ordinarily some figure between 14.4 and 16.7 pounds per square inch; (5) 460 plus the base temperature (ordinarily 60° F),divided by 460 plus the reservoir temperature in degrees F; (6) 1/Z, where Z is the compressibility factor at reservoir pressure. Z can be determined by laboratory measurements or estimated from published curves. The concluding step is to multiply the figure obtained through the application of this formula by the area of the gas accumulation in acres and by the average thickness of the reservoir in feet. The accuracy of the final result depends largely upon the accuracy of these dimension figures, especially the average thickness estimate.

[50]D. L. Katz, "High Pressure Gas Measurement," Part 2, "A Suggested Standard Method for Calculation of High Pressure Gas Measurement," *Refiner and Natural Gasoline Manufacturer* (June, 1942), pp. 64–69.

RESERVOIR SEALS

The reservoir rock is the container in which oil and gas are stored. Containment is only possible if the walls of the container are effectively sealed. Because of the importance of *reservoir seals* to the accumulation of hydrocarbons, this subject is given the dignity of a separate chapter, even though it is a brief one.

Reservoir seals confine not only gas and oil, but also the underlying body of water which is almost invariably present and through which the hydrocarbons move in passing upward into the trap.

Reservoir seals can be classified into two groups, depending upon their structural relationship with the reservoir rock. Where the sealing surface is parallel to the bedding planes of the reservoir rock, the seal is a *parallel seal.* If this surface, no matter how irregular, crosses the stratification of the reservoir rock it constitutes a *transverse seal.*

QUALIFICATIONS FOR SEALS

A seal is any material or combination of materials in the earth's crust which is impervious to the passage of fluids in any volume. Nothing is completely impervious, and reservoir seals are no exception. The sealing is imperfect in many gas and oil fields, notably Kirkuk in Iraq, where hydrocarbons seep to the surface from scores of vents.

To be relatively impervious there can be virtually no interconnecting fractures or supercapillary pores. It has already been noted (in Chapter 6) that all brittle rocks are vulnerable to fracturing, and fracture systems are in themselves oil and gas reservoirs. Therefore seals, to be widely effective, must have a certain degree of *plasticity*, which permits them to give

241

or flow rather than fracture during movements of the earth's crust. In all probability many "dry" (water-filled) traps owe their deplorable condition to the fact that the assumed seal has fracture permeability.

Parallel Seals. The conformably overlying, relatively impervious rock stratum that prevents fluids from moving vertically upward out of the reservoir rock is usually referred to as the "cap rock." This term has a quite different meaning when it is used to designate the disk of anhydrite and perhaps other minerals at the top of an intrusive mass of salt (Chapter 12). Furthermore, reservoir rocks have to be sealed by an underlying impervious stratum as well, to which the term cap rock is not applicable. For these reasons "parallel seals" is the preferred term here, with the overlying impervious stratum heretofore called the cap rock referred to as the "overseal."

Wherever the occurrence of oil within a reservoir is vertically erratic, as in fracture-controlled porosity, the oil accumulation at any one point may have both an *immediate overseal* and an *ultimate overseal*. The immediate overseal is a tight phase at the top of the reservoir rock through which wells must be drilled before reaching the "pay." The ultimate overseal is the consistently tight overlying rock which stops further upward movement of hydrocarbons. Best examples of immediate and ultimate overseals are in carbonate rock reservoirs.

Parallel seals are the sole seals in anticlinal traps (Chapter 11); they are also a part of the sealing system in most types of varying permeability ("stratigraphic") traps (Chapter 14).

Transverse Seals. A transverse seal supplies the dam across the stratification of a reservoir rock below which hydrocarbons may be impounded. They are essential to both fault and varying permeability traps. The shape and geological history of transverse seals are discussed in Chapters 13 and 14. In this chapter we shall confine ourselves to the rock materials which make sealing possible. With few exceptions these are the same whether the seals are parallel or transverse.

TYPES OF SEALS

Shale. Included here are the non-laminated as well as the laminated, fine-grained, clastic rocks. Shale is the most abundant sedimentary rock in the earth's crust. It is commonly interbedded with either sandstones or carbonate rocks, or both. Therefore the chances are that a reservoir rock will be both overlain and underlain by a bed of shale.

According to Grim,[1] the degree of impermeability of shales depends upon both the *texture* and the *types* of minerals present. "Many shales

[1] Ralph E. Grim, *letter* dated October 3, 1958.

are composed of clay mineral particles which are flat flake-shaped units with these units arranged in a parallel fashion, so that the flake-shaped particles overlap like the shingles on a roof." The orientation of the aggregate is much less uniform in other shaley materials; no doubt the sealing efficiency is also less. Regarding mineral content Grim states:

. . . if the montmorillonite clay mineral is present in a shale it would be likely to have the maximum degree of imperviousness, because of its fine-grained character, because it tends to adsorb a modest amount of water and then very little more, and because of its relatively high degree of plasticity. At the other end of the scale, one would expect shales containing kaolinite in which the particles of kaolinite would be relatively less flake-shaped and more granular, larger in particle size and with much less water adsorbing capacity.

Presumably the less plastic kaolinite shales are more prone to fracture under stress. Mud shales, consisting mainly of finely divided quartz grains, are probably the most brittle of the shale type rocks, and therefore the least trustworthy as seals. Calcareous shales are mixtures of finely divided calcite or dolomite with clay or mud particles. Their degree of plasticity depends upon the clay mineral content.

Swann [2] has observed that some of the shales overlying oil reservoirs in the Illinois basin leak and others do not: "Most lower Chester sandstone reservoirs are not effectively sealed from the next overlying sand units by 20–50 feet of shale unless the shale contains at least 5–10 feet of limestone. The limestone itself is probably not the seal but is rather an indicator of the type of shale most effective as a seal." Fissured shales overlying sandstone reservoir rock at Salt Creek, Wyoming, are oil-bearing. Other examples of commercial production from fractures in shale is given in Chapter 6 under "Fracture Reservoirs."

Fine clastic particles, especially clay, also actually provide the seals in other types of rocks. Argillaceous sandstones consisting of poorly sorted quartz and clay grains are important seals in Tertiary accumulations in many parts of the world. Many permeability wedge-outs are due to the up-dip "shaling" of sandstone or carbonate rock reservoirs.

It is also possible in some instances that shale beds higher in the section above an oil accumulation provide the actual seal, rather than the allegedly tight non-shale rock immediately overlying the reservoir which gets the credit.

Carbonate Rocks. In terms of petroleum accumulation, the carbonate rocks are the most versatile. Without doubt some have, in the past, functioned as source rocks, and carbonate rocks are extremely important reservoirs in various parts of the world. Carbonate rocks under certain conditions are also seals. The most widespread limestone and dolomite

[2] David H. Swann, "Effective Seals Above Chester Sandstones of the Eastern Interior (Illinois) Basin," *Program Am. Assoc. Petrol. Geol. Annual Meeting*, 1956, pp. 31–32.

overseals are those which are either: (1) argillaceous, grading into calcareous shales; or (2) excessively fine-grained and to some degree plastic, like marls and certain chalks; or (3) anhydritic, containing disseminated particles of anhydrite. The relatively pure dense limestone or dolomite is brittle and hence vulnerable to cracking through earth movements. As discussed in Chapter 6 this type of carbonate rock is an important hydrocarbonate reservoir, with the oil and gas occurring in fissures. Whether or not a brittle carbonate rock *is* fractured depends upon several factors, including topography, load, and especially earth movements. Presumably a limestone in a stable area could be a seal, whereas the same rock in a disturbed belt could be a fissure reservoir. The next section describes limestones that changed from seals to channelways with increasing intensity of folding.

Evaporites. Although it does not ordinarily act as a seal, salt is an ideal material for the purpose. It is probably the most impermeable of sedimentary minerals. Mines in thick salt beds are invariably dry, except for water entering through the shaft or other man made channelway. The "flowability" of salt under pressure is well known. In most places, however, as beneath an evaporite series or on the flank of a salt plug, where salt could act as a seal, a bed or sheath of anhydrite intervenes and does the actual sealing.

For this reason, and because anhydrite is more widespread in its occurrence than salt and other saline precipitates, it is by far the most important seal among the evaporite minerals. It also possesses impermeability and flowability, but to a somewhat lesser degree than salt. Three different types of anhydrite sealing in the northeastern Williston Basin (North Dakota-Saskatchewan-Manitoba) have been described by Johnson;[3]

(1) Secondary anhydrite along the unconformity at the top of the Madison forms an impervious zone of variable thickness by filling the fractures and pores of associated rocks.

(2) A seal is created by primary anhydrites interbedded with porous, commonly vuggy, fragmental limestones. Since the angle between the unconformity and bedding is small, the primary anhydrites may help to form the seal along the weathered area.

(3) A regional facies change, in which porous reservoir beds grade to dense anhydrites in an easterly direction as the basin edge is approached, creates the seal.

During Jurassic time in what is now the Persian Gulf area a succession

[3] Walter Johnson, "Mississippian Oil Fields of Northeastern Williston Basin," *Program Am. Assoc. Petrol. Geol. Annual Meeting* (1956), pp. 25–26.

of interbedded shallow-water limestones and anhydrites were deposited. The limestones have functioned as source and reservoir rocks and the anhydrites have provided the necessary seals.[4] At Kirkuk in northern Iraq the present overseal for this great hydrocarbon accumulation is the Tertiary Fars evaporite section. Dunnington believes, however, that the first overseals were thin-but-impervious pre-Fars limestone layers interbedded with permeable limestone reservoirs. He also believes that renewed folding of the Kirkuk anticline cracked these tight limestones and permitted the hydrocarbons to ascend across the bedding until stopped by the Fars evaporites.[5] Even this overseal is not perfect, as is shown by the leakage mentioned earlier in this chapter and in Chapter 3 under "Seeps."[6]

A recent study of 12 of the more important Michigan oil and gas fields showed that in six of these fields anhydrite plays an important role in sealing in the hydrocarbons. In some instances the overseal is a fairly pure bed of anhydrite, but in most fields the anhydrite is disseminated through a shale or carbonate rock overseal.[6]

Gypsum may also seal in hydrocarbons.[7] This mineral is exceptionally tight.

Miscellaneous Seals. Solid or semisolid hydrocarbons, such as natural *asphalt*, are known to seal in underlying liquid hydrocarbons in a few places (Chapter 14). Some fault sealing (Chapter 13) is accomplished by the occurrence of impermeable *fault gouge* along the fault plane rather than by the presence of an impervious material across the fault. Furthermore, fault gouge may prevent leakage up the fault plane.

Many siltstones are so low in permeability that they could seal in liquid hydrocarbons, and some do. Most "tight" sandstones are impervious because they are argillaceous. It is the clay in the sandstone pores and the shaley interbeds that provide the actual seal. Sandstones in which the cementation is virtually complete are also impervious, but such sandstones are very unusual. Both siltstones and sandstones are brittle rocks and are vulnerable to fracturing brought about by movements of the earth's crust.

[4] R. M. Ramsden, "Some Problems Concerning Sealing Factors by Jurassic Anhydrite in the Persian Gulf," *Program Am. Assoc. Petrol. Geol. Annual Meeting* (1956), p. 39.

[5] H. V. Dunnington, "Generation, Migration, and Accumulation of Oil in Northern Iraq," *Habitat of Oil* (American Association of Petroleum Geology Special Vol., 1958), pp. 1194–1251.

[6] *Seminar in Advanced Petroleum Geology*, University of Michigan, 1959.

[7] Sherman A. Wengerd, "Sealing Factors for Non-Structural Paleozoic Oil in the Paradox Basin of the Four Corners Region," *Program Am. Assoc. Petrol. Geol. Annual Meeting* (1956), pp. 26–27.

POSTACCUMULATION SEALS

According to Heald[8] there are many examples of oil accumulations becoming "frozen" in place by subsequent complete cementation of the reservoir rock at the oil-water interface. This freezing may be caused by precipitation of anhydrite in the reservoir-rock pores due to inter-action between calcium ions in the edge water and sulfate ions in the oil, or it may be due to contact asphaltization.

The geological implications of this sealing in of a petroleum deposit are considerable. It would mean that oil would not be spilled out by subsequent regional tilting. It would provide another explanation for inclined oil-water interfaces. Furthermore this freezing in place of hydro-carbons would cut off all hydrostatic connections, so that a water drive would not be possible. Actually, the best evidence for postaccumulation sealing is the fact that some oil reservoirs, which from evidence obtained both from within and beyond oil field borders, appear to have the fluid pressure and the porosity and permeability adequate for water drive do not have one.

[8]K. C. Heald, *lecture*, Geology-Mineralogy Journal Club, University of Michigan (December 5, 1951).

MIGRATION OF OIL
AND GAS[1]

Because oil and gas ordinarily do not occur in commercial deposits in the same rocks in which they probably originated, *migration* of these hydrocarbons from source rock to reservoir rock is postulated. In addition, most students of petroleum geology subscribe to the belief that further migration can and does take place through the reservoir rock until the hydrocarbons either escape or are caught in some type of natural trap. Therefore migration is a probable chapter in the history of an oil or gas deposit, falling between generation and accumulation.

Owing to the extreme mobility of natural gas, there is little, if any, dissent to the concept of its migration. Gas under pressure moves through all but the tightest rocks in the direction of lesser pressure; usually, but not necessarily, this direction is upward. We have already seen that in the deeper oil reservoirs gas and oil merge into a homogeneous vapor phase, therefore deep-seated oils have a mobility comparable with that of natural gas. Moody[2] believes that long distance migration is possible under these circumstances. However, we also have petroleums that never experienced the conditions necessary for vaporization. Most of the dis-

[1] F. M. Van Tuyl, Ben H. Parker, and W. W. Skeeters, "The Migration and Accumulation of Petroleum and Natural Gas," *Quarterly Colo. School of Mines*, Vol. 40 (January, 1945), 111 pp. (112 titles in bibliography); William B. Heroy, "Petroleum Geology," *Geol. Soc. Am., 50th Anniversary Vol.* (1941), pp. 512–548; V. C. Illing, "The Migration of Oil," *Science of Petroleum* (Oxford University Press, 1938), Vol. 1, pp. 209–215; James Frost, "Oil Migration," *Jour. Inst. Petrol. Technologists*, Vol. 31 (December, 1945), pp. 486–493.

[2] C. L. Moody, "Petroleum Geology, 1951," *Bull. Am. Assoc. Petrol. Geol.*, Vol. 35 (July, 1951), pp. 1491–1504.

cussion that follows is concerned with the migration of these liquid hydrocarbons.

Does Oil Migrate? As stated in Chapter 7, oil underground has mobility approaching and even exceeding that of water at the surface because of the presence of gas in solution and because of the higher temperatures that exist beneath the surface. This fact is of extreme importance in any discussion on the pros and cons of oil migration.

Evidence opposed to oil migration. The disagreement about oil migration is one of *degree* of mobility. No one actually believes that each drop of oil in the reservoir rock is the relic of an organism that once occupied that exact spot. The opponents of migration believe, rather, that the oil comes from the enclosing and immediately surrounding rocks and has not moved any great distance.

Perhaps the best argument for relatively slight migration is the occurrence of oil in lenticular sand bodies completely surrounded by dense shale. As in every other instance, the origin of the oil in the sand is not known, but it probably was squeezed out of the enclosing shale during compaction. That it came from a more distant source is most unlikely. Another argument opposing migration, at least across stratification planes, is the presence in different superimposed reservoirs in a single field of different types of petroleum, showing that there has been no intermingling of crude oils. This rule is general but by no means universal; exceptions that tend to prove migration will be cited subsequently. Additional arguments, advanced by Clark,[3] are that if oil accumulates by migration from far-flung source materials, then (1) every trap should contain oil, and (2) the water-filled reservoir rock below oil pools should contain traces of the oil that once passed through. However, there are other explanations for barren traps, and it is questionable that oil passing through a wet sand would leave any trace.

Evidence in support of oil migration. Oil underground is a fluid with considerable mobility. To deny categorically that it can migrate, or that it can migrate through any great distance, is also to deny that underground water has migratory possibilities. Under the influence of a pressure differential of considerable magnitude brought about by poking a well into a reservoir, the oil migrates with rapidity from surrounding rock into the borehole.

Every water-drive field that has been developed under engineering control yields unequivocal, visual evidence of migration. East Texas is perhaps the best example.

[3] Frank R. Clark, "Origin and Accumulation of Oil," *Problems of Petroleum Geology* (American Association of Petroleum Geologists, 1934), pp. 309–335.

On the west side of this great field square mile after square mile of the area that was formerly a forest of derricks and a bedlam of pumping wells has now returned to pastoral and sylvan quietude. It won't be long until it may be said that, like the mining boom towns of the west, "her tanks are rust; her men are dust; it's forty years since she went bust." I personally am a victim of migration here; I am a royalty owner under a square mile of ex-oil-land. Many millions of barrels of oil have migrated from Woodbine sands underlying this area into wells, through tubing to surface separators, then through pipelines to refineries; thence to commerce via the filling station, and, through tailpipes of automobiles, to the atmosphere, where presumably these now changed hydrocarbons are again jockeying for position in a new carbon-cycle that may eventually take them back underground in yet unmade rocks to form new oil deposits . . . which probably won't be needed by that time! And where these millions of barrels used to be are now equal volumes of briny salt water. Such is man's role in underground fluid migration in the hills of east Texas.

Proof that migration in the abandoned zone of this huge field is complete has been established by much coring, both sidewall and conventional; by re-surveys of old wells by the electric logging devices; and by many bottom-hole-pressure and fluid-level measurements. Pressure decline in the producing part of the field has been controlled by limiting oil output and injecting into the Woodbine sands salt water in down-dip locations. Plenty of underground fluid migration here!

Incidentally the superstition that migrating oil must leave tell-tale evidence of its former occupancy is pretty well dissipated by examination of cores from the Woodbine in the abandoned area. Many core-analyses and many electric logs show zero saturation for oil and 100 per cent saturation for salt water in sands.

It seems quite safe to say that all the natural water drive fields of the world, and also those that have been treated to artificial drives through engineering activity, prove without qualification that oil, as we know it today, *does* migrate. Therefore, uniformitarianly speaking, it *has* migrated in the past.[4]

Points of evidence that oil has migrated *in the geological past* follow:

(1) *The presence of oil seeps.* The mere existence of an oil seep is evidence of the natural movement of oil today. The petroleum that emerges at the surface has migrated from a buried reservoir.

(2) *Accumulations in inorganic rocks.* Most commercial petroleum deposits occur in rocks that in all probability never contained the source organisms from whence the oil came. Therefore the oil must have migrated from source to reservoir. The most common oil reservoir, sandstone, is deposited under strand-line conditions that are inimical to the growth and preservation of organisms. Carbonate rocks, second only to sandstone in importance as oil reservoirs, may be organic in origin, but in

[4]Clarence L. Moody, *letter* dated October 13, 1958. Although not intended for publication the scientific and literary quality of these paragraphs are such that permission to incorporate was requested and granted, albeit with reluctance.

most cases, at least, it is very unlikely that the oil is indigenous. Much carbonate rock porosity (Chapter 6) is the result of solution-leaching, which does not take place until after lithifaction and emergence. It is difficult to picture indigenous oil waiting around until leaching has produced cavities. Even where the cavities are primary, as is common in a reef, the oil may have migrated into the porous zone from without.

A strong argument for migration into reservoirs exists when oil deposits occur in intrusive and crystalline basement rocks. It is impossible to conceive of such oil as indigenous. A similar situation exists when oil has accumulated in the porous tops of buried hills and ridges. This oil, like that in the crystalline rocks, must have migrated into place after the porosity was developed, which was considerably later than the formation of the rock itself.

(3) *Correlation between oils in reservoirs and residual oils found in source rocks.* A research team has succeeded in identifying four types of petroleum and asphalt in different reservoir rocks in the Uinta basin of northern Utah. These same four types were found as residua in different source beds in the same area.[5] It is concluded that each type migrated from source rock to reservoir rock.

(4) *Chemically similar oils in a series of superimposed reservoirs.* Although the general rule is to the contrary, there are examples of migration of oil from reservoir to reservoir in a multiple-zone field. At Garber, Oklahoma, the same type of oil has been obtained in 14 distinct reservoirs, lying one above the other and ranging in age from Ordovician to Permian. Obviously trans-formational channelways have permitted migration between reservoirs at Garber. At Oklahoma City the oils in Ordovician reservoirs are alike, but the opportunities for transverse migration do not extend upward across the unconformity into the Pennsylvanian rocks, for there the oil is different. Other examples of migration between reservoirs could be cited, but they are much less common than examples of no migration between superimposed reservoirs.

(5) *Structural adjustment of hydrocarbons in a reservoir.* The crust of the earth is constantly yielding to diastrophic forces which fold and tilt the rock strata, and yet, regardless of the recency of such activity, the gas, oil, and water are usually in adjustment with the structure. Regional tilting changes markedly the contour pattern of a dome. Not only the highly mobile gas of a gas cap but also the oil and underlying water migrate relatively quickly into the new positions called for by the structural change. In the great oil districts in the San Joaquin Valley and in the Los Angeles Basin in California, no traps with adequate sedi-

---

[5] John M. Hunt, Francis Stewart, and Parke A. Dickey, "Origin of Hydrocarbons of Uinta Basin, Utah," *Bull. Am. Assoc. Petrol. Geol.,* Vol. 38 (August, 1954), pp. 1671–1698.

mentary section have been found barren as yet, in spite of their extreme youth.[6]

(6) *Quantitative considerations.* Probably the strongest argument for large-scale migration is the presence of oil in such enormous quantities in fields like those of the salt domes of the Gulf Coast, of East Texas, of Burghan, Kuwait, and of Leduc, Alberta, that it is quantitatively inconceivable that a local, or even near-local, accumulation of organic debris would have been great enough to yield all the oil.

The conclusion would seem to be that even though migration of oil may have been minor in some fields, such as the "shoestring sand" pools, still, in the great majority of fields, large-scale migration is not only possible but probable.

Primary and Secondary Migration. The first migration that oil experiences after generation is from fine-grained source rock into porous and permeable reservoir rock. Except when the reservoir rock is an isolated lens, the *primary* migration is followed by *secondary* migrations through the reservoir rock until the oil either escapes or is trapped.

The primary and secondary migration of gas, and of oil-gas mixtures in a vapor phase due to high pressures, is much easier than that of liquid petroleum. The gas merely streams in the direction of lesser pressure. Water-filled pores present minimum resistance to this movement. It should be remembered that gas, and oil-gas mixtures in vapor phase, can travel not only through the same interconnecting pores that are followed by migrating petroleum, but also they can move through micropores and micro-fractures as well.[7]

Causes of Movement of Oil.[8] The motivation for oil movement through rock depends upon whether the migration is primary or secondary. In primary migration hydraulic forces are dominant; in secondary migration the buoyancy of the oil with respect to the associated water is probably most important.

Hydraulic forces.[9] Flowing water is capable of carrying oil along

[6] Harold W. Hoots, "Origin, Migration, and Accumulation of Oil in California," *Calif. Div. Mines, Bull.* 118 (August, 1941), pp. 261–262.

[7] M. A. Kapelushnikov, "Migration of Petroleum and Secondary Recovery," reviewed by George V. Chilingar in *Bull. Am. Assoc. Petrol. Geol.*, Vol. 40 (July, 1956), pp. 1725–1726.

[8] E. DeGolyer and Harold Vance, "Bibliography of the Petroleum Industry," *Bull. Agr. and Mech. Coll. Texas*, 83 (1944), pp. 343–345 (57 titles).

[9] Malcolm J. Munn, "Studies in the Application of the Anticlinal Theory of Oil and Gas Accumulation," *Econ. Geol.*, Vol. 4 (1909), pp. 141–157; John L. Rich, "Problems of the Origin, Migration, and Accumulation of Oil," *Problems of Petroleum Geology* (American Association of Petroleum Geologists, 1934), pp. 337–345, "Moving Underground Water as a Primary Cause of the Migration and Accumulation of Oil and Gas," *Econ. Geol.*, Vol. 16 (September-October, 1921), pp. 347–371, "Further Notes on the Hydraulic Theory of Oil Migration and Accumulation," *Bull. Am. Assoc. Petrol. Geol.*, Vol. 7 (May-June, 1923), pp. 213–225.

with it. There is no doubt an outflow of water during the *compaction*[10] of fine-grained sediments such as source rock. Initially the source rock is a clay, mud, or calcareous ooze with porosity as high as 90 per cent. As it is squeezed by the weight of overlying sediment, or by lateral pressures accompanying diastrophism, the fine-grained material is compacted, with a reduction in porosity down to 35 per cent or less. Obviously the fluids occupying the pore space that is obliterated by the compaction are driven out. They move in the direction of least resistance into non-compacting porous formations such as sandstones and permeable limestones. Water is always the more abundant fluid, but if any oil is present in the interstices between the water-wet grains of the source rock it is pushed out along with the non-wetting water.

Studies made by coring recent sediments now undergoing compaction in eastern Venezuela have shown the existence of a pressure gradient in the muds upward toward laterally continuous sands and downward toward the unconformity at the top of the Pleistocene. Both the sands overlying and underlying the compacting muds are apparently acting as conduits which permit the escape of the fluids to the surface where they are dissipated. "However, any of the fluid that enters an enclosed sand body has to re-enter the overlying finer sediments to reach the surface. The water phase of the fluid can pass from a sand back into a water wet mud, but any entrained oil will be filtered out at the first sand-mud interface."[11]

Even sandstones are not completely incompressible, and the compaction of such a rock may be a contributing force to the migration of oil through sandstone reservoirs.[12]

---

[10] Hollis D. Hedberg, "Gravitational Compaction of Clays and Shales," *Am. Jour. Science*, 5th Series, Vol. 31 (April, 1936), pp. 241–287; L. F. Athy, "Density, Porosity and Compaction of Sedimentary Rocks" and "Compaction and Oil Migration," *Bull. Am. Assoc. Petrol. Geol.*, Vol. 14 (January, 1930), pp. 1–36; R. C. Beckstrom and F. M. Van Tuyl, "Compaction as a Cause of the Migration of Petroleum," *Bull. Am. Assoc. Petrol. Geol.*, Vol. 12 (November, 1928), pp. 1049–1055; H. D. Hedberg, L. C. Sass, and H. J. Funkhouser, "Oil Fields of Greater Oficina Area, Central Anzoategui, Venezuela," *Bull. Am. Assoc. Petrol. Geol.*, Vol. 31 (December, 1947), p. 2137 and footnote, p. 2138; William C. Gussow, "Migration of Oil," *World Oil* (Aug. 1, 1956), pp. 79–83; M. G. Cheney, "Geology of North-Central Texas," *Bull. Am. Assoc. Petrol. Geol.*, Vol. 24 (January, 1940), pp. 113–116.

[11] Albert L. Kidwell and John M. Hunt: "Oil Migration in Recent Sediments," *World Oil* (July, 1958), pp. 79 et seq; "Migration of Oil in Recent Sediments of Pedernales, Venezuela," *Habitat of Oil* (American Association of Petroleum Geologists Special Vol., 1958), pp. 790–817.

[12] David Donoghue, "Elasticity of Reservoir Rocks and Fluids with Special Reference to East Texas Oil Field," *Bull. Am. Assoc. of Petrol. Geol.*, Vol. 28 (July, 1944), pp. 1032–1035.

Limestones may have two quite different pressure expulsions. If a limestone was originally a calcareous ooze, it experienced an early pre-lithifaction squeezing with concurrent fluid expulsion as just described. After lithifaction overburden weight may cause pressure solution and a reduction of volume of as much as 40 per cent. Stylolites are a visible result of this shrinkage by solution. Ramsden [13] points out that a 40 per cent reduction in volume is accompanied by the expulsion of all fluids occupying that space. Dunnington [14] adds that the actual pore-volume reduction may be much greater than the percentage of rock-volume shrinkage, because of precipitation of carbonate minerals within the pores owing to super-saturation of the interstitial water brought about by stylolite formation.

Some geologists also believe in the efficacy of hydraulic forces to move oil through reservoir rock, with accumulation the result of density strati-fication of water and hydrocarbons when a trap is reached. To many, however, the hydraulic theory is unsatisfactory unless one is willing to concede a velocity to the ground-water circulation in excess of the speed of upward (vertical or up-dip) movement of oil through water. Actually, at the depths we are considering, the water in the reservoir is virtually stationary. It is only in motion in the near vicinity of an outlet, such as a spring or well. Confined aquifers are similar to surface lakes and reservoirs in this regard; no appreciable current is present except close to the outlet stream or spillway.

There is, however, considerable evidence to the effect that where there is a tilted potentiometric (piezometric) surface due to a confined aquifer having an outlet at lower elevation than the intake, the gas-water and oil-water interfaces in bodies of trapped hydrocarbons are also tilted. This is the hydrodynamic concept developed by Hubbert. [15] As its application is principally in the emplacement of oil or gas in the trap, it is discussed in more detail under "Hydrodynamic Traps" in Chapter 13.

There is no dispute but that hydraulic forces are responsible for the movement of oil through the reservoir during the exploitation of a water-drive field. In this situation the reservoir water is in motion, because of nearness to outlet wells.

Buoyancy. Although there have been many attempts to minimize and even deny its importance, buoyancy (sometimes called "gravity" or

[13] R. M. Ramsden, "Stylolites and Oil Migration," *Bull. Am. Assoc. Petrol. Geol.*, Vol. 36 (November, 1952), pp. 2185–2186.

[14] H. V. Dunnington, "Stylolite Development Post-Dates Rock Induration," *Jour. Sedimentary Petrology*, Vol. 24 (March, 1954), pp. 27–49.

[15] M. King Hubbert, "Entrapment of Petroleum under Hydrodynamic Conditions," *Bull. Am. Assoc. Petrol. Geol.*, Vol. 37 (August, 1953), pp. 1954–2026.

"flotation") remains the primary cause for movement of oil through reservoir rocks (secondary migration). There are only two basic requirements: (1) immiscible fluids, and (2) fluids with different densities. It is as simple as the accumulation of cream at the top of a bottle of whole (non-homogenized) milk. It has already been pointed out that water and crude oil are immiscible, and that crude oil is always lighter than the reservoir water (even though that oil may be heavier than pure water). Natural gas if of course the lightest of all. It is also immiscible with the other two reservoir fluids after they have become gas-saturated for the pressures and temperatures in effect.

Therefore, wherever water and oil, or water and gas, or water, oil, and separate gas are present in a container of adequate permeability, they assume a density stratification. Because these containers are almost always water-filled, the oil and gas are crowded beneath a relatively impermeable roof (overseal) in a hydrocarbon trap (Chapters 10 to 14). In the rare instances when the container is not full of water the oil still overlies the water, wherever it may be, or if no water is present the oil lies at the bottom of the container. There is no repeal of the law of gravity.

In order to reach the zone of density stratification the hydrocarbons must have migrated through rock pores. There is, of course, a lower limit in pore dimension below which the forces that tend to impede oil movement are greater than the buoyant force that causes oil to rise through water to the top. The usual source rock is so fine of grain that water-oil stratification within the rock is impossible. On the other hand, since reservoir rocks are, by definition, rocks of permeability as well as porosity, density adjustment is possible within.

The buoyancy theory therefore becomes operative when oil arrives in a reservoir rock, following primary migration from a source rock. The pores in the reservoir rock are water-filled, and the lighter oil moves upward in a manner to be described subsequently. Rarely is the reservoir vertical, so after the overseal ("hanging wall") is reached further movement must be up an inclined plane. The minimum angle of dip up which oil can move depends upon the viscosity of the oil, the volume involved, the specific gravities of both the oil and the reservoir water, the viscosity of the reservoir water, and other factors. Many great oil accumulations have taken place where the reservoir dips do not now, and probably did not in the past, exceed 40 to 50 feet to the mile, and the Trenton oil accumulations in the Lima-Indiana district of the United States appear to have taken place with dips as low as 10 feet to the mile.

It is possible (if not probable) that young, relatively sluggish oils are squeezed into reservoirs in which the dips are less than the critical point for migration for that degree of mobility.   In this event the oil remains, for the time being, more or less dispersed at the top of the reservoir rock. This situation can be changed by: (1) increased dips, due to earth movements; (2) decreased viscosity of the oil due to increased temperature with burial, and the natural evolution of the petroleum itself; or (3) a combination of both increased dip and increased mobility.   Eventually the oil ends up in an up-dip trap, or it may even reach the surface, where it tends to become dissipated by weathering and erosion.

The secondary migration through the reservoir does not necessarily end with the first trapping.   Later tiltings of the regional dip, caused by intermittent sagging of the sedimentary basin or trough floor, result in "spilling" of the hydrocarbons and renewed migration up-dip (Figs. 9–1 and 9–2) as brought out by Gussow and Walters.[16]

[16]W. C. Gussow, "Differential Entrapment of Oil and Gas: a Fundamental Principle, *Bull. Am. Assoc. Petrol. Geol.*, Vol. 38 (May, 1954), pp. 816–855; Robert F. Walters, "Differential Entrapment of Oil and Gas in Arbuckle Dolomite of Central Kansas," *Bull. Am. Assoc. Petrol. Geol.*, Vol. 42 (September, 1958), pp. 2133–2173.

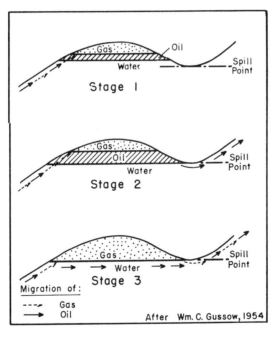

Fig. 9–1. *Stage* 1: The stratification of gas, oil, and water above the trap spill point. *Stage* 2: Hydrocarbons now fill the trap to the spill point; oil is spilling out and migrating farther up-dip. *Stage* 3: Trap is now filled with gas; gas moving up from below enters the trap, but a like volume spills out at the same time; oil bypasses the trap entirely.

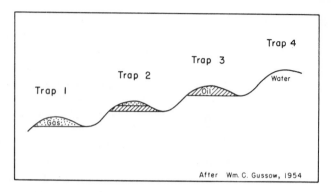

Fig. 9–2. Four traps in tandem; the lowest is filled with gas; the next highest is filling with gas, but some oil remains; the third trap up-dip is filled with oil which will now begin to spill and enter the highest, water-filled, trap.

Every oil pool is evidence of migration caused by the action of buoyancy. Almost invariably (1) the oil is underlain by water or is in contact with water at the down-dip edge, and (2) the oil occupies the highest point beneath an impervious roof or beneath a cap of still lighter natural gas. Even if other means are advanced to move the oil great distances through the reservoir rock, it is necessary to use differences in density for the final arrangement of gas, oil, and water.

Because of its early application to oil finding, the buoyancy theory was first called the "anticlinal" theory. However, since buoyancy explains the presence and position of oil in all other types of traps as well, the more restrictive term has been dropped.

Some doubt developed at one time regarding the adequacy of buoyancy because of the failure of early migration experiments. But these experiments themselves tended to be inadequate because in most of them were used either surface conditions of temperature, pressure, and oil-gas saturation, or dry (rather than water-wet) sand, or both. It follows from the work of Schilthuis [17] on the presence of interstitial ("connate") water in oil and gas reservoir rocks that experiments using sand which is not water-wet are of doubtful value.

Other possible forces. Dilatancy may be responsible for the movement of oil from overlying reservoir rocks into fractured basement rocks. Dilatancy is the volume increase which is brought about by the fracturing. It causes a sudden pressure decrease with the result that fluids in the overlying rock are sucked into the just-formed cracks.

[17] R. J. Schilthuis, "Connate Water in Oil and Gas Sands," *Trans. Am. Inst. Min. Met. Engrs.*, Vol. 127 (1938), pp. 199–214.

A frequently listed cause of oil migration is *capillarity*. This force may be important in assisting in the expulsion of oil from source rocks, but in reservoirs capillarity may serve more to retard than to promote oil movement.[18] The confinement of oil to the coarser zones in thick sandstones is probably due to the fact that the envelopes of "connate" water surrounding the grains of finer sediment impinge upon each other to such an extent that oil simply cannot enter rocks of this type. This latter explanation is preferred to the theory that confinement is due to differential capillarity between oil and water.[19]

Miscellaneous possible causes for oil migration include the expansion of associated water, gas streaming ("effervescence") upon release of pressure, and pressures created by the precipitation of mineral cements within the reservoir. Since these were discussed under causes of reservoir pressure in Chapter 7, further elaboration here is not necessary.

Character of Movement. Roof and Rutherford[20] postulate four mechanisms by means of which petroleum can migrate. Mechanism I involves the simultaneous flow of both water and petroleum phases. Only the petroleum phase flows in Mechanism II; the water remains stationary. In Mechanisms III and IV petroleum migrates on a molecular scale; water flows in III but not in IV.

Mechanism I is undoubtedly applicable to the primary migration of oil out of the source rock. The compaction of clay or ooze containing liquid hydrocarbons causes the expulsion of both oil and water. This fluid flow can be in any direction, including downward. The flow of both water and petroleum phases through reservoir rock (secondary migration) is the hydraulic theory of migration; objections to this concept were given in the preceding section.

Mechanism II, flow of the petroleum phase only, with water stationary, is difficult to apply to primary migration, but is by far the best of the four mechanisms to explain secondary migration. "After the droplets of oil have by some means been removed from the source bed, they may then move upward by buoyancy. For droplets smaller than the fine pores such upward migration appears to be possible . . . If the path has at all points an upward component and if the droplets do not wet the porous medium, they will move continuously through the formation and accumulate at the top."[21]

[18] R. Van A. Mills, "Experimental Studies of Subsurface Relationships in Oil and Gas Fields," *Econ. Geol.*, Vol. 15 (July-August, 1920), pp. 398–421.

[19] Alex W. McCoy and W. Ross Keyte, "Present Interpretations of the Structural Theory for Oil and Gas Migration and Accumulation," *Problems of Petroleum Geology* (American Association of Petroleum Geologists, 1934), pp. 253–307.

[20] J. G. Roof and W. M. Rutherford, "Rate of Migration of Petroleum by Proposed Mechanisms," *Bull. Am. Assoc. Petrol. Geol.*, Vol. 42 (May, 1958), pp. 963–980.

[21] J. G. Roof and W. M. Rutherford, *op cit.*, p. 968.

Secondary migration through reservoir rock is not confined to petroleum droplets. These can be expected to meet and unite into increasingly larger and more buoyant masses as they progress upward beneath an inclined overseal toward a trap. Furthermore, once accumulation has taken place, the oil may resume its migration in large "slugs", if it is spilled out of the trap by tilting or some other means. These have even more buoyancy, and probably they move up-dip with relative expedition.

Roof and Rutherford conclude, in regard to Mechanisms III and IV, that there is probably no substantial migration on the molecular scale alone. This type of movement could, however, be accessary to one or both of the other migration mechanisms. The direction of molecular movement cannot be other than upward.

Disposal of displaced water. Oil entering a water-filled reservoir can make room for itself only by displacing a like volume of water. In an open system this displacement does not create a problem, for the volume is kept constant by outflow at the water table beneath the outcrop. But in a closed reservoir the additional volume can be accommodated only by (1) compression of the reservoir fluids, or (2) forcible penetration of the overlying strata by the displaced water. Water has greater penetrability than oil, and a seal may permit the passage of water and yet retain the oil.

A greater mystery is the phenomenon of oil sands that contain little or no free (other than "connate") water. Some sandstone lenses, including "shoestring" sands, are of this type. They are subaqueous deposits and obviously were filled with water initially. Several explanations have been attempted for the disappearance of this water. Perhaps the best one is that the water was forcibly displaced by oil but the subsequent shrinkage of the oil owing to escape of gas has left some void space unoccupied by either liquid hydrocarbons or water. Other suggested explanations are that the water has been absorbed by hydrating minerals in the surrounding rocks or that the water has been evaporated by natural gas.

Direction of Oil Migration.[22] Migration directions are considered in terms of the stratification planes; the oil migrates either *parallel* or *transverse* to the stratification. Parallel migration is ordinarily referred to as "lateral" and transverse as "vertical," but these terms are unfortunate, especially when the sedimentary layers are steeply inclined. Parallel migration ordinarily takes place through the reservoir rock, whereas transverse migration requires the presence of permeable zones that cross

[22] F. H. Lahee, "A Study of the Evidences for Lateral and Vertical Migration of Oil," *Problems of Petroleum Geology* (American Association for Petroleum Geology, 1934), pp. 399–427; F. M. Van Tuyl and Ben H. Parker, "The Time of Origin and Accumulation of Petroleum," *Quarterly Colo. School of Mines*, Vol. 36 (April, 1941), Chapter 17, pp. 116–123.

the stratification planes. As a general, but not invariable, rule, the primary migration from source rock to reservoir rock is transverse, and the secondary migration through the reservoir to the trap is parallel.

Transverse migration. Transverse migration can be downward or upward. If movement is taking place because of differences in density of oil and water, the migration direction of the oil is obviously upward. But if the oil is being squeezed from a compacting rock, it moves in the direction of least resistance, whether that direction is downward, upward, or sideways. Oil from different parts of a compacting rock probably moves in different directions, like water from a squeezed sponge. The sole prerequisite is that a receptive layer be present to receive the fluids. A receptive rock is one with porosity and permeability and with a fluid pressure less than that of the liquids being driven from the compacting rock. The latter pressure is due to the weight of a column of mud. It is greater than the hydrostatic pressures existing even in the subjacent rock, so oil can be squeezed into underlying as well as overlying reservoirs.

The channelways that are used by fluids squeezed from a compacting rock are the interconnecting pores between the grains, which become closed to further migration when compaction is complete. Subsequent transverse migration must be by way of secondary channelways that cut across the bedding. Outstanding examples are joint fractures, which are especially prevalent in brittle rocks such as limestone. Pervious fault planes may also provide channelways for migratory hydrocarbons, but in Chapter 13 we shall see that in sedimentary rocks faults are more often dams than conduits. Sandstone dikes have been suggested as channelways for the transverse migration of oil through shales in California,[23] Romania,[24] and elsewhere.

One excellent example of transverse migration is the previously cited Garber, Oklahoma, field, where like oil occurs in superimposed reservoirs.[25] Bailey[26] suggests that the oil in the Sespe redbeds in California got there by migrating upward from Eocene shales through "several hundred to a few thousand feet" of intervening strata that consist of sandstones with shaly interbeds cut by "countless minor joints and cracks." Upward transverse migration is the most reasonable explanation for oil that occurs in the Khaur field of West Pakistan in fissures in tightly cemented fresh water sandstones at all levels from the surface

[23] Olaf P. Jenkins, "Sandstone Dikes as Conduits for Oil Migration through Shales," *Bull. Am. Assoc. Petrol. Geol.*, Vol. 14 (April, 1930), pp. 411–421.

[24] W. A. J. M. van Waterschoot van der Gracht, "The Stratigraphical Distribution of Petroleum," *Science of Petroleum* (Oxford University Press, 1938), Vol. 1, p. 60.

[25] James H. Gardner, "Vertical Source in Oil and Gas Accumulation," *Bull. Am. Assoc. Petrol. Geol.*, Vol. 29 (September, 1945), pp. 1349–1351.

[26] Thomas L. Bailey, "Origin and Migration of Oil into Sespe Redbeds, California," *Bull. Am. Assoc. Petrol. Geol.*, Vol. 31 (November, 1947), pp. 1913–1935.

(where it seeps) to a depth of 5400 feet. The underlying Eocene rocks are the probable source.[27]

Where the usual dense Arbuckle "cap rock" is absent at Chetopa, Kansas, Arbuckle oil has migrated upward into the overlying Chattanooga shale.[28] Some wells at Salt Creek, Wyoming, produced commercially from fracture zones in the shale overlying the major sandstone reservoir; it is logically concluded that this oil migrated upward across the lithologic boundary. Near Toyah in Reeves County of trans-Pecos Texas, wells produced "from a few gallons to a few barrels per day" at depths of less than 100 feet in Pleistocene or Recent alluvium.[29] Obviously this oil rose from beneath through vertical channelways that cross the nearly horizontal bed-rock strata. Transformational seeps, as on the island of Trinidad, illustrate transverse migration, but seeps can also result from parallel migration through an inclined reservoir rock.

Northern Iraq appears to be the outstanding province for transverse movement of oil in large volume. According to interpretations by Dunnington[30] "over 95 per cent . . . of the known oil field reserves of the region have come into place by substantial migration across the bedding" (p. 1250). This migration has been due primarily to the brittleness of the limestones which make up the major part of the stratigraphic section involved. Oil caught in earlier traps, both structural and stratigraphic, has escaped upward when subsequent diastrophism has ruptured the seals. The oldest and deepest seal successfully to resist destruction by fracturing is the Lower Fars evaporite section, therefore the major oil accumulations of northern Iraq are below the Fars. Furthermore, where fracture-resistant seals have not been present or have been breached by subsequent erosion, large accumulations of oil have migrated to the surface and become dissipated. Dunnington (pp. 1241–1245) considers the volume of oil lost in this manner to be very large, in one area alone measurable in the "hundreds of millions of tons."

Examples of transverse migration downward from source rock to reservoir rock are some of the occurrences of oil in crystalline basement rocks

[27] E. S. Pinfold, "Oil Production from Upper Tertiary Fresh Water Deposits of West Pakistan," *Bull. Am. Assoc. Petrol. Geol.*, Vol. 38 (August, 1954), pp. 1653–1660.

[28] G. E. Abernathy, "Migration of Oil from Arbuckle Limestone into Chattanooga Shale in Chetopa Oil Pool, Labette County, Kansas," *Bull. Am. Assoc. Petrol. Geol.*, Vol. 25 (October, 1941), pp. 1934–1937.

[29] Ronald K. DeFord, in F. M. Van Tuyl and Ben. H. Parker, "The Time of Origin and Accumulation of Petroleum," *Quarterly Colo. School of Mines*, Vol. 36 (April, 1941), Chapter 17, p. 120.

[30] H. V. Dunnington, "Generation, Migration, Accumulation, and Dissipation of Oil in Northern Iraq," *Habitat of Oil* (American Association of Petroleum Geology Special Volume, 1958), pp. 1194–1251.

(Chapter 6) and in buried hills. Oil also may move transversely downward from overlying reservoir rocks into basement rocks at the time the latter are fractured, as described under causes of migration (dilatancy). Other possible examples of downward transverse migration are the oil accumulations beneath unconformities, and especially those in the leached upper surfaces of thick limestones.[31] Most of the movement of oil from source shales into lenticular sandstones is transverse, as described earlier in this chapter.

Parallel migration. Migration parallel to the stratification is possible when a porous and permeable rock layer occurs in the sedimentary section. Most numerous examples are sandstones and porous carbonate rocks. Parallel migration is by no means confined to widespread ("sheet") sandstones or regional porous limestones, however. Sand-filled channels and bars in thick shale sections also may be utilized as conduits for migrating hydrocarbons.[32] Before compaction makes the muds and oozes impervious, parallel movement is the preferred direction for fluids passing through these materials because of the lamellar characteristics of the minerals and their orientation parallel to the sea floor on which they were deposited.[33]

Parallel migration is so common as to be almost universal. Only a relatively insignificant part of each reservoir contains hydrocarbons. Unless it is assumed that by strange coincidence oil entered the reservoir only where there were traps, it must be concluded that oil entering where traps were absent must have migrated laterally until trapped. The confinement of oil accumulations to the highest levels in the reservoir rock is presumptive evidence that oil moved through the rock until those levels were attained. The presence of extensive deposits in a given formation and none in higher potential reservoirs is also evidence that oil travels and accumulates parallel to the bedding far more often than across the bedding. This conclusion is supported by the fact that like oils may occur in a single reservoir formation in fields scattered over an area extending for scores of miles, whereas other formations, separated vertically by but a few feet, will have unlike oils.

Practically any oil field can be used to illustrate parallel migration.

[31] Roy L. Ginter, "Exercise on Amount of Source Bed Required to Furnish Oklahoma City Oil Pool," *Bull. Am. Assoc. Petrol. Geol.*, Vol. 25 (September, 1941), pp. 1706–1712; Robert F. Walters and Arthur S. Price, "Kraft-Prusa Oil Field, Barton County, Kansas," *Structure of Typical American Oil Fields* (American Association of Petroleum Geologists, 1948), Vol. 3, p. 268.

[32] W. C. Krumbein and L. T. Caldwell, "Areal Variation of Organic Carbon Content of Barataria Bay Sediments, Louisiana," *Bull. Am. Assoc. Petrol. Geol.*, Vol. 23 (April, 1939), pp. 582–594.

[33] D. A. Greig, *memorandum* dated Dec. 14, 1948.

The enormous accumulation at East Texas (Fig. 14–14) has been at the upper end of a great sheet sand body where it has been truncated by erosion and overlapped by younger, impervious formations. The best explanation for this great concentration of oil is that it entered the sandstone at an infinite number of points down the flank of the east Texas (Tyler) basin, whence it drained upward through the water-filled pores of the reservoir until it was impounded below the tightly sealed unconformity where the sandstone wedges out. Lateral migration in a radial direction away from the center of the eastern Venezuela basin is indicated by the fact that practically all accumulations of oil "are found on the basinward side of the barriers to such migration."[34]

Some seeps occur at the outcrop of reservoir rocks, which is evidence of parallel migration from a deeper, down-dip source. The Bartlesville sandstone, reservoir rock for many rich, relatively shallow, oil fields in southeastern Kansas and northeastern Oklahoma, is quarried at its outcrop in southwestern Missouri because of the presence of seep oil, which makes the crushed rock valuable as a road dressing.

Parallel migration is also a strong possibility for the primary movement from source rock to reservoir in the "shoestring" sand fields (Fig. 14–10). The belief has been expressed that the source material is the organic accumulations which were being deposited in the adjacent lagoons at the same time that the sand bars were being built.[35] The oil, as it formed, migrated laterally into the sand. Some transverse movement from younger, overlying organic muds is also a possibility.

Without doubt, much of the migration from younger to older rocks is actually parallel rather than transverse. For example, the accumulation of oil in the basement rock schists in the Edison field of California has probably taken place through movements up-dip through sedimentary reservoirs into the fractured crystalline rock against which the sedimentaries abut: "The oil originated in the westerly extending Tertiary sedimentary basin and migrated into the pore and fracture spaces of the structurally higher schist."[36]

It can be concluded that parallel and transverse migration are not mutually exclusive and that many if not most oils, in journeying from

[34] H. D. Hedberg, L. C. Sass, and H. J. Funkhouser, "Oil Fields of Greater Oficina Area, Central Anzoategui, Venezuela," *Bull. Am. Assoc. Petrol. Geol.*, Vol. 31 (December, 1947), p. 2138.

[35] L. M. Neumann et al. (Research Committee, Tulsa Geological Society), "Relationship of Crude Oils and Stratigraphy in Parts of Oklahoma and Kansas," *Bull. Am. Assoc. Petrol. Geol.*, Vol. 31 (January, 1947), pp. 92–148.

[36] J. C. May and R. L. Hewitt, "The Nature of the Basement Complex Oil Reservoir, Edison Oil Field, California," *Bull. Am. Assoc. Petrol. Geol.*, Vol. 31 (December, 1947), p. 2240.

source to trap, travel paths both transverse and parallel to the stratification planes.

Distance of Oil Migration. The distance through which oil can migrate, and has migrated in the geological past, is a function of time, if continuity of permeability and gradient is assumed. If oil can migrate an inch, it can migrate a mile. A movement of but a foot a year becomes 190 miles in a million years.

Obviously the opportunities for migration over long distances are much greater by parallel than by transverse migration. In the latter movement, the distance is limited to the thickness of the sedimentary column (or a somewhat greater distance where migration is on the bias), whereas the potential migration range for parallel movement is the distance from bottom to rim of a structural basin. The "gathering area" for a trap is considered to be the down-dip extension of the reservoir rock. If the trap is high on the flank of a basin and the reservoir rock has sheet porosity, the trap can impound the upward drainage from over an enormous area. This statement may explain why some of the greatest accumulations have just such a setting. Examples are the oil fields of Lake Maracaibo in Venezuela, the San Joaquin Valley fields of California, East Texas, Oklahoma City, and many others.

Definite figures of distance of migration are difficult to obtain. Brauchli[37] believes that the original accumulation at Oklahoma City may well be the result of drainage from far out in the Anadarko basin, a potential distance of more than 100 miles. Hoots[38] states that it is a matter of "several miles" between some of the California accumulations in nonmarine and inorganic sediments and their down-dip marine organic facies from whence the oil probably came. From "a few" to 15 miles is the range of migration given for some of the oil that subsequently devolatilized to form the asphalt deposits of western Kentucky.[39] The only logical source rocks for the oil (and salt water) in continental sands in the Velasquez field, Colombia, are marine shales which unconformably underlie the fresh water sediments "several miles" to the east.[40]

Examples of minimum distance of migration are the often-cited accumulations of oil in sand bodies completely surrounded by shale. The main input probably takes place during the compaction of the enclosing

[37] R. W. Brauchli, "Migration of Oil in Oklahoma City Field," *Bull. Am. Assoc. Petrol. Geol.*, Vol. 19 (May, 1935), pp. 699–701.

[38] H. W. Hoots, "Origin, Migration, and Accumulation of Oil in California," *Calif. Div. Mines Bull.* 108 (August, 1941), pp. 261–262.

[39] W. L. Russell, "Origin of the Asphalt Deposits of Western Kentucky," *Econ. Geol.*, Vol. 28 (September-October, 1933), pp. 571–586.

[40] W. S. Olson, "Source-Bed Problem in Velasquez Field, Colombia," *Bull. Am. Assoc. Petrol. Geol.*, Vol. 38 (August, 1954), pp. 1645–1652.

shale. It is difficult to visualize further migration after compaction and lithifaction are complete, unless the shale is extensively fractured.

Carrier beds. Rich, the foremost proponent of migration of oil over great distances, has coined the term "carrier beds"[41] for deep, highly porous, and permeable rocks through which oil can migrate. He cites as a possible example of the use of carrier beds the movement of oil from the geosynclinal basins of southern Oklahoma to the anticlinal traps of the central Kansas uplift. Possible carrier beds include sheet sandstones, cavernous limestones, and weathered surfaces beneath widespread unconformities. The oil works its way upward at every opportunity from the carrier beds to the reservoir rocks by means of fissures. Once the reservoir bed is reached, the oil may be immediately trapped, as in a "shoestring" sand, or it may continue parallel migration until a trap is found. Rich confines the term "reservoir rock" to the stratum or strata in which accumulation occurs.

Time of Migration.[42] The accumulation of oil or gas in a commercial deposit is only possible through a combination of physical events taking place in proper sequence. If the sequence is out of order, accumulation is not possible. For example, a trap formed *after* the hydrocarbons have gone by will be barren. Therefore to our list of prerequisites to an oil or gas deposit (source rock, reservoir rock, reservoir seals, migration, and trap) we have to add the factor of *timing*. It is fully as important to commercial accumulation as any of the other prerequisites.

The time of migration can be broken down into three categories: (1) time of primary migration from source rock to reservoir rock; (2) time of the first of the secondary migrations into a trap, transient or otherwise; and (3) time of subsequent migrations into higher traps following spillage. All three of these migrations are in turn dependent upon the timing of other activities. Thus, before oil can be squeezed from a source rock into a reservoir rock the latter has to come into being. Some reservoirs are not concurrent sedimentary deposits, like sandstones, but are the result of postsedimentation activity. Examples are rocks with porosities resulting from postlithifaction solution (carbonate rocks) or fracturing. Migrations into and out of traps are dependent upon the timing of the trap formation, and of subsequent wrigglings and writhings of the earth's crust.

Assuming proper sequence of timing of the prerequisites, we have then to consider the time during which the three categories of migration take

[41] John L. Rich, "Function of Carrier Beds in Long-Distance Migration of Oil," *Bull. Am. Assoc. of Petrol. Geol.*, Vol. 15 (August, 1931), pp. 911–924; "Distribution of Oil Pools in Kansas in Relation to Pre-Mississippian Structure and Areal Geology, *ibid.*, Vol. 17 (July, 1933), pp. 798–815.

[42] William Carruthers Gussow, "Time of Migration of Oil and Gas," *Bull. Am. Assoc. Petrol. Geol.*, Vol. 39 (May, 1955), pp. 547–574.

place. Primary migration may begin even before source-bed deposition is complete, because of the weight of the upper sediment on the lower. This migration is, of course, based on the assumption that there is oil in the sediment at this stage. The primary migration continues until compaction is virtually complete. That in turn is a function of the weight of the overlying load. Actually, compaction and expulsion are probably never really complete, but there comes a time when the rate-of-expulsion curve flattens sharply, and that is the end of quantitatively significant primary migration.

Carbon 14 measurements of age have shown that oil apparently about 14,000 years old is being squeezed out of muds in eastern Venezuela. Some of this oil is accumulating in a lenticular sandstone which was deposited about 5000 years ago.[43]

Secondary migration probably gets off to a very slow start owing to the sluggish character of the immature oil and the low dip of the reservoir seal. This topic was discussed under the buoyancy theory of oil migration. Once secondary migration does start, it is subject to temporary interruptions. Irregularities in the hanging wall surface, and anticlines too small to contain commercial deposits, may act as *"microtraps"* by holding the oil in a pocket until either the pocket is filled (so that subsequent migrating oil underpasses the accumulation), or as *"transient traps"* which hold the oil until regional tilting eliminates the closure. Microtraps are of very common occurrence, and many "shows of oil" reported during the drilling of a well are actually accumulations of hydrocarbons caught in such a trap. Oil underpassing a filled microtrap, or spilled out of a transient trap, eventually may enter a trap of such a type, or of such magnitude of closure, that later tiltings do not spill all of it out. However, the oil that is spilled out resumes its secondary migration through the reservoir rock.

We can conclude that although primary migration takes place early in the history of an oil deposit and is of relatively short duration, secondary migration is a discontinuous affair which may take place over a long span of geologic time.

Natural gas rides along in solution in the oil as long as that oil remains under-saturated. However, once an oil reaches the saturation point ("bubble point") gas emerges and goes its own way. If it reaches and fills a trap ahead of the oil, the latter underpasses the trap and keeps on going. Because of its high mobility and small pore penetrability, gas moves up gradients and through rocks that oil cannot. Therefore, gas may end up in a trap some distance away from its parent.

[43] Albert L. Kidwell and John M. Hunt, "Oil Migration in Recent Sediments," *World Oil* (July, 1958), pp. 79 et seq.

Further discussion of timing is given in Chapter 10 under "Time of Accumulation."

Changes in Oil During Migration. The act of migration should not interfere with the natural evolution of oil. In fact, this evolution might even be accelerated because of increased contact (through a wetting-water film) with catalysts. As the oil moves from places of higher pressure to lower (*never* the reverse), some change in character may take place owing to the escape of a part of the dissolved gas. Temperature changes occurring en route may also affect the amount of gas held in solution. Contamination of the oil by minerals forming the pore walls is no longer considered possible because of the intervening water film now believed to be universally present. However, chemical changes due to contact with water through which the migrating oil passes may be possible. Some believe that asphaltization of crude oil near the surface is due to chemical interaction between the upward migrating oil and downward percolating surface waters.

It is a well-known fact, utilized in refining both petroleum and vegetable oils, that some clays (such as fuller's earth) can be used as filters to remove discoloring or foul-smelling compounds from oils or to separate the heavy viscous asphaltic hydrocarbons from the lighter, more mobile compounds. Without doubt, the passage of crude oil underground through rocks of varying permeability also has a filtering effect that results, at the end of the migration, in an oil somewhat different from what it was when it started on its journeys. A light oil occurring in relatively minor amounts across the Whittier fault from the Brea-Olinda field may possibly be due to the filtering effect of the fault gouge, which allows only the lightest of the liquid hydrocarbons to pass through.

If the oil and water are migrating together, intimately intermixed, the ability of the water, but not the oil, to enter the finer pores (already water-wet) of an argillaceous zone results in a gradual enrichment of oil in the rock in front of the barrier.[44]

It is also possible, but difficult of proof, that gas spilled out of deep retrograde condensate fields would condense while migrating up-dip owing to the effect of lower pressure. Moody[45] points out that some of the oil fields of southern Arkansas produce from the same reservoirs that contain the gas-condensate accumulations down-dip in Louisiana.

Summary. Strong evidence makes it probable that oil can and does migrate through considerable distances. It is also probable that for some accumulations the distance of migration has been relatively short. The

---

[44] V. C. Illing, "The Migration of Oil," *Science of Petroleum* (Oxford University Press, 1938), Vol. 1, pp. 209–215.

[45] C. L. Moody, *letter* dated October 13, 1958.

first movement is upward, downward or laterally out of the source rock into a reservoir rock. This oil is flushed out by water which is expelled from the source rock by compaction. Once the reservoir rock is reached, the oil probably lies dormant for a considerable period, until natural evolution and diastrophism gives it the necessary fluidity and gradient for up-dip migration induced by buoyancy. *The probability that it takes time, perhaps even geological time, for oil to develop the fluidity necessary for secondary migration may be the reason for the virtual absence of commercial deposits of oil today in formations younger than the Pliocene.*

Once secondary migration starts it is expedited by the ingathering of the droplets which give them increased mobility. Migration is not confined to the planes of stratification; where the overseal is breached the oil migrates across the bedding.

Migration ceases when the oil (1) reaches a trap, or (2) reaches the surface (or the water table below the surface). When the oil is trapped it can resume migration (1) through tilting of the trap with consequent spilling of the oil, (2) through being pushed out by gas accumulating above it, or (3) through cracking of the overseal with escape upward.

Primary migration starts with the deposition of the source sediments and secondary migration is possible as long as there is any oil buried in the rocks of the earth's crust. In its travels oil continues to evolve. In fact, evolution may be accelerated by greater contact with catalysts. Other changes may take place during migration because of natural filtering and lesser fluid pressures.

There is no controversy over the migratory habits of natural gas. It has maximum mobility and moves through finer openings than oil. It is probably formed as a concurrent product during the evolution of oil. Once the source oil becomes saturated, any new gas generated is free to move out in the direction of least resistance, which is upward or at least up-dip. It, too, travels until it either escapes or is trapped. However, it is also subject to capture by any undersaturated oil it may meet enroute. Gas can escape a trap as does oil by spilling or·through the overseal.

10

TRAPS: INTRODUCTION AND DRY SYNCLINE TYPE

The final stage in the natural history of an oil or gas field is the trapping of the hydrocarbons into an accumulation of exploitable size. The first prerequisite to accumulation, a reservoir rock, was discussed in Chapter 6. The second prerequisite is that the reservoir be *closed* so that the hydrocarbons are trapped within. A closed reservoir, hereafter referred to as the *trap*, is a body of reservoir rock completely surrounded above a certain level by impervious rock. The amount of *closure* is the vertical distance between the "certain level" and the highest point reached by the reservoir rock. All traps must have closure, but the concept is best illustrated in elliptical anticlines and domes, which can be mapped readily by structure contours. Closure begins at the level of the lowest *possible* closed contour and extends upward to the apex of the structure. The approximate amount of closure can be calculated by multiplying the number of closed contours by the contour interval. The actual closure lies within one contour interval of this figure.

The amount of closure is also the *maximum* vertical distance through which hydrocarbons can accumulate in the reservoir rock. Any additional oil would flow out of the structure beneath the inverted lip of the trap. Rarely are traps completely filled with oil. More commonly the oil-water contact lies well above the lowest closed contour.

The "gathering area" is the maximum expanse down-dip from a trap. In the case of a basin flank trap the gathering area may extend to the bottom of the basin. Other things being equal, the greater the gathering area the larger the oil accumulation. Therefore the position of a trap in

268

respect to the regional structure is of great importance. This aspect of accumulation is developed more fully in Chapter 15.

TIME OF ACCUMULATION

The time of accumulation [1] of hydrocarbons in traps in the earth's crust is a matter not only of considerable interest but also of great practical importance in oil finding. Unfortunately our knowledge is rarely sufficient to permit the exact dating of accumulation. However, one can always start with the general rule that the accumulation is later than the establishment of the trap.

Post-dating accumulation until after trap formation does not mean that a long period of time between the two is necessary. There are many examples of probable early accumulation, in which the word "early" is used in a relative sense. For example, it is quite likely that the oil trapped in the shoestring bodies entered those bodies shortly after deposition during the compacting of the surrounding shale. It is also possible that some anticlinal accumulations may have taken place long before the last folding. Subsurface studies show that most of the folds in the older rocks in the earth's crust have been formed through a series of recurrent movements, and adequate closure for the trapping of oil may have existed several periods previous to the major diastrophism ordinarily credited with the creation of the present picture. To some degree, the same situation exists in regard to faulting. Many faults have been recurrent, and the offset may have been sufficient, after the initial movement, to create a subsurface dam for the impounding of oil. Therefore, some structural accumulations may have occurred at a relatively early date and subsequent earth movements may merely have accentuated the traps.

We do know, however, of anticlines containing oil which were formed by a single, relatively late, period of diastrophism, and many faults have had but one period of movement of any magnitude. Furthermore, in the unconformity traps a relatively late date of the accumulation can scarcely be denied. Examples of late accumulation are many. Bartram,[2] finding

[1] William C. Gussow, "Time of Accumulation and Evolution of Hydrocarbons," *World Oil* (September, 1956), pp. 113–115; Robert H. Paschall, "How the Time Element Affects Oil Traps," *Oil and Gas Jour.* (March 5, 1956), pp. 151–153; A. I. Levorsen, "Time of Petroleum Accumulation," *Econ. Geol. 50th Anniversary Vol.* (1955), Part 2, pp. 748–756; F. M. Van Tuyl and Ben H. Parker, "Time of Origin and Accumulation of Petroleum," *Quarterly Colo. School of Mines,* Vol. 36 (April, 1941); William B. Heroy, "Petroleum Geology," *Geol. Soc. Am. 50th Anniversary Vol.* (1941), pp. 512–548; Stanley C. Herold, "Criteria for Determining the Time of Accumulation under Special Circumstances," *Bull. Am. Assoc. Petrol. Geol.,* Vol. 22 (July, 1938), pp. 834–851.

[2] John G. Bartram, in F. M. Van Tuyl and Ben H. Parker, *op. cit.,* p. 153.

no evidence of incipient folding in the Rocky Mountain fields producing from the Embar and Tensleep sands, does not believe that the oil found in these Pennsylvanian and Permian reservoirs could have accumulated until the end of the Cretaceous or early Eocene, when folding took place. Hoots [3] notes that the Ventura anticline in California, which produces from Pliocene and Pleistocene sediments, was not folded until middle Pleistocene time. The tilted and truncated Ordovician reservoirs at Oklahoma City contain great deposits of hydrocarbons immediately beneath a seal of Cherokee shale which was not deposited until Pennsylvanian time. Furthermore, Levorsen, [4] from a study of reservoir pressures at Oklahoma City, believes that the accumulation did not reach its maximum until after the overburden above the truncated edges of the reservoirs had reached somewhere near its present thickness. Walters and Price [5] state their belief that migration of oil into the Arbuckle below the Pennsylvanian unconformity at Kraft-Prusa, Kansas, took place in late or post-Pennsylvanian time. The reasons of Walters and Price are: (1) solution porosity in the Arbuckle was not developed until the early Pennsylvanian; (2) Arbuckle dolomites probably would not have retained indigenous oil during exposure on early Pennsylvanian surface; (3) overlying and oil-bearing Pennsylvanian sediments constitute an adequate source; (4) there was probably no closure on this unconformity until after Lansing-Kansas City time; (5) reservoir structure is completely filled, indicating late adjustment to reservoir capacity.

Gussow [6] deduced that the accumulation of oil and gas in the upper Devonian reservoirs of the Nisku and Leduc fields of Alberta did not take place until Colorado (Upper Cretaceous) time. He reached that conclusion by analyzing the effects on timing of (1) compaction of the source material, (2) date of regional tilting, (3) date of trap formation, (4) hydrostatic pressure, (5) saturation or bubble-point pressure (a function of depth of burial at time of migration), and (6) history of secondary lithologic development such as cementation and solution.

The principal faulting which caused the accumulation in the Oficina area in Venezuela did not take place until after the beginning of Las Piedras (Miocene-Pliocene) time, although the reservoirs are of Oligocene-

[3] H. W. Hoots, ibid., p. 153.

[4] A. I. Levorsen, "Time of Oil and Gas Accumulation," Bull. Am. Assoc. Petrol. Geol., Vol. 29 (August, 1945), pp. 1189–1194.

[5] Robert F. Walters and Arthur S. Price, "Kraft-Prusa Oil Field, Barton County, Kansas," Structure of Typical American Oil Fields (American Association of Petroleum Geologists, 1948), Vol. 3, p. 268.

[6] William Carruthers Gussow, "Time of Migration of Oil and Gas," Bull. Am. Assoc. Petrol. Geol., Vol. 39 (May, 1955), pp. 547–574.

Miocene age.[7]   An extreme situation has been noted by Reed[8] in the San
Pedro and other oil fields in northern Argentina.   There the only possible
source rock is Devonian, and the structure-forming movement which created the trap in which the oil is found did not take place until the Pleistocene.   In the fractured shale pools, such as the Florence field of Colorado,
it is obvious that the oil could not have accumulated until after the fissures were formed, and these could not appear until the shale had attained
some degree of lithifaction.   It is also true that the oil which has accumulated in leached and porous zones in limestone could not have done so
until after the period of emergence and subareal erosion that created the
cavernous situation, and this may not have taken place until long after the
original deposition of the carbonate rock.   The ultimate in late accumulation is illustrated by the occurrence of oil beneath surficial asphalt deposits formed during the present cycle of erosion.   The seeps that occur
along the outcrop of some monoclinal reservoir rocks at the present time
point to the possibility that accumulation is going on today where traps
lie across the pathway of ascending oil.

Every great unconformity creates an almost infinite number of varying-
permeability traps over many thousands of square miles beneath the plane
of the unconformity.   However, the magnificent opportunity during the
long time of emergence and erosion for the escape of hydrocarbons from
the older rocks should also be noted.   Heald[9] points out that in pre-
Pennsylvanian time erosion across the Barton arch in Kansas stripped off
a great volume of Mississippian, Devonian, Silurian, and Ordovician
rocks, exposing the Precambrian complex along the arch axis.   During
that period any oil within those rocks must have had ample opportunity
to escape.   Yet today many great oil pools occur immediately beneath
this unconformity, sealed there by the overlapping Pennsylvanian sediments, with accumulation in porous rocks which must have been at or
very close to the surface before being covered by the Pennsylvanian sea.
A similar situation exists at Oklahoma City, but there some asphalt has
been found at the top of the truncated sands which implies the presence
of a seep at the time this surface was exposed.   The presence of asphaltic
material here can be taken as evidence that Ordovician oil was escaping
during the pre-Pennsylvanian emergence, but the quantity of asphaltic

[7] H. D. Hedberg, L. C. Sass, and H. J. Funkhouser, "Oil Fields of Greater Oficina Area,
Central Anzoategui, Venezuela," *Bull. Am. Assoc. Petrol. Geol.*, Vol. 31 (December, 1947),
p. 2138.

[8] Lyman C. Reed, "San Pedro Oil Field, Province of Salta, Northern Argentina," *Bull.
Am. Assoc. Petrol. Geol.*, Vol. 30 (April, 1946), p. 605.

[9] K. C. Heald, "Essentials for Oil Pools," *Elements of the Petroleum Industry* (American
Institute of Mining and Metallurgical Engineers, 1940), pp. 38–39.

material is too insignificant to warrant the postulation of a tar seal adequate to contain beneath it the great oil deposits now found there. Howard describes a similar problem in the northern Rockies: "The Madison limestone of Wyoming and Alberta contains oil but this formation is overlain directly by deposits which were laid down in the Jurassic, possibly 100,000,000 years later. It is difficult to conceive of oil forming in the Madison, staying there for that length of time and then rising into the reservoir only after a seal was laid down over it."[10]

At least six possible solutions of the enigma of "old" oil caught in "young" traps have been suggested. One obvious possible explanation is that the oil found beneath unconformities came from the overlying strata which functioned both as seal and source rock. The overlying material must then have been a highly organic mud or calcareous ooze. The compaction which it underwent while lithifying into shale or limestone would cause some of the contained fluids to be squeezed downward into porous carbonate rock or sandstones immediately underlying the unconformity. Though this simple explanation appears applicable in many areas, it is difficult to apply at Oklahoma City, because there the oil in the Ordovician reservoirs is quite dissimilar to that found in the Pennsylvanian reservoirs. The two oils are more nearly similar in the Barton arch area of Kansas, but there the Pennsylvanian reservoirs contain considerably more gas.

Another suggested explanation is the recurrent generation hypothesis advocated by Stadnichenko and described in Chapter 5. According to this concept, the Oklahoma City oil was being generated from source materials in pre-Pennsylvanian times. This oil, leaving occasionally some asphaltic residue, migrated up the inclined strata and escaped at the surface. Subsequently, after sealing by overlapping Pennsylvanian sediments, elevated earth temperatures and pressures which were deeply buried brought about the harvesting of later crops from the same source material. This time the oil accumulated beneath the plane of the unconformity. However, the previously discussed studies of Brooks and others cast a doubt upon the validity of any process of oil generation that involves appreciable increase in temperature.

A third idea, advanced by Dorsey,[11] is that the oil accumulated in whatever traps were in existence before the uplift and truncation. With the uplift the water table was lowered and the oil followed the water down the dip, withdrawing from the near-surface zone where it was in

[10] W. V. Howard, "The Derivation of Reservoir Rocks," Oil and Gas Jour., Vol. 42 (June 24, 1943), Part 1, p. 158.

[11] George Edwin Dorsey, "Preservation of Oil During Erosion of Reservoir Rocks," Bull. Am. Assoc. Petrol. Geol., Vol. 17 (July, 1933), pp. 827–842.

greatest danger of escape or destruction. Subsequently, after resubmergence and the deposition of younger sediments above the unconformity, the water table rose to levels above the unconformity, permitting the oil to move back up-dip and become impounded below the plane of the unconformity. A major objection to this hypothesis is that the upper surface of the oil accumulation, perched on top of the water table during the time of uplift and erosion, was in constant contact with the atmosphere, even though the water table may have been several hundred feet below the topographic surface. Through this exposure the more volatile petroleum compounds would escape and the remaining hydrocarbons would tend to combine with oxygen. Both of these activities produce a tarry or asphaltic residue which would in all probability not be able to move up to the plane of the unconformity after the overlying sediments were deposited.

Van Tuyl and Parker[12] suggest that perhaps the oil "may be stored for considerable periods after generation, either in a disseminated condition or as pools, which may undergo renewed migration as a result of changes in geologic structure." A somewhat similar concept has been expressed by Hedberg, Sass, and Funkhouser: "Since the principal faulting causing the accumulation did not occur until after the beginning of Las Piedras time, it is evident that while oil migration may have been taking place previously, it must have continued well into the time of Las Piedras deposition. It may be that the extensive migration could not proceed until a certain dip gradient had been established in the Oficina formation by the combined results of compaction and basin subsidence,"[13] Davies[14] believes that the Mesozoic oil of the Rocky Mountain province was lying dormant in the deeper parts of the sedimentary basins until Laramide orogeny produced traps for its concentration.

The fifth possibility which is to a considerable extent an elaboration of the Van Tuyl-Parker-Hedberg et al. idea, is that the oil was formed early and was temporarily stored in "transient traps," which are very minor irregularities in the structure of the reservoir rock. These minor wrinkles, which are rather universally present in stratified rocks, would yield oil by spillage with every slight tilt. Therefore, the sinking of a regional basin, which takes place intermittently, would release oil from time to time which would drain up the flanks of the basin. The oil stored be-

[12] F. M. Van Tuyl and Ben H. Parker, "Time of Origin and Accumulation of Petroleum," *Quarterly Colo. School of Mines*, Vol. 36 (April, 1941), p. 164.

[13] H. D. Hedberg, L. C. Sass, and H. J. Funkhouser, "Oil Fields of Greater Oficina Area, Central Anzoategui, Venezuela," *Bull. Am. Assoc. Petrol. Geol.*, Vol. 31 (December, 1947), p. 2138.

[14] H. F. Davies, "Structural History and Its Relation to the Accumulation of Oil and Gas in the Rocky Mountain District," *Problems of Petroleum Geology* (American Association of Petroleum Geology, 1934), p. 690.

neath the unconformities in the Mid-Continent and elsewhere could have accumulated after basinward tilting, which took place after (perhaps long after) the deposition of the overlapping younger sediment.

Finally there is the idea, expressed in the preceding chapter, that young oil may be too sluggish to migrate, so that it remains dispersed in the reservoir rock, after being squeezed out of the source rock, until it acquires adequate mobility to work its way up-dip. In other words, it may take millions of years of geologic time for an oil to achieve, by natural evolution, the fluidity necessary to migrate up very gentle slopes. In the meantime up-dip traps may have come into existence.

BARREN TRAPS

In many areas, barren (water-filled) traps tend to be the rule rather than the exception. A number of possible explanations for this situation, most of which have been given before, are assembled and summarized at this point. It is assumed that the trap has been tested thoroughly,[15] that it is completely closed, and of adequate porosity and permeability, with water filling the reservoir to the top.

(1) NO SOURCE MATERIAL. The absence of potential source material in the sedimentary section may be due to unfavorable climate, too rapid sedimentation, wrong depositional environment[16] (which includes a topographically high position on the sea floor), or the destruction of once-present source material during the biochemical stage, or by erosion subsequent to deposition.

(2) NO GENERATION OF OIL. Inasmuch as we do not know for certain the oil-producing processes, we can only guess at what may be responsible for lack of transformation of organic solid into liquid or gaseous hydrocarbons. Suggested possibilities in this regard include absence of the proper bacteria, absence of the necessary catalysts, inadequate time, and insufficient cover.

(3) OIL FAILED TO REACH TRAP. This may be due to down-dip impounding, or to diversion. Damming could be caused by lenticular permeability, or by faults. A large anticline lying athwart the migration pathway would prevent oil from reaching a smaller up-dip anticline until and unless it was filled to the spill point. Diversion would have the same effect as impounding. According to Gussow[17] hydrocarbons migrate

[15] "Why Was My Well Dry? (Or Was It?)," *Oil and Gas Jour.* (Nov. 14, 1955), pp. 143–159.

[16] Frank R. Clark, "Origin and Accumulation of Oil," *Problems of Petroleum Geology* (American Association of Petroleum Geologists, 1934), pp. 334–335.

[17] William Carruthers Gussow, "Differential Entrapment of Oil and Gas: A Fundamental Principle," *Bull. Am. Assoc. Petrol. Geol.*, Vol. 38 (May, 1954), pp. 816–853.

along definite migration paths which are normal to the structure contours. Traps along these paths accumulate hydrocarbons and those off the path do not. Diversion could also take place through selectivity of gradients. Hydrocarbons migrate preferentially up the steeper slopes, just as water at the surface seeks out and moves down the steeper gradients.

(4) THE OIL HAS ESCAPED. Regional tilting may have caused the oil to spill out of the trap, after which renewed folding reproduced the trap. Fissures and faults may have allowed the hydrocarbons to escape upward to the surface, where they become dissipated. The preceding chapters (8 and 9) describe how renewed folding of the Kirkuk anticline in Iraq caused fracturing of overseals and upward drainage of oil; perhaps this type of escape has been more common in the past than heretofore realized. Because no rock is absolutely impervious, a great enough span of geologic time might permit the more volatile hydrocarbons, at least, to disperse through the confining rocks without benefit of any fracture system.[18] Illing[19] states that as a general rule less oil is stored in the first 1000 feet of depth than in succeeding 1000-foot units, and he believes that the lesser accumulation at shallow depth has been due to the more active water circulation which is found there. He suggests that the oil and gas are dissipated long before the actual uncovering of the reservoir strata by erosion. Furthermore the reservoir, after losing its hydro-carbons, may never actually be uncovered by erosion. Instead the rock section may be resubmerged and covered by younger sediments. This theory perhaps explains the absence of oil in some major anticlines that have experienced recurrent folding. These anticlines were more vulnerable to hydrocarbon escape in the geologic past than were neighboring lower anticlines. The greater the vertical distance between reservoir rocks at the anticlinal crest and the next overlying unconformity, the better the chance that trapped oil was retained during the emergence indicated.

The thesis of oil escape upward when a trap is tilted or punctured by a fracture is based upon the presence of water. The water-saturated zone extends only to the water table. If the water table is high, as in humid belts, so that it intersects the surface in the topographically low spots, the oil escapes and joins the surface runoff. If the water table is deep, as in arid environments, the oil may remain perched at the water-table surface until evaporation of the more volatile constituents has reduced it to a tarry mass impregnating the rocks at that level. In arid regions of high elevation cut by deep valleys, the valley-floor elevation controls the water-

---

[18] William B. Heroy, "Petroleum Geology," *Geol. Soc. Am. 50th Anniversary Vol.* (1941), pp. 512–548.

[19] Vincent C. Illing, "Role of Stratigraphy in Oil Discovery," *Bull. Am. Assoc. Petrol. Geol.*, Vol. 29 (July, 1945), p. 880.

table level except in confined aquifers. Under these conditions it is of no avail to look for oil in free (as opposed to confined) reservoirs in the anticlines above drainage level, for there is nothing to hold the oil up there. Furthermore, any oil once stored in these traps could have been drained out, along with the underlying water, by the downcutting of the canyons. Even the confined reservoirs would be drained of their fluids above creek level if a valley were cut through the impervious cap rock into the underlying aquifer. Storm[20] describes a number of wells drilled on anticlines near the Colorado River in southeastern Utah in which any oil that was trapped probably drained down the river in geologically recent time.

(5) OIL DESTROYED. Hydrocarbons in rock can be destroyed in at least three ways: (a) by relatively intense diastrophism; (b) by weathering, and (c) by the activity of hydrocarbon-consuming bacteria. For the last two methods of destruction to become operative, enough cover must be stripped so that the oil accumulation lies within the zone of oxidation. Heald[21] cites the Moorcroft field of Wyoming as an example of the exposure of an oil pool by erosion during the present cycle. No doubt many accumulations have been exposed and dissipated in the geologic past. It is equally probable that hydrocarbon deposits have been destroyed (or at least expelled) by diastrophism where the sedimentary rocks show any degree of metamorphism. However, with these exceptions, it is probably most difficult to destroy oil in the earth's crust. As the unstable crust teeters and twists, the oil in laterally persistent reservoirs moves around like the bubble in a carpenter's level. At times some of the oil may slip by the traps in its path and reach the surface, where it is lost. It should always be remembered that most of the movements of the earth's crust are infinitesimally slow, and as a general rule the oil can migrate with sufficient rapidity to remain in adjustment with the changing tectonic pattern.[22]

(6) TRAP FORMED TOO LATE. The timing of oil accumulation is of utmost importance. Obviously a trap is of no value if it does not come into existence until the hydrocarbons have ceased movement through the rocks. Van Tuyl and Parker[23] point out that in some areas the barren anticlines contain a greater thickness of the formations lying above the

[20] L. W. Storm, "Oil and Ground Water in High Rocky Mountain Structures," *Mines Mag.*, Vol. 39 (December, 1949), pp. 62–64.

[21] K. C. Heald, "Essentials for Oil Pools," *Elements of the Petroleum Industry* (American Institute of Mining and Metallurgical Engineers, 1940), p. 36.

[22] T. C. Hiestand, "Regional Investigations, Oklahoma and Kansas," *Bull. Am. Assoc. Petrol. Geol.*, Vol. 19 (July, 1935), pp. 948–970.

[23] F. M. Van Tuyl and Ben H. Parker, "The Time of Origin and Accumulation of Petroleum," *Quarterly Colo. School of Mines*, Vol. 36 (April, 1941), pp. 161–162.

reservoir rocks than in adjacent productive anticlines. Presumably the anticlines overlain by thinner stratigraphic sections have been growing sporadically through a considerable span of geological time, whereas the anticlines containing normal sections have been formed by but one relatively recent period of folding. In the case of the anticline overlain by thin stratigraphic sections, the oil is trapped in an incipient phase of the upfold. It stays there during the subsequent periods of diastrophism and serves to increase the relief of the anticline. Structures that do not have any closure in that much earlier period catch no oil. Often cited as an example of a trap formed too late to catch oil is the Kelsey dome, a prominent closed structure 18 miles northwest of the East Texas field and 15 miles east of the Van field.[24] The Kelsey dome has more closure in the surface rocks than the Van structure, but it has been tested by five dry holes drilled through the Woodbine formation, which is the reservoir rock in both the Van and East Texas fields. There is no apparent difference between the dry Kelsey dome and the producing Van dome in regard to the nature of the associated organic rocks which presumably contained the oil source materials. A major difference does exist, however, in regard to the structural history of these two domes. The Van dome was a closed structure by the end of Woodbine time and has continued to be a closed structure to the present. The Kelsey dome, on the other hand, did not come into being until the Eocene. It is quite possible that the Van dome, by being in existence at an earlier date, trapped the oil going by, whereas the Kelsey dome arrived too late to partake of this accumulation.

The Grenville dome in central Wyoming has been tested and found barren by ten wells. Structural studies have shown that this fold was a plunging nose until late Tertiary time when it acquired closure.[25] Most of the oil in the Rocky Mountains is produced from traps that formed in the late Cretaceous or early Tertiary. In all probability the oil had gone by the Grenville structure before it became closed. A similar analysis of the age of deformation of barren anticlines in California indicates that "regional migration had passed their sites" before folding had produced closure.[26]

---

[24] Alex W. McCoy and W. Ross Keyte, "Present Interpretations of the Structural Theory for Oil and Gas Migration and Accumulation," *Problems of Petroleum Geology* (American Association of Petroleum Geologists, 1934), p. 302.

[25] James A. Barlow, "Time of Deformation in Structural Evaluation, Grenville Dome, Wyoming," *Am. Assoc. Petrol. Geol., Program Annual Meeting* (April, 1957), p. 32.

[26] Robert H. Paschall, "Fourth Dimension in Oil-Trap Analysis," *Bull. Am. Assoc. Petrol. Geol.*, Vol. 40 (February, 1956), p. 432.

CLASSIFICATION OF TRAPS

Several classifications of traps have been proposed. Wilson,[27] before presenting his own classification in a paper published in 1934, reviewed those put forth by others. Since then several other classifications have appeared.[28] The classification used in this book, evolved from that originally submitted by Wilson[29] and modified by Heald,[30] is:

Classification of Oil and Gas Traps
I. Structural traps.
    a. Dry synclines
    b. Anticlines
    c. Salt cored structures
    d. Hydrodynamic
    e. Fault
II. Varying permeability traps.
    a. Varying permeability caused by sedimentation.
    b. Varying permeability caused by ground water.
    c. Varying permeability caused by truncation and sealing.

Structural traps are the result of movements of the earth's crust. Upfolds are by far the most important of the structural traps; dry synclines are included in the classification largely in order to make it complete. The second class (II) contains all traps in which closure is due to more or less abrupt termination of permeability in a direction parallel to the bedding without the intervention of a fault. The term most often used for this class is "stratigraphic" trap, but Wilson[31] points out that it is

[27] W. B. Wilson, "Proposed Classification of Oil and Gas Reservoirs," *Problems of Petroleum Geology* (American Association of Petroleum Geologists, 1934), pp. 433–445.
[28] Willaim B. Heroy, "Petroleum Geology," *Geol. Soc. Am. 50th Anniversary Vol.* (1941), pp. 512–548; C. W. Sanders, "Stratigraphic Type Oil Fields and Proposed New Classification of Reservoir Traps," *Bull. Am. Assoc. Petrol. Geol.*, Vol. 27 (April, 1943), pp. 539–550; O. Wilhelm, "Classification of Petroleum Reservoirs," *Bull. Am. Assoc. Petrol. Geol.*, Vol. 29 (November, 1945), pp. 1537–1580; I. O. Brod, "Geological Terminology in Classification of Oil and Gas Accumulation," *Bull. Am. .Assoc. Petrol. Geol.*, Vol. 29 (December, 1945), pp. 1738–1755; Sylvain J. Pirson, "Genetic and Morphologic Classification of Reservoirs," *Oil Weekly*, Vol. 118 (June 18, 1945), pp. 54–59; Alex W. McCoy III, Robert L. Sielaff, George R. Downs, N. Wood Bass, and John H. Maxson, "Types of Oil and Gas Traps in Rocky Mountain Region," *Bull. Am. Assoc. Petrol. Geol.*, Vol. 35 (May, 1951), pp. 1000–1037.
[29] W. B. Wilson, *op. cit.*, p. 442.
[30] K. C. Heald, "Essentials for Oil Pools," *Elements of the Petroleum Industry* (American Institute of Mining and Metallurgical Engineers, 1940), pp. 26–62.
[31] W. B. Wilson, "Classification of Oil Reservoirs," *Bull. Am. Assoc. Petrol. Geol.*, Vol. 26 (July, 1942), pp. 1291–1292; H. R. Lovely, "Classification of Oil Reservoirs," *ibid.*, Vol. 27 (February, 1943), p. 224.

not exactly applicable, for the strata may continue even though the permeability does not.

Oil accumulations may result from *single traps, multiple traps,* or *combination traps.* A single trap is illustrated by the accumulation of oil in a reservoir rock across the top of a structural dome. An example of multiple trapping is the occurrence of oil beneath an asphalt seal in one part of a field and beneath an anticlinal axis in another part. Another example is presented by traps all of the same nature, as a series of small domes superimposed upon a large anticline. Combination trapping is very common. It differs from multiple trapping in that the traps are mutually dependent in closing the reservoir. *Most of the accumulations credited to varying permeability are actually due to a combination of erratic permeability and structural position.* Some of the fields so credited are entirely structurally controlled, as pointed out by Sanders.[32] Examples are accumulations lying across anticlines and containing local barren spots due to erratic permeability. In these places the trapping is entirely anticlinal, but the distribution within the trap is controlled by the local porosity.

Some great oil fields are due to structural trapping; others are the result, in part at least, of varying permeability. Several attempts have been made to evaluate the relative importance of these two types of accumulation. Wilson[33] and Hornbaker[34] each made a census of the published descriptions of oil fields. Wilson found that approximately half the fields that were described produced from anticlines or domes without complications. Hornbaker noted that 57 per cent of the descriptions covered fields with structural traps. Varying permeability had been an important factor in oil accumulation in the remaining 43 per cent. Of course a census of published descriptions tends to minimize the importance of the commonplace, in this case anticlinal accumulation, because of the greater interest in the unusual. A more recent study by Knebel and Rodriguez-Eraso[35] tends to confirm this fact. They prepared detailed statistics for 236 major oil fields, all of the then-known fields, outside of Soviet Russia and affiliated countries, with an ultimate anticipated recovery exceeding 100 million barrels. Eighty per cent of the oil stored in these fields is in anticlines, 13 per cent in stratigraphic traps, 6 per cent in "combination" traps and slightly over one per cent in fault traps. It should

[32] C. W. Sanders, "Stratigraphic Type Oil Fields and Proposed New Classification of Reservoir Traps," *Bull. Am. Assoc. Petrol. Geol.,* Vol. 27 (April, 1943), p. 540.

[33] W. B. Wilson, "Proposed Classification of Oil and Gas Reservoirs," *Problems of Petroleum Geology* (American Association of Petroleum Geologists, 1934), p. 445.

[34] A. L. Hornbaker, "Structural and Stratigraphic Oil Traps," *unpublished thesis* (University of Michigan, 1947), pp. 57–58.

[35] G. M. Knebel and Guillermo Rodriguez-Eraso, "Habitat of Some Oil," *Bull. Am. Assoc. Petrol. Geol.,* Vol. 40 (April, 1956), pp. 547–561.

be noted, however, that anticlinal traps are much more readily discovered than the varying permeability type, and therefore it is highly probable that a greater proportion of the total number of varying-permeability traps remains to be discovered than of structural traps. Jacobsen[36] emphasizes that the known occurrences of oil are inevitably tied in with the direction and degree of exploration effort. He asks if the 80 per cent of oil that has been found in anticlines is there because of actual habitat, or because most wildcats have been drilled on anticlines? Time, no doubt, will supply the answer to this question. It is to be hoped that these statistical analyses of known oil occurrence will be repeated periodically.

The conclusion appears justified that structural position or attitude is of utmost importance and that it is the sole factor in many accumulations and a contributing factor in most others. It is also concluded that varying porosity has played an important role in oil trapping in many fields, and its *relative* importance is likely to increase in the future.

DRY SYNCLINE TRAP

As a trap, the closed syncline, or basin, is possible only in the absence of water, which is a very unusual situation. Some geologists question the existence of an actual synclinal trap. According to Heald,[37] ". . . investigation of synclinal fields has failed to discover a single one in which both water and gas were absent." The Griffithsville, West Virginia, oil pool is illustrated in Fig. 10-1, as an example of synclinal accumulation. Because it has a large gas cap, it can be argued that the gas pushed the oil to the bottom of the structure and the oil pushed the water farther down the plunging synclinal axis. But it may also be possible that since there was very little water in the reservoir rock in the first place, the oil collected at the bottom of the fold and the gas collected close to the top.

Perhaps a better example of dry synclinal accumulation is the shale-fissure accumulation in the Florence pool in the Cañon City embayment, Colorado.[38] This field is the oldest in Colorado, with enough production in 1887 to support a refinery. Over 1000 wells were drilled, and the field has produced about 14 million barrels of oil. The oil occurs in near-

[36] Lynn Jacobsen, "Habitat of Oil," *Bull. Am. Assoc. Petrol. Geol.*, Vol. 40 (October, 1956), pp. 2514–2522.

[37] K. C. Heald, "Essentials for Oil Pools," *Elements of the Petroleum Industry* (American Institute of Mining and Metallurgical Engineers, 1940), p. 46.

[38] Ronald K. DeFord, "Surface Structure, Florence Oil Field, Fremont County, Colorado," *Structure of Typical American Oil Fields* (American Association of Petroleum Geologists, 1929), Vol. 2, pp. 75–92.

vertical fissures in the black, organic Pierre (Cretaceous) shale in a belt about 3 miles wide along the eastern side of a regional syncline. The fissures apparently failed to reach the surface, and because the fissure walls pinch together with depth, neither surface water nor water from an underlying aquifer can penetrate to these openings. As a result they are water-free. Therefore, the oil lies in the lower part of the fissures, and as it is pumped, the fluid level in the fissures is lowered.

Some alleged examples of synclinal trapping are in reality porosity traps. The only rock sufficiently permeable for use as a reservoir happened to be at or near the bottom of a syncline.

There is also a possible type of accumulation in between the dry syncline and the fluid-filled anticline. When water is present in the reservoir, but under insufficient head to rise to the top of the anticline, oil may overlie the water down the anticlinal flank (or up the synclinal flank). This situation may be the explanation for the down-flank fractured shale production on the Caddo anticline in Carter County, Oklahoma, although no water has been found beneath the oil as yet.[39] There does not appear to have been a thorough search for flank production in areas in which the potentiometric surfaces of potential reservoirs lie below the anticlinal crests.

[39] Frank J. Gardner, "Exploration Mystery in Carter County," *Oil and Gas Jour.* (Oct. 18, 1954), pp. 165–166.

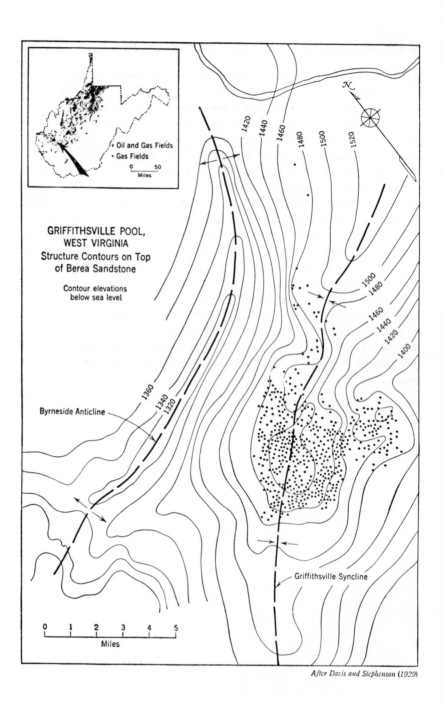

GRIFFITHSVILLE POOL,
WEST VIRGINIA
Structure Contours on Top
of Berea Sandstone

Contour elevations
below sea level

Byrneside Anticline

Griffithsville Syncline

• Oil and Gas Fields
• Gas Fields

0   50
Miles

0   1   2   3   4   5
Miles

1420 1440 1460 1480 1500 1520

1500 1480 1460 1440 1420 1400

1360 1340 1320

After Davis and Stephenson (1929)

282

TRAP CASE STUDIES   Synclinal Accumulation

Fig. 10-1. Griffithsville Pool, West Virginia.

The Griffithsville oil field° is in Lincoln County, West Virginia, about 18 miles southwest of Charleston. It was discovered in 1908. Development was rapid, and before long the producing area had spread over 20 square miles. The Berea sandstone is the main oil-producing stratum. It is between 20 and 25 feet thick and is fine-grained, hard, and tightly cemented. All the oil wells have been small but remarkably long-lived.

The structure map on the opposite page is contoured on the Berea. The greatest accumulation of oil occurs across the axis of the northeastward-plunging Griffithsville syncline between elevations −1480 and −1400. Above the −1400 contour and surrounding the oil pool on three sides is one of the major gas fields of West Virginia. The gas accumulation extends to, or close to, the crests of the adjacent anticlines lying to the southeast and to the southwest. Practically no water has been produced in connection with the development of this field. The gas trapping in this field is anticlinal, but the oil is gathered in local basins along the synclinal axis. Water may be present beneath the oil down the axis of the syncline to the north, but none has been found in wells drilled in the pool.

Similar synclinal accumulations in other parts of southern West Virginia are described by Davis and Stephenson in the same article. The Tanner Creek field in Gilmer County produces from the Maxon sand along the axis and up the flanks of the Robinson syncline. Oil is especially abundant in several small structural basins along the synclinal axis. No water has been found beneath the oil. The synclinal axis plunges downward both to the northeast and southwest from the Tanner Creek field. The Granny's Creek field of Clay County, West Virginia, is also cited as an illustration of synclinal accumulation. The structural map of this field, however, shows that not only is the synclinal axis much deeper to the northeast but also that the production is largely confined to one flank of the syncline. This condition indicates probable erratic permeability which has controlled the distribution of the oil. *Courtesy American Association of Petroleum Geologists.*

° Ralph E. Davis and Eugene A. Stephenson, "Synclinal Oil Fields in Southern West Virginia," *Structure of Typical American Oil Fields* (American Association of Petroleum Geologists, 1929), Vol. 2, pp. 571–576.

11

*T*RAPS: ANTICLINAL

Most of the world's known oil deposits are trapped in anticlines. It was noted in the preceding chapter that 80 per cent of the oil in the 236 major oil fields of the free world is in anticlines.

This chapter is concerned with anticlinal traps other than those which are (or probably are) salt-cored. The latter traps will be discussed in Chapter 12.

Origin of Folds. Anticlines and synclines can result from either vertical or horizontal movements in the earth's crust; pseudofolds can result from initial dips.

There are two types of vertical movement: (1) movement up or down due to diastrophic and perhaps even igneous activity in the earth's crust, and (2) settling due to either compaction or to leaching. As a general rule, the locus of the geologic activity that causes vertical movement is in the unexplored crystalline basement rocks.[1] Consequently, the nature of the deep-seated activity that results in folds in the sedimentary rock veneer is not known. Faulting,[2] igneous intrusion, isostatic adjustment, and rock flowage are possibilities that have been suggested at one time or another. There is abundant seismological evidence of deep-seated faulting, and there are many examples in the visible parts of the earth's crust of faults damping out upward into folds. That igneous intrusions also may arch overlying rock is well known from the occurrences in Utah,[3]

[1] Clyde G. Strachan, "The Interpretation of Structure in the Exploration for Oil and Gas," *Tulsa Geol. Soc. Digest*, Vol. 25 (1957), pp. 132–134.

[2] Alex W. McCoy, "An Interpretation of Local Structural Development in Mid-Continent Areas Associated with Deposits of Petroleum," *Problems of Petroleum Geology* (American Association of Petroleum Geologists, 1934), pp. 581–627.

[3] Charles B. Hunt, "New Interpretation of Some Laccolithic Mountains and Its Possible Bearing on Structural Traps for Oil and Gas," *Bull. Am. Assoc. Petrol. Geol.*, Vol. 26 (February, 1942), pp. 197–203.

around the periphery of the Black Hills, and elsewhere.[4] An unusually symmetrical dome (Fig. 11-1), thought to be the result of laccolithic intrusion, contains oil in Gallatin County, Illinois.[5] Evidence favoring the hypothesis of an underlying intrusive is the presence of dikes and sills of peridotite up to 50 feet in thickness cutting the sedimentary section, including the reservoir strata. It is of course not necessary that the laccolithic-like intrusions reach the sedimentary shell. They can be quite effective in producing differential vertical movements even though they stop in the crystalline basement rocks.

The vertical movements that produce folds in the sedimentary rocks do not have to be upward movements. Settling or subsidence may produce folds, providing that it is differential in character. Areas of lesser settling, if the sedimentary rocks bend and do not break, become anticlinal in structure; the greater settling produces synclines. The most common cause for settling is compaction.[6] Clay and mud, as originally deposited on the sea floor, have porosities as high as 90 per cent. With the piling on of younger sediment, the weight of the overlying rock squeezes and compacts this material until it eventually becomes shale. The possible significance of the liquids escaping during this squeezing was discussed in previous chapters. If the compacting material overlies a smooth floor, and is homogeneous throughout, the settling is not differential and no folds result. But if the floor contains hills or monadnocks or reefs[7] of older rock which were surrounded by clay or mud before being covered by the same material, there is less settling over these topographic features, and the structure of the overlying rock reflects to some extent the buried topography (Fig. 11-2). The "draping" of sediments over reefs by differential compaction is an important clue in finding buried reefs today. Likewise, if the compacting material contains lenses of sand which are

---

[4] G. L. Knight and Kenneth K. Landes, "Kansas Laccoliths," *Jour. Geol.*, Vol. 40 (January-February, 1932), pp. 1–15.

[5] R. M. English and R. M. Grogan, "Omaha Pool and Mica-Peridotite Intrusives, Gallatin County, Illinois," *Structure of Typical American Oil Fields* (American Association Petroleum Geologists, 1948), Vol. 3, pp. 189–212.

[6] Hollis D. Hedberg, "The Effect of Gravitational Compaction on the Structure of Sedimentary Rocks," *Bull. Am. Assoc. Petrol. Geol.*, Vol. 10 (November, 1926), pp. 1035–1072; L. F. Athy, "Density, Porosity, and Compaction of Sedimentary Rocks," *Bull. Am. Assoc. Petrol. Geol.*, Vol. 14 (January, 1930), pp. 1–24; L. F. Athy, "Compaction and Its Effect on Local Structure," *Problems of Petroleum Geology* (American Association of Petroleum Geologists, 1934), pp. 811–823; John L. Rich, "Application of Principle of Differential Settling to Tracing of Lenticular Sand Bodies," *Bull. Am. Assoc. Petrol. Geol.*, Vol. 22 (July, 1938), pp. 823–833; G. D. Hobson, "Compaction and Some Oil Field Features," *Jour. Inst. Petrol.*, Vol. 29 (February, 1943) pp. 37–54.

[7] Gerson H. Brodie, "Structure Forming Role of Limestone Reefs in Eastern Platform Area of West Texas," *Am. Assoc. Petrol. Geol., Program Annual Meeting* (1950), p. 14.

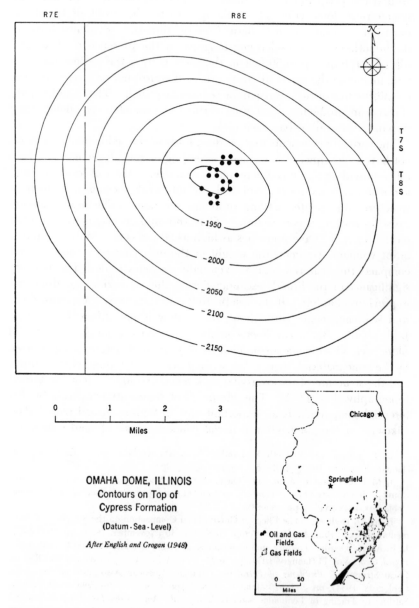

OMAHA DOME, ILLINOIS
Contours on Top of
Cypress Formation

(Datum - Sea - Level)

After English and Grogan (1948)

Fig. 11-1. Producing dome possibly formed by laccolithic intrusion. Omaha field, Illinois. *Courtesy American Association of Petroleum Geologists.*

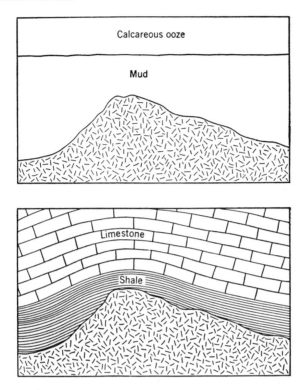

Fig. 11-2. Differential compaction over hill. *Upper:* mud deposited around and over hill of crystalline rock. *Lower and later:* mud has compacted into shale; both the shale and overlying rock are draped over unyielding hill as a result of greater compaction where the mud was thicker. *Drawing by John Jesse Hayes.*

relatively non-compactible, the rocks above such lenses settle through a lower vertical interval and anticlinal bulges result (Fig. 11-3).

When soluble rocks occur in the section, leaching followed by slumping may produce sagging in the overlying strata, with arching over the blocks that have not collapsed. These folds do not, of course, extend below the soluble rock zone. If the potential reservoir rocks are below this zone, the structure maps based on higher datum planes are valueless. The soluble rocks are limestone and dolomite, salt, and gypsum. Salt collapse structures are especially prevalent around the leached edges of evaporite deposits, as in the Michigan Basin.[8] They can be recognized by (1) their

[8]Kenneth K. Landes, G. M. Ehlers, and George M. Stanley, "Geology of Mackinac Straits Area," *Mich. Geol. Survey, Publ.* 44 (1945), pp. 123–153.

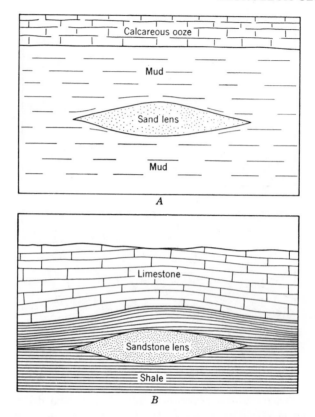

Fig. 11-3. Compaction over sand lens. *Upper:* lens of sand surrounded by mud. *Lower and later:* mud has compacted into shale; strata are arched over sand lens because of its relative incompressibility. *Drawing by John Jesse Hayes.*

position in respect to the periphery of the salt deposit; (2) the presence of small faults, slickensides, and breccia; (3) the prominence of depression contours; and (4) the irregularity of slope and lack of parallelism of the structures.

Residual structural highs may also possibly result from the plastic flow of salt. As will be described later, some salt columns are surrounded by so-called rim synclines, which are most easily explained as downwarps of beds above the salt that are a result of the flow of the salt into domes. Presumably an area surrounded by these synclines would be structurally higher than the immediate surroundings. It has been suggested that the presence of such an isolated body of undisturbed sedimentary salt may be responsible for the doming of the younger sediments and the trapping of oil in the Katy field of Texas.[9]

Horizontal movements are at their maximum in orogenic belts where the earth's crust is under compression. The rock strata are buckled into folds and may even be broken and overthrust. The result is a shortening of the earth's crust, by a matter of several miles in some folded mountain belts.

The folds in the sedimentary basins outside of the orogenic belts have relatively low dips and are less faulted, but otherwise they have the same appearance as the orogenic belt folds. They extend into the basement rock. The surface of the basement conforms structurally with the overlying folded sediments.

The simplest explanation for the folding of basin sediments is that given by Dallmus.[10] He demonstrated that the floors of sedimentary basins, when plotted to the same vertical and horizontal scale, are still convex upward, although less so than the curvature of the earth. The compressional (wedging) forces created by the subsidence of these basins are entirely adequate to explain the folding of the basin sediments. Hudson[11] believes that weakening of the basement rocks by crushing, fracturing, and faulting prior to the deposition of the sediments assists in translating the tangential forces in the basement into folds in the sediments.

Whatever the cause for the folding, it must be, in most instances, an activity that is subject to repeat performances, for recurrent folding[12] is the rule rather than the exception in the Mid-Continent and similar areas. Subsurface isopach mapping has shown that most structures are the result of several periods of folding. "Once an anticline, always an anticline"[13] is a maxim with wide applicability.

Oil can be and is trapped in upfolds, whether they be formed by horizontal or by vertical forces. Oil is even found in regions of the overthrust faulting. The Rose Hill field in Virginia[14] is an example of this, as is the

[9] A. P. Allison et al., "Geology of the Katy Field, Waller, Harris, and Fort Bend Counties, Texas," *Bull. Am. Assoc. Petrol. Geol.*, Vol. 30 (February, 1946), p. 169.

[10] K. F. Dallmus, "Mechanics of Basin Evolution and its Relation to the Habitat of Oil in the Basin," *Habitat of Oil* (American Association of Petroleum Geologists, Special Vol., 1958), pp. 833–931.

[11] Frank S. Hudson, "Folding of Unmetamorphosed Strata Superjacent to Massive Basement Rocks," *Bull. Am. Assoc. Petrol. Geol.*, Vol. 39 (October, 1955), pp. 2038–2052.

[12] F. M. Van Tuyl and Ben H. Parker, "The Time of Origin and Accumulation of Petroleum," *Quarterly Colo. School of Mines*, Vol. 36 (April, 1941), Chapter 11 (Recurrent Folding and Accumulation), pp. 83–89.

[13] Credited to E. O. Ulrich by G. M. Ehlers.

[14] Ralph L. Miller, "Rose Hill Oil Field, Lee County, Virginia," *Structure of Typical American Oil Fields* (American Association of Petroleum Geologists, 1948), Vol. 3, pp. 452–479.

older Turner Valley, Alberta field, and the Polish oil fields in front of the Carpathian Mountains. [15]

Pseudo-folds, so called because they look like folds but are not the result of any movement, are formed by deposition parallel to an uneven ocean floor. If the deposition is from suspension, and the slopes are not steeper than the angle of repose of the unconsolidated sediment, the layers may be deposited with initial dips which conform to the topography of the submerged surface. [16] Submergence must be fairly rapid, so that deposition takes place concurrently on hilltop and valley floor. If the low areas fill with fine sediment while the hills are still emergent, the stage is set for compaction but not for initial dip. These two methods of obtaining a reflection of buried topography in the structure of the overlying rocks are mutually antagonistic. Initial-dip structures tend to diminish in relief upward and to become lost by flattening a few hundred feet above the erosional unconformity.

Hills in the Permian surface appear to have been reflected upward by initial dips in the overlying Cretaceous sediments in western Kansas. [17] Limestone reefs may provide arched floors for the conformable deposition of overlying sediment. [18] Initial-dip folds extend downward only to the unconformity, and considerable money has been spent in drilling wells on surface anticlines that were not anticlines at the depths of the potential reservoirs.

Types of Anticlines. There are two types of anticlines in nature, but neither one has a horizontal axis like the ridgepole of a roof, for that structure exists only in textbook drawings. Invariably the axis is a vertical arc, as seen in a longitudinal section, commonly plunging in both directions from the highest point on the fold. As the axis reaches lower elevations at both ends, the structure contours loop around from one flank to the other. This description is the closed anticline, hereinafter referred to merely as *anticline*. Oil and gas are trapped beneath the overlying seal at the tops of such folds because they have reached that point by traveling up one flank or the other, owing to their inherent buoyancy. To con-

[15] W. A. J. M. van Waterschoot van der Gracht, "Oil Fields in Folded Rocks," *Science of Petroleum* (Oxford University Press, 1938), pp. 247–251.

[16] Josiah Bridge and C. L. Dake, "Initial Dips Peripheral to Resurrected Hills," *Mo. Bur. Geol. and Mines, Bienn. Repts. State Geologist,* 1927–1928 (1929), pp. 93–99; "Buried and Resurrected Hills of Central Ozarks," *Bull. Am. Assoc. Petrol. Geol.,* Vol. 16 (July, 1932), pp. 629–652.

[17] Kenneth K. Landes and J. W. Ockerman, "Origin of Domes in Lincoln and Mitchell Counties, Kansas," *Bull. Geol. Soc. Am.,* Vol. 44 (June 30, 1933), pp. 529–540.

[18] Heinz A. Lowenstam, "Marine Pool, Madison County, Illinois, Silurian Reef Producer," *Structure of Typical American Oil Fields* (American Association of Petroleum Geologists, 1948), Vol. 3, pp. 153–188.

tinue down the other side would be to move in the direction of greater pressure, which is not possible.

The other type of anticline is superimposed upon the flank of a much larger monocline, with the anticlinal axis normal to the regional strike (parallel to the monoclinal dip). The axis of this type of anticline is also a vertical arc, but because of the regional structure, the dip on the "uphill" side may be relative rather than actual, and the fold may not close in that direction. The structure contours note the presence of a possible trap by an outward bulge in the direction of the regional dip. The closest imitative topographic feature is a fan, or delta. Folds of this type are sometimes referred to as "structural noses," but this expression is incongruous for a feature that is usually over one hundred times wider than it is high.

The plunging anticline is of utmost importance in fault-trap and combination-trap accumulations. It supplies closure on three sides; the fourth side may be closed by strike fault, up-dip facies change, unconformity overlap, or other type of permeability barrier. Without the presence of a plunging anticline, these local dams athwart the regional dip are valueless because the would-be trap lacks closure at the sides.

Anticlinal trapping could be illustrated by literally hundreds of oil and gas fields. The selected examples that are shown in the trap case studies at the end of this chapter are subdivided into three arbitrary classes: *elongate anticlines*, which are over four times as long as wide; *anticlines*, which are from two to four times longer than wide; and *domes*, which are less than twice as long as wide. Anticlines that are approximately circular in plan are sometimes referred to as *quaquaversal domes*.

Besides coming in all shapes, anticlines also come in all sizes. The length varies from less than a mile to many miles and the amount of closure ranges from tens of feet to thousands of feet. As a general rule, the larger anticlines do not carry oil the entire length. Instead, the superimposed domes or nodes are oil-bearing and the intervening saddles contain only water. The large anticlines are also less likely to be filled with hydrocarbons to the spill point.

In earlier days of oil finding, many pools were considered to be terrace, or "nose," accumulations because no closure was evident in the structure contours drawn on the outcropping formations. However, with the development of subsurface structure contouring it was found that most of these structures were closed in the producing formations. This discovery is strikingly illustrated in two maps published by W. B. Wilson[19] (Fig.

[19] W. B. Wilson, "Proposed Classification of Oil and Gas Reservoirs," *Problems of Petroleum Geology* (American Association of Petroleum Geologists, 1934), pp. 438 and 439.

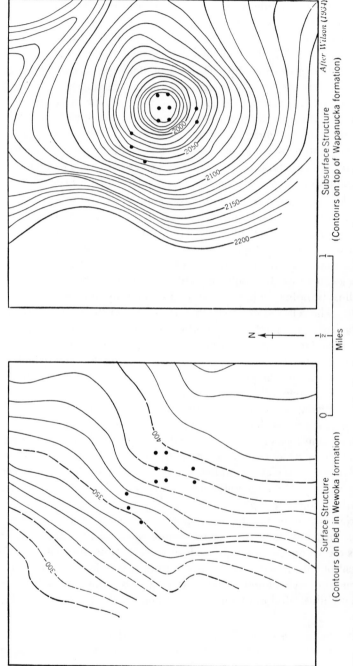

Fig. 11-4. Development of closure on terrace with depth. *Left:* structure at surface. *Right:* structure over same area at deeper level. *Courtesy American Association of Petroleum Geologists.*

11–4). Exceptions to structural closure with depth are hydrodynamic (Chapter 13) and varying permeability (Chapter 14) traps.

Anticline Hunting. Almost all the oil-finding techniques have been developed for the purpose of, and are most effective in, finding anticlinal traps. Anticlines can be discovered and identified by the use of surface geologic methods where outcrops exist, by core drilling, by geophysical methods, and by subsurface geology. These hunting procedures are followed in all parts of the world, both onshore and offshore.[20] Their success in finding anticlines explains why 80 per cent of the known major oil supplies in the free world are in such structures. No doubt the actual percentage of anticline-trapped oil is less.

[20] Pierre A. A. H. Masson and F. J. Agnich, "Seismic Survey of Sinai and the Gulf of Suez," *Geophysics*, Vol. 23 (April, 1958), pp. 329–342.

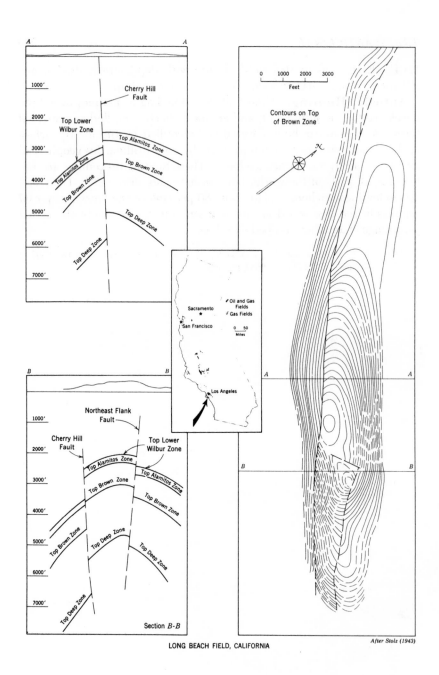

A A

1000'

2000' Cherry Hill
Fault

Top Lower
Wilbur Zone

3000' Top Alamitos Zone

Top Alamitos Zone Top Brown Zone

4000' Top Brown Zone

5000' Top Deep Zone

6000' Top Deep Zone

7000'

0   1000 2000 3000
Feet

Contours on Top
of Brown Zone

N

Oil and Gas
Fields
Gas Fields

Sacramento

San Francisco

0   50
Miles

Los Angeles

A A

B B

B B

1000' Northeast Flank
Fault

Cherry Hill Top Lower
Fault Wilbur Zone

2000' Top Alamitos Zone

Top Brown Zone Top Alamitos Zone

3000' Top Brown Zone

4000'

5000' Top Brown Zone Top Deep Zone

Top Deep Zone

6000'

7000' Top Deep Zone

Section B-B

LONG BEACH FIELD, CALIFORNIA

After Stolz (1943)

294

Fig. 11–5. Long Beach Field, California.

The Long Beach, California, oil field,° also known as the Signal Hill field, lies 20 miles south of Los Angeles in the Los Angeles structural basin. It is but one of a number of oil fields located along a line of uplift which crosses the Los Angeles basin from northwest to southeast. This uplift appears on the surface as a series of low hills, of which Signal Hill is one. The anticline was recognized many years before the discovery well was drilled in 1921. Signal Hill and the surrounding territory had been subdivided and streets laid out, and although very few residences had been built the lots were individually owned and the discovery of oil was followed by an inefficient and chaotic town-lot drilling campaign.

A subsurface structure contour map of the Long Beach anticline, and two cross sections, are shown. This anticline is 1 by 4 miles in approximate dimension and trends northwest-southeast. The dips on the southwest flank range from 40° to 45° as against 15° on the northeast flank. The closure is over 1200 feet. At least two recent faults, which are shown on the structure map, cut the anticline. The longer one of these, the Cherry Hill fault, has a maximum displacement of 1000 feet. It is so recent in age that it makes a prominent scarp (Signal Hill) at the surface. Accumulation is anticlinal, however, rather than fault controlled.

Below the Quaternary sands and gravels exposed at the surface are from 1000 to 1500 feet of Pleistocene sediments. The underlying Pliocene is 3600 feet thick. It is divided into the Upper and the Lower Pliocene. The lower sands of the Upper Pliocene have been found productive here and there along the top of the anticline and on the southwest flank. The most prolific reservoirs are the Alamitos zone and the underlying Brown zone of the Lower Pliocene. The Alamitos zone is about 650 feet thick and consists of four sands up to 175 feet in thickness with interbedded shales. The Brown zone is about 1450 feet thick and extends into the Upper Miocene. It likewise consists of a series of thick sandstones separated by shales. Some production has also been obtained from two deeper Miocene sand zones. The total thickness of the Miocene sediment is not known but may exceed 15,000 feet. *Courtesy California Division of Mines.*

°Harry P. Stolz, "Long Beach Oil Field," *California Div. Mines, Bull.* 118 (1943), pp. 320–324.

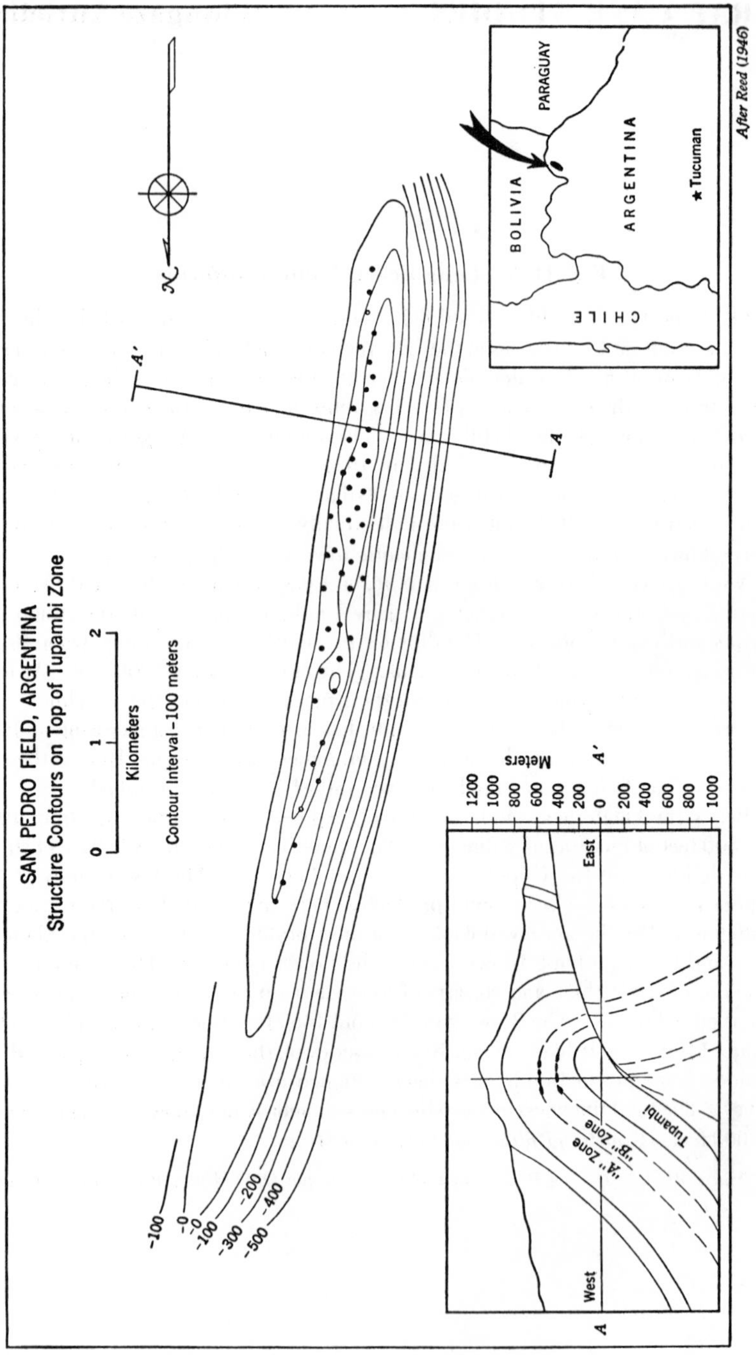

SAN PEDRO FIELD, ARGENTINA
Structure Contours on Top of Tupambi Zone

Contour Interval –100 meters

Kilometers
0   1   2

-100
-0
-100
-200
-300
-400
-500

A
A'

West
East

Tupambi
"A" Zone
"B" Zone

Meters
1200
1000
800
600
400
200
0
200
400
600
800
1000

A
A'

BOLIVIA
CHILE
PARAGUAY
ARGENTINA
★ Tucuman

After Reed (1946)

296

Fig. 11-6. San Pedro Field, Argentina.

The San Pedro oil field° is in Salta province in northern Argentina. It is the most important of the seven fields that lie in the Tarija sedimentary basin. The San Pedro field occurs along the crest of a partly eroded, tightly folded anticline which makes a ridge in the San Antonio range. The structure was mapped by several oil company geologists beginning as early as 1921. The discovery well for this particular field was completed in 1928, but the first commercial production in the district had been obtained two years earlier.

The structure of the San Pedro anticline as mapped at the top of the main producing sand is shown on the opposite page. The anticline is many times longer than wide, and the production is confined to the uppermost 300 feet of a total closure far in excess of that figure. This anticline is a more or less superficial wrinkle on a great overriding thrust sheet.

Oil occurs in the San Pedro anticline in three zones, of which the upper two are relatively unimportant. These are sandstones in a thick "clay grit" section in the Tarija formation of Permo-Triassic age. The major oil-producing zone is the C zone in the underlying Tupambi formation, which is also of Permo-Triassic age. The actual reservoirs are sandstones occurring in lenses and layers in a siltstone section 300 feet thick. Apparently enough oil was available at San Pedro to fill only the very top of the trap. The major overthrust was a deep low-angle thrust from the west; the San Pedro anticline, and the relatively shallow fault shown in the cross section on the opposite page, were formed during that eastward push. San Pedro is but one of five or more parallel thrust-faulted anticlinal ranges in the Tarija basin, but it is the only field in which the oil occurs at the highest point of the anticline. The other Salta fields produce from lenticular sands along the axis, but down the plunge a considerable distance from the highest point.

The Tarija basin is filled with sediments which may exceed 32,000 feet in thickness. These sediments consist mostly of Devonian shales, Permo-Carboniferous clastics of glacial origin, and Pliocene and Pleistocene continental sediments. The thrusting with concurrent folding did not take place until Pleistocene, an unusually recent date for a hydrocarbon trap. *Courtesy American Association of Petroleum Geologists.*

° Lyman C. Reed, "San Pedro Oil Field, Province of Salta, Northern Argentina," *Bull. Am. Assoc. Petrol. Geol.,* Vol. 30 (April, 1946), pp. 591–605.

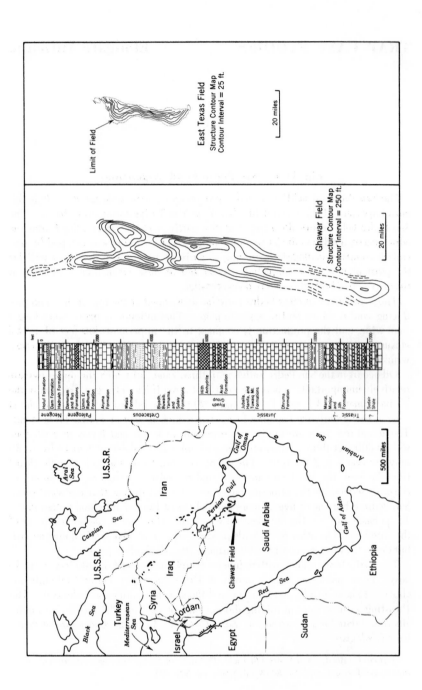

East Texas Field
Structure Contour Map
Contour Interval = 25 ft.

Limit of Field

20 miles

Ghawar Field
Structure Contour Map
Contour Interval = 250 ft.

20 miles

feet
0

2000

4000

6000

8000

10000

12000

Neogene	Heluf Formation
	Dam Formation
	Hadrukh Formation
Paleogene	Dammam and Rus Formations
	Umm Er Radhuma Formation
Cretaceous	Aruma Formation
	Wasia Formation
	Biyadh, Buwaib, Yamama, and Sulaiy Formations
	Hith Anhydrite
Riyadh Group	Arab Formation
Jurassic	Jubaila, Hanifa, and Tuwaiq Mt. Formations
	Dhruma Formation
Triassic	Marrat, Minjur, and Jilh Formations
?	Sudair Shale

U.S.S.R.

Aral
Sea

U.S.S.R.

Black Sea

Turkey

Caspian Sea

Iran

Mediterranean Sea

Israel

Syria

Jordan

Iraq

Egypt

Sudan

Ghawar Field

Saudi Arabia

Persian Gulf

Gulf of Oman

Arabian Sea

Red Sea

Gulf of Aden

Ethiopia

500 miles

Fig. 11-7. Ghawar Oil Field.

The Ghawar° field is one of the largest oil fields in the world. It is the result of the merger of five oil fields along a great anticlinal arch. Currently (1958) the length is about 145 miles and the width 13 miles. The discovery well was completed in the Ain Dar area near the north end of the field in 1948. Subsequently 90 producers had been drilled by the end of 1957 for an average well depth of 7300 feet and for an average potential of 11,400 barrels per day. The estimated production for the Ghawar field during its first decade is approximately 1¼ billion barrels.

The trap at Ghawar is the elongate north-south anticline shown on the opposite page. The dips on the flanks range from 3° to 5°. Ghawar is a part of the even longer En Nala anticlinal axis which probably extends north at least as far as the Fadhili field which is 85 miles beyond the northern tip of Ghawar. So far, seven nodes have been found along the En Nala axis in the Ghawar field. The maximum oil-filled closure is about 1300 feet. The oil-water contact rises to the south and at the same time the salinity of the underlying water decreases from 240,000 parts per million of total solids in the north to 36,000 parts per million in the south.

The reservoir rock is the Arab D member which overlaps the upper Jubaila and the lower Arab formations. The average thickness of the reservoir is 260 feet. The Arab D is a limestone with primary porosity due to the presence of voids between (1) oölites, (2) organic detritus, (3) carbonate-rock sand grains, and (4) calcareous pellets. The seal is the lowest of three anhydrite members of the Arab formation which immediately overlies the D member.

The huge accumulation is due to the large gathering area, especially to the east, which has been present ever since the structure began to form in post-Jurassic time.

The discovery of the different original producing areas in what is now the Ghawar field was primarily due to structure drilling. However, the initial clues to the existence of the En Nala axis came from surface mapping. *Courtesy Arabian-American Oil Company.*

° Max Steineke, R. A. Bramkamp, N. J. Sander, "Stratigraphic Relations of Arabian Jurassic Oil," *Habitat of Oil* (American Association of Petroleum Geologists Special Vol. (1958), pp. 1294–1329; Geological Staff of Arabian American Oil Company, "Ghawar Oil Field," *Program Annual Meeting* (American Association of Petroleum Geologists (1958), pp. 42–43; Warren H. Thralls and R. C. Hasson, "Basic Pattern of Exploration in Saudi Arabia Remains Unchanged," *Oil and Gas Jour.* (July 15 and July 22, 1957); W. H. Thralls and R. C. Hasson, "Geology and Oil Resources of Eastern Saudi Arabia," *Symposium on Deposits of Petroleum and Gas*, Vol. 2, XX International Geol. Congress (1956), pp. 9–32.

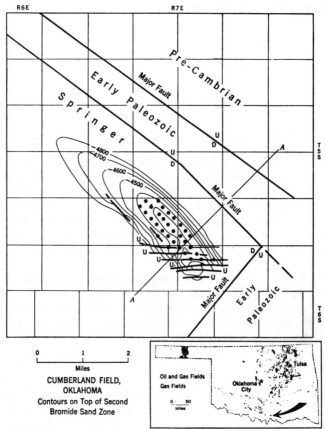

Miles

**CUMBERLAND FIELD,
OKLAHOMA**

Contours on Top of Second
Bromide Sand Zone

After Cram (1948)

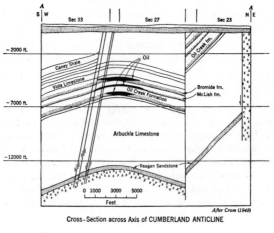

After Cram (1948)

Cross-Section across Axis of CUMBERLAND ANTICLINE

Fig. 11-8. Cumberland Oil Field, Oklahoma.

The Cumberland oil field° lies on the south flank of the Arbuckle Mountain system in southern Oklahoma. The Precambrian core of the Arbuckles is exposed only 12 miles to the northwest. Previous to the drilling of the discovery well in 1940, the Cumberland area had been covered by surface, subsurface, magnetic, and seismograph surveys, each one of which increased the interest in the possibilities of the area.

The Cumberland structure is a closed faulted anticline which lies within a down-faulted block of sedimentary rock. The structure map is contoured on top of the second Bromide (Ordovician) sand zone. The contour interval is 100 feet. The map indicates a closure of approximately 550 feet, but it may actually be as much as 1000 feet. The faults that cross the southeast end of the anticline have displacements as great as 300 feet. The major faults at the sides of the graben which contains the Cumberland anticline have displacements somewhere between 7000 and 13,000 feet.

The right figure shows a cross section along line *AA'* drawn from southwest to northeast across the Cumberland anticline in the left drawing. The dips are of such magnitude that no exaggeration of the vertical scale is necessary. The section shows the graben character of the block containing the Cumberland anticline, the Comanchean Trinity sand blanketing everything, and the anticlinal control on the trapping of the oil.

The reservoirs are sandstones belonging to the Bromide, McLish, and Oil Creek formations of the Simpson group of Ordovician age. There are three producing sandstones in the lower part of the Bromide, each one over 100 feet in thickness. At the base of the underlying McLish formation is another 100-foot sandstone which is one of the main producing zones in the field. The still deeper First Oil Creek sand averages 185 feet in thickness, and the Second Oil Creek sand, a large producer, is 210 feet thick. *Courtesy American Association of Petroleum Geologists.*

° Ira H. Cram, "Cumberland Oil Field, Bryan and Marshall Counties, Oklahoma," *Structure of Typical American Oil Fields* (American Association of Petroleum Geologists, 1948), Vol. 3, pp. 341–358.

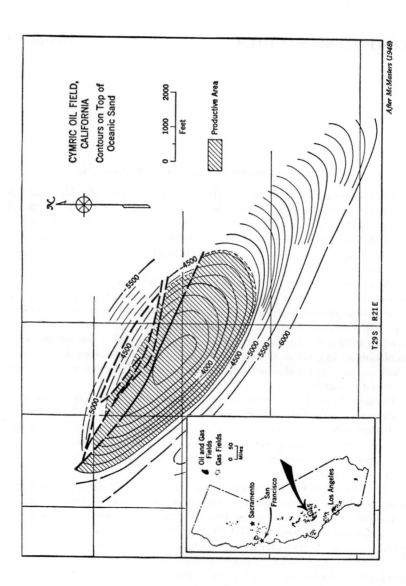

CYMRIC OIL FIELD, CALIFORNIA

Contours on Top of Oceanic Sand

Feet
0 1000 2000

Productive Area

Oil and Gas Fields

Gas Fields

Miles
0 50

Sacramento

San Francisco

Los Angeles

T 29 S   R 21 E

After McMasters (1948)

302

Fig. 11-9. Cymric Oil Field, California.

The Cymric oil field° is in Kern County, California, on the west side of the San Joaquin Valley.   The first discovery was made in 1916 in a shallow reservoir; the deeper zone was not discovered until 1945.

On the opposite page is shown the Cymric structure contoured on top of the Oligocene Oceanic sand, which is the leading producing formation out of eleven distinct superimposed reservoirs.   The anticline is elliptical in plan and unusually symmetrical.   The oil in the Oceanic sand lies above the −4400-foot contour. Faulting on the northeast flank does not contribute to the trapping, but it does limit production by dropping the down-dropped reservoir rock below the general level of the oil-water contact in the Oceanic formation.

The shallow reservoirs, which were the sole producers at Cymric between 1916 and 1945, are the Cymric sands and the Amnicola sand, both belonging in the Tulare formation of Pleistocene-Pliocene age.   These sediments are of continental origin, and the reservoir sands disappear to the north and west.   The oil is black and tarry with a gravity ranging from 11° to 13°.   The deeper reservoirs include one in the Pliocene, six in the Miocene, one in the Oligocene, and one in the Upper Eocene.   The Pliocene reservoir is a fine, silty sand.   The six Miocene producers occur in four Lower Miocene formations.   The prolific Oceanic reservoir of Oligocene age is a uniform, medium-grained, friable, and somewhat silty sandstone. The contained oil is very much lighter than that found in the shallow sands; the gravity varies between 37° and 38°.   The eleventh and deepest reservoir is the Point of Rocks sand in the Upper Eocene.   This is a hard, fine-grained sandstone containing primarily gas.   Structurally high wells in this zone have tested as much as 30,000,000 cubic feet of gas per day with some 52° gravity condensate.   Lower wells produce less gas and more oil; the deeper oil runs about 36° in gravity.

Accumulation in the Miocene and older sands is anticline controlled, but distribution of oil varies somewhat from one zone to the next owing to faulting, variable permeability, and truncation beneath unconformities.   The anticline is not present in the post-Miocene reservoirs, and trapping there has been the result of wedging-out of the reservoir sands on the flanks of a monocline.   *Courtesy American Association of Petroleum Geologists.*

° J. H. McMasters, "Cymric Oil Field, Kern County, California," *Structure of Typical American Oil Fields* (American Association of Petroleum Geologists, 1948), Vol. 3, pp. 38–57.

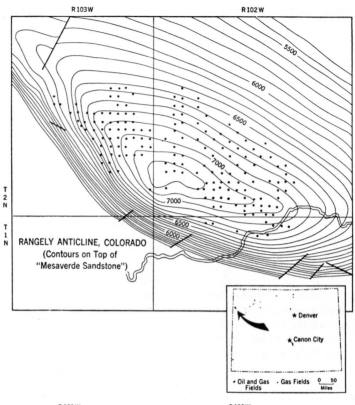

R 103 W   R 102 W

RANGELY ANTICLINE, COLORADO
(Contours on Top of
"Mesaverde Sandstone")

★ Denver

★ Canon City

• Oil and Gas   · Gas Fields   0   50
  Fields   Miles

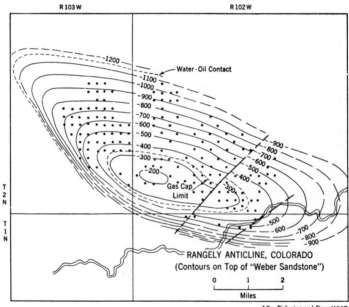

R 103 W   R 102 W

Water · Oil Contact

Gas Cap
Limit

RANGELY ANTICLINE, COLORADO
(Contours on Top of "Weber Sandstone")

0   1   2
         Miles

After Pickering and Dorn (1948)

Fig. 11-10. Rangely Field, Colorado.

The Rangely field° lies in northwestern Colorado on the northeastern edge of the Uinta basin. The Cretaceous Mesa Verde sandstone is exposed throughout this area and was used to construct the structure contour map shown on the left. The Rangely anticline at the surface has a closure of 1900 feet. It was recognized by federal geologists as early as 1878. Oil was discovered in shallow-fissured shale belonging to the Mancos (Upper Cretaceous) formation in 1902. A large accumulation in the Weber sandstone of Pennsylvanian age was discovered by the drilling of a 7000-foot well in 1931. Exploitation of the field, however, did not become active until 1943, when the demand for oil overcame problems of inaccessibility.

The right map is the Rangely anticline contoured on the producing Weber formation. This anticline, both at the surface and in the subsurface, is decidedly asymmetrical with the steeper dip on the southwest side. It is approximately 20 miles long and nearly 8 miles wide in the central part. The structure is crossed by faults toward the eastern end, but these have had little or no effect in the trapping of oil. The oil-water contact plane lies at an elevation of −1160 feet, about 1000 feet below the top of the anticline. However, a gas cap is present in the uppermost 170 feet of the structure, so that the thickness of the oil-bearing section is about 830 feet.

The Weber is a fine-grained, calcareous, and somewhat tight sandstone. The total thickness of the formation is about 550 feet, of which about 30 per cent is calculated to be effective oil sand. In spite of the relatively tight character of the sandstone, initial productions range up to 1000 barrels a day. Development and production procedures at Rangely are controlled by the Rangely Engineering Committee.

The Rangely field is notable for two reasons in addition to its size. It was the scene of one of the few authentic cases (popular opinion to the contrary notwithstanding) of the temporary capping of a successful wildcat because of remoteness from market. Secondly, its discovery was delayed because earlier tests stopped short of the Weber for the reason that it was popularly, but mistakenly, referred to as the Weber "quartzite," a most unlikely reservoir rock. *Courtesy American Association of Petroleum Geologists.*

° W. Y. Pickering and C. L. Dorn, "Rangely Oil Field, Rio Blanco County, Colorado," *Structure of Typical American Oil Fields* (American Association of Petroleum Geologists, 1948), Vol. 3, pp. 132–152.

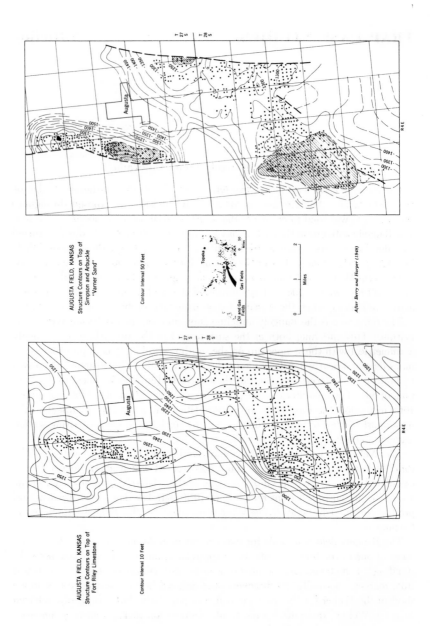

AUGUSTA FIELD, KANSAS
Structure Contours on Top of
Simpson and Arbuckle
"Warner Sand"

Contour Interval 50 Feet

After Berry and Harper (1948)

AUGUSTA FIELD, KANSAS
Structure Contours on Top of
Fort Riley Limestone

Contour Interval 10 Feet

306

Fig. 11-11. Augusta Field, Kansas.

The Augusta field[°] is in Butler County, Kansas, extending from 3 miles north of the city of Augusta to 7 miles south of it.   It overlies the Nemaha Granite Ridge, and the local reversal of the general westerly dip can be recognized in the outcropping cherty limestones.   The field was discovered in 1914 as a result of surface structure mapping.

The left figure is the structural map of the Augusta anticline contoured on the Permian surface rocks.   The anticline is divided by a saddle into the north Augusta and south Augusta domes.   The latter in turn is divided by a saddle into two nodes.   The maximum closure is only about 60 feet.

The right figure shows the Augusta anticline with the structure contoured on the Varner "sand," the pay zone at the top of the truncated Ordovician section.   Each of the structural domes overlies and reflects a hill on the crystalline rock surface. In post-Mississippian pre-Cherokee (Pennsylvanian) time, uplift along the Granite Ridge folded and faulted the overlying sediments and erosion truncated the tilted beds.   The shaded area on the right map shows where the Pennsylvanian rocks directly overlie the Ordovician Arbuckle "Siliceous Lime."   Subsequent uplifts have arched the Pennsylvanian and Permian sediments with the result shown in the left figure.   The maximum closure in the Ordovician is about 350 feet.

Most of the oil has come from the truncated Ordovician formations.   These include the Simpson sandstone and shale formation which is 60 feet thick on the flanks of the Augusta anticline and missing altogether across the top, and the Arbuckle cherty dolomite which ranges in thickness from 1000 feet on the flanks to 400 feet on the axis.   Some oil and all the gas have come from four Pennsylvanian reservoirs.

Trapping in the Pennsylvanian formations above the unconformity has been due entirely to the anticlinal structure.   In the older formations the oil has moved up through the Ordovician strata flanking the Nemaha Ridge and has been trapped by the unconformity lying between the truncated Ordovician formations and the Cherokee shale of Pennsylvanian age.   It can therefore be said that the Ordovician oil has been trapped by varying permeability, but it migrated into its present position as the result of anticlinal structure below the plane of the unconformity. *Courtesy American Association of Petroleum Geologists.*

[°] George F. Berry, Jr., and Paul A. Harper, "Augusta Field, Butler County, Kansas," *Structure of Typical American Oil Fields* (American Association of Petroleum Geologists, 1948), Vol. 3, pp. 213–224.

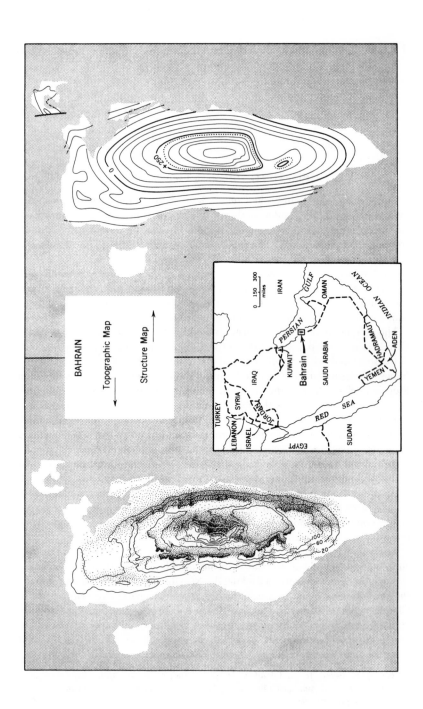

BAHRAIN

Topographic Map

↓

Structure Map

↑

Fig. 11-12. Bahrain Field.

The Bahrain° field occupies the central part of Bahrain Island, which lies at the mouth of the bay between Qatar Peninsula and the mainland of Saudi Arabia. This field, the first oil discovery in the southern Persian Gulf, commenced production in 1932. The Bahrain anticline is one of the most obvious structures in the Middle East. It is elliptical in plan, and an outcropping resistant limestone of Eocene age has been removed by erosion over the crestal part of the anticline with the result that the center and oil producing part of the island consists of a topographic basin completely rimmed by an outward-dipping limestone hogback. Most of the island outside of the central depression consists of the dip slope of this Eocene limestone. The Bahrain anticline is about 12 miles long and 4 miles wide. The two maps on the opposite page show the close relationship between the surface topography (left) and the geological structure (right).

Exploration for oil on this anticline was encouraged by the presence of an asphalt seep in the topographic basin.

The greater part of the Bahrain oil comes from a porous limestone ("Bahrain zone") which lies in the Wasia formation of Cretaceous age. This section is a thick sandstone on the Arabian mainland but has very little sand in it on Bahrain Island. Deeper drilling so far has not resulted in the discovery of important oil reserves below the Wasia formation, but a thick limestone in the Upper Jurassic Arab formation, which corresponds to the Arab C member in Saudi Arabia, carries a dry gas under a pressure of 2238 pounds per square inch absolute at elevation −4245. The Bahrain reservoir did not have a gas cap; the oil is undersaturated. The initial reservoir pressure was due to water drive. It was found that the pressure declined as oil withdrawal took place because the water drive did not keep up with oil removal. Therefore it was decided to maintain reservoir pressure by injecting gas from the deeper Jurassic reservoir into the Wasia oil reservoir without further compression. This injection has created an artificial gas cap and has effectively offset the drop in reservoir pressure.

The Bahrain field produced at the average rate of about 32,000 barrels a day during 1957. It is a small field by Middle East standards, but nonetheless it is a major oil field with cumulative production already in excess of 200 million barrels.

An interesting fact about the Bahrain oil accumulation is that the water beneath the oil contains only 8000 parts per million of dissolved solids, only about one-fifth the salinity of Persian Gulf water.

° C. T. Barber, "Review of Middle East Oil," *Petroleum Times* (June, 1948), pp. 69–72.

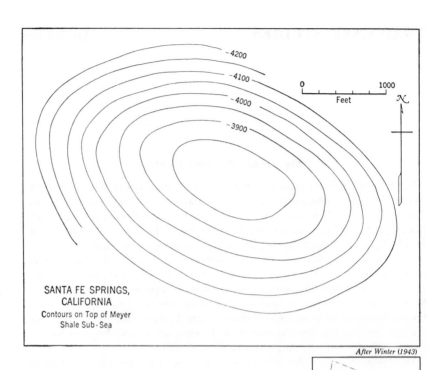

SANTA FE SPRINGS,
CALIFORNIA
Contours on Top of Meyer
Shale Sub·Sea

After Winter (1943)

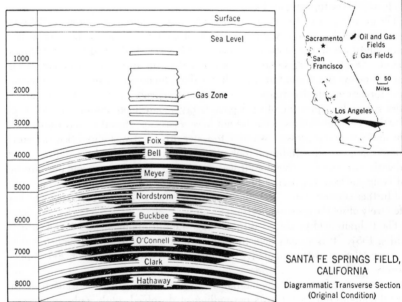

SANTA FE SPRINGS FIELD,
CALIFORNIA
Diagrammatic Transverse Section
(Original Condition)

310

Fig. 11-13.   Santa Fe Springs Oil Field, California.

The Santa Fe Springs oil field° lies in Los Angeles County, about 12 miles southeast of the center of the city of Los Angeles.   Although the productive area never exceeded 1500 acres, the total output of the field has been nearly a half billion barrels, making Santa Fe Springs one of the great oil fields of the world.   Because of lack of outcrops the anticline could not be mapped by surface surveys, but its presence was suspected from the topography and the appearance of gas in water wells.   The discovery well was drilled in 1919.

The subsurface structure map on the opposite page, contoured on one of the oil zones, shows the anticline to be an unusually symmetrical elliptical dome.   The dip is quite uniform, and no faults have ever been discovered.   The field was originally 2½ miles long and 1 mile wide at the widest point.

The reason for the prolific production at Santa Fe Springs is the presence of eight distinct superimposed oil zones, with from 100 to 550 feet of initially productive sand in each zone.   In addition there is a shallow gas zone which lies above the oil sands in the upper part of the Pliocene section.   Below are six Pliocene-producing zones, of which the third one from the top, the Meyer, has yielded nearly half of the total production from the Santa Fe Springs field.   It lies beneath a thick shale at a depth of 4150 feet.   The zone consists of 650 feet of sediment, of which two-thirds was initially oil-filled sand.   The original area over which this zone was productive was 1450 acres, three times larger than any other Pliocene-productive zone.

The two lowest productive zones, the Clark and the Hathaway, are Miocene in age.   They produce from depths between 7250 and 8000 feet.   The Miocene sands are harder and less porous than the Pliocene sands.   The Clark-Hathaway sands underlie 900 acres and contained initially 375 feet (out of twice that thickness) of oil-saturated sand, but they have only produced about 10 per cent of the field's total.

Many of the wells were completed as multiple-zone producers.   Three of the Pliocene zones have had per acre recoveries in excess of 150,000 barrels.   The initial productions per well ranged from 1000 to 10,000 barrels per day.

Trapping is entirely anticlinal at Santa Fe Springs, with water underlying the oil around the periphery of each zonal accumulation.   *Courtesy California Division of Mines.*

° H. E. Winter, "Santa Fe Springs Oil Field," *Calif. Div. Mines, Bull.* 118 (March, 1943), pp. 343–346.

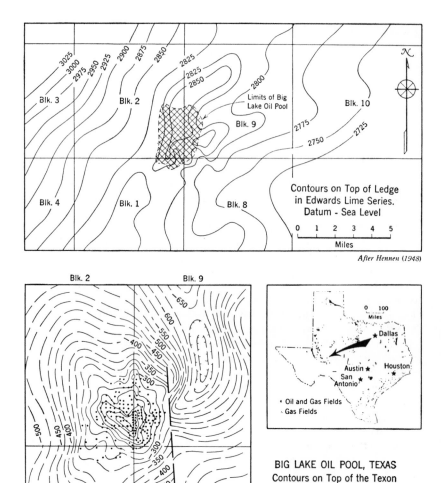

Blk. 3   Blk. 2   Blk. 9   Blk. 10

Limits of Big Lake Oil Pool

Blk. 4   Blk. 1   Blk. 8

Contours on Top of Ledge
in Edwards Lime Series.
Datum - Sea Level

0   1   2   3   4   5

Miles

After Hennen (1948)

Blk. 2   Blk. 9

Blk. 1   Blk. 8

0   100
Miles

Dallas

Austin   Houston
San
Antonio

Oil and Gas Fields
Gas Fields

BIG LAKE OIL POOL, TEXAS
Contours on Top of the Texon
(Discovery) Oil Zone
(Datum · Sea · Level; Contour
Interval · 25 Feet)

0   1   2   3   4

Miles

312

Fig. 11-14. Big Lake Pool, Texas.

The Big Lake oil pool° is in the southwestern corner of Reagan County, in west Texas. It was the first major oil field to be discovered in what is known now as the west Texas or Permian basin petroliferous province. The discovery well was completed in 1923. The exploration was based on surface geologic studies; the Edwards limestone and other Lower Cretaceous formations crop out in this area.

The upper figure shows the structure of the Big Lake area as contoured on top of a ledge within the Edwards limestone formation. The regional dip is to the northwest into the Permian basin, and the Big Lake fold is one of several anticlines which lie on the southeast flank of this basin. As can be seen from the figure, closure in the surface formations is only 30 feet and the anticline is decidedly elongate in a northeast-southwest direction. The lower figure shows the Big Lake dome as contoured on the producing Texon (Permian) zone. At this level the Big Lake structure is a quaquaversal dome with 250 feet of closure. This is more than eight times the closure in the surface rocks and double that present in the intervening salt measures. The fault shown on the east flank of the dome does not appear in the surface rocks and was probably formed before Cretaceous deposition. It has had no effect on trapping.

Four permeable zones—two in Permian rocks, one in the Pennsylvanian, and one in the Ordovician—carry oil. The two Permian producers are the Shallow and the Texon. The Shallow zone is a silty sand from 30 to 50 feet in thickness lying at a depth of about 2450 feet. Much more important in size and yield is the Texon zone, which lies between 2800 and 3100 feet below the surface. The Texon is a highly porous oölitic dolomite, averaging 22 feet in thickness. The third oil zone is 6300 feet deep. It is a sandstone, 15 feet thick, of probable Pennsylvanian age. The Ordovician reservoir is the deep, but prolific, Ellenberger dolomite. Oil accumulation has been brought about by oil rising through the water-filled reservoirs to the top of the anticline. The oil may have traveled a considerable distance up the flank of the Permian basin from the northwest. *Courtesy American Association of Petroleum Geologists.*

° Ray V. Hennen, "Big Lake Oil Pool, Reagan County, Texas," *Structure of Typical American Oil Fields* (American Association of Petroleum Geologists, 1929), Vol. 2, pp. 500–541.

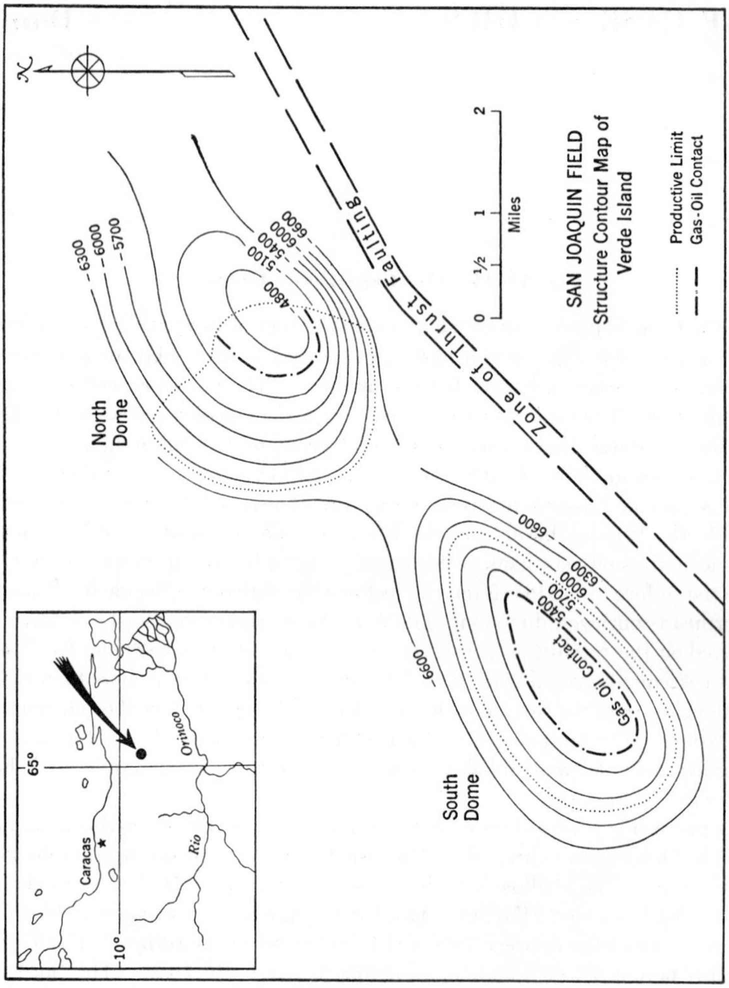
SAN JOAQUIN FIELD
Structure Contour Map of
Verde Island
Productive Limit
Gas-Oil Contact
Miles
0
1/2
1
2
North
Dome
South
Dome
Zone of Thrust Faulting
Gas-Oil Contact
-6300
-6000
-5700
-6600
-6000
-5400
-5100
-4800
-6600
-6300
-6000
-5700
-5400
-6600
Caracas
Orinoco
Rio
65°
10°

After Funkhouser, Sass, and Hedburg (1948)

Fig. 11–15. San Joaquin Field, Venezuela.

The San Joaquin field is one of a group of fields lying in the central part of the state of Anzoategui, in eastern Venezuela.° The San Joaquin and the neighboring Santa Ana domes were first outlined as the result of air-photograph study and surface geological reconnaissance in 1934. This study was followed by a seismograph check and the drilling of the discovery well for the Santa Ana field in 1936. The discovery well for the San Joaquin field was completed in 1939.

The map on the opposite page shows by means of contours the structure of the two domes that constitute the San Joaquin field. The datum for the contours is the Verde I sand, one of many oil-bearing sandstones which lie within the Oficina formation of Oligocene-Miocene age. The domes are believed to be the result of drag over a northwest-dipping zone of thrust faulting; the fault planes crop out to the southeast as shown on the map.

The Oficina formation is from 7500 to 10,000 feet thick. It contains twenty-eight sands which have produced oil, and at least fifteen more are known to contain gas. Within the productive section of the Oficina, sandstone beds constitute but 8 per cent. Only the uppermost 1900 feet of the underlying Merecure formation (upper Eocene-lower Oligocene) have been explored as yet. This section is about 50 per cent sandstone and 50 per cent shale. Oil and gas have been found within the sandstones wherever tested. The sandstones are probably interconnecting and so constitute a single reservoir.

The south dome has a closure of about 1200 feet. In the Verde I sand the oil-water contact lies about 600 feet above the saddle between the north and the south domes. The top of the dome contains a gas cap. The oil-water contact is several hundred feet lower on the north dome; a gas cap is present there also. As the Verde I sand lenses out to the east and north, only the west end of the north dome is productive at this stratigraphic level.

The San Joaquin field illustrates almost perfectly the trapping of oil at the top of a dome-shaped structure where a pervious reservoir rock is present. Some gas has separated out from the oil and has accumulated below the very top of the trap. *Courtesy American Association of Petroleum Geologists.*

° H. J. Funkhouser, L. C. Sass, and H. D. Hedberg, "Santa Ana, San Joaquin, Guario, and Santa Rosa Oil Fields (Anaco Fields), Central Anzoátegui, Venezuela," *Bull. Am. Assoc. Petrol. Geol.*, Vol. 32 (October, 1948), pp. 1851–1908.

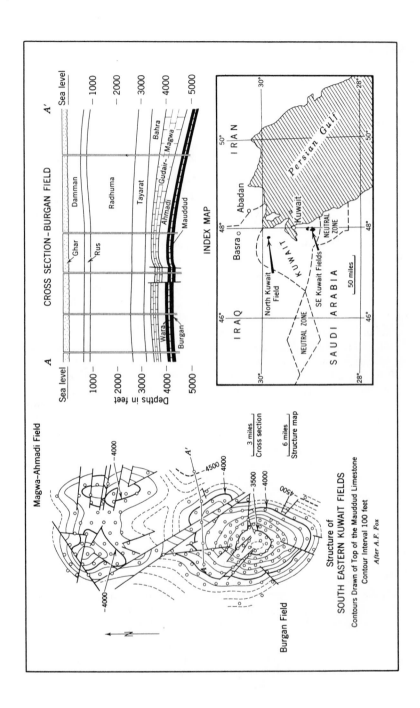

CROSS SECTION-BURGAN FIELD

INDEX MAP

Structure of
SOUTH EASTERN KUWAIT FIELDS
Contours Drawn of Top of the Mauddud Limestone
Contour Interval 100 feet
After A.F. Fox

316

Fig. 11-16. Burgan Field, Kuwait.

The Burgan° field, along with the adjoining Magwa and Ahmadi fields, has for some years been the largest source of petroleum in the world, with a daily output of over one million barrels. The discovery of oil and gas seeps by a geological surface party led to the seeking of a concession which was granted in 1932. The discovery well was drilled in 1938, but because of the war, major development and production did not get under way until 1945.

As can be seen on the opposite page, the Burgan field occupies the southern half of an anticlinal complex. The Burgan production comes from a faulted elliptical dome with the major axis trending almost north-south. So far the field has proved dimensions of about 15 by 10 miles, and the structural closure is at least 1200 feet. The average dip is about 3°. The cause of the upward bulge may be basement block faulting and possibly some flowage of deep Cambrian salt.

The reservoirs are mid-Cretaceous sandstones with an aggregate thickness of over 1300 feet. The upper one is Cenomanian age and the lower group are probably Albian. The top of the intervening formation, the Mauddu limestone, is the datum for the structure contour map. The seal above the reservoir sand series is a gray shale and marly limestone formation.

The cross section on the opposite page shows: (1) the graben character of some of the faults; (2) the confinement of the faults to the Middle Cretaceous; and (3) the thickening of the highest Middle Cretaceous formation (Magwa) and the lowest Upper Cretaceous formation (Gudair) off the Burgan anticline. The maximum known displacement in the graben cut by the cross section is 300 feet. Wells drilled into the graben penetrated abnormal thicknesses of the Magwa, Gudair, and Bahra formations. The thickening of the Magwa off the anticline is due to lesser post-Magwa pre-Gudair erosion away from the crest of the structure. The thinning of the Gudair toward the anticline is a result of overlap; the dome was an island in the early Gudair sea.

The minimum depth of the first reservoir sand is about 3550 feet. A deep test drilled on the Burgan anticline went through a Jurassic evaporite series, including thick salt beds, between 8956 and 10,072 feet. This is the probable time equivalent of the Arab limestone, the prolific oil reservoir rock in Saudi Arabia.

°A. F. Fox, "Oil Occurrences in Kuwait," *Symposium on Deposits of Petroleum and Gas,* XX International Geol. Congress, Vol. 2 (1956), pp. 131–158; R. M. S. Owen and Sami N. Nasr, "Stratigraphy of the Kuwait-Basra Area," *Habitat of Oil* (American Association of Petroleum Geologists Special Vol., 1958), pp. 1252–1278; Staff Kuwait Oil Co., "Kuwait," *Science of Petroleum,* Vol. 6, Part 1 (1953), pp. 99–100.

TRAPS: SALT-CORED

A salt-cored [1] structure is one in which pressures in the earth's crust have caused normally bedded salt deposits to flow laterally and upward plastically, first bulging and then in many places rupturing the overlying sediments.

Among the many types of oil and gas traps, those that result from the flowage of salt are the most bizarre. In spite of their odd nature and origin, however, a substantial percentage of the world's petroleum comes from salt-cored structures. Outstanding in this regard are those of the Gulf Coast of Louisiana and Texas (both on land and offshore), and salt-cored traps in Romania. Mexico, Germany, and Russia are other, but relatively minor, sources of oil that has accumulated in salt structures. Some of the Middle East anticlines probably have salt cores. In addition to oil and gas, the greater part of the annual world yield of sulfur, and some of the salt and potash production, come from salt deposits of this type.

[1] A voluminous bibliography has resulted from the widespread interest in structures cored (or probably cored) with salt. The DeGolyer and Vance *Petroleum Bibliography* (College Station, Texas, 1944) lists 94 titles on this subject. The American Association of Petroleum Geologists has published a symposium, *Gulf Coast Oil Fields* (Tulsa, Oklahoma, 1936) containing 36 papers dealing with salt-cored and probable salt-cored structures; several of these papers include extensive bibliographies. There is also the older symposium, *Geology of Salt Dome Oil Fields* (Tulsa, Oklahoma, 1926). Another symposium appeared in the *Journal of the Institute of Petroleum Technology*, Vol. 17 (1931), pp. 252–371. Additional bibliographies are appended to the following articles on this subject: M. A. Hanna, "Geology of Gulf Coast Salt Domes," *Problems of Petroleum Geology* (American Association of Petroleum Geologists, 1934), pp. 629–678; G. D. Hobson, "Salt Structures; Their Form, Origin, and Relationship to Oil Accumulation," *Science of Petroleum* (Oxford University Press, 1938), Vol. 1, pp. 255–260; P. I. Bediz, "Salt Core Structures and Their Importance in Petroleum Geology," *Mines Mag.*, Vol. 32 (May, 1942; June, 1942), pp. 215 et seq.

Classification. Salt-cored structures may be classified into two main types: non-piercement and piercement (Fig. 12–1). The first includes the *salt anticline*, a laccolith-like structure in which the bedded salt has thickened locally by flowage but has not ruptured the overlying strata. Salt anticlines occur in the Utah-Colorado salt basin and in many other parts of the world. To date no oil deposits have been found associated with these structures. The term non-piercement has also been applied, somewhat questionably, to deep-seated probable salt masses not yet reached by the drill. Under the piercement classification are all the salt intrusions which transect the stratification of the intruded sediments. A few of these injected salt masses followed fissures and so are dike-like in shape. Much more abundant are the vertical, or near-vertical, pipe-like intrusions, which are similar to igneous stocks, plugs, and necks. In some of these, the flowing salt has reached the surface through these conduits.

Examples of salt extrusion have been found in the Colorado-Utah district, on the Isthmus of Tehuantepec, on islands in the Canadian Arctic, in the Maritime Provinces, in northern Spain and southwestern France, in Morocco, and in central and southern Iran.[2] Because of the extreme aridity in Iran, the salt that has broken through the overlying sediments makes a dome-shaped hill, whence it flows down the flanks in tongues or "glaciers" for distances as great as three miles.[3] In some areas leaching of the salt has been followed by surface collapse producing topographic depressions.

Salt-cored structures of the piercement type are ordinarily referred to as salt domes in the United States and diapiric folds in the Eastern Hemisphere. Wide variation exists between these structures in size, shape, and degree of tectonic disturbance of the associated sediments. In the north German salt basin, all stages are present between undisturbed salt layers, salt anticlines, and salt stocks extending upward into the overlying ruptured sediments. A moderate degree of diastrophism has taken place in this basin, as shown by the folding of the rock strata. The diastrophism in Romania has been much more intense, and the salt has been "squirted" upward through the apices of the anticlines into bodies of irregular shape. Both the German and Romanian salt intrusions are small compared with those of the Gulf Coast.

Non-Gulf Coast Salt Structures. Eby[4] has published maps showing not only the salt basins but also the salt-dome areas throughout the

---

[2] C. A. E. O'Brien, "Salt Diapirism in South Persia," *Geologie en Munbouw* (September, 1957), pp. 357–376.

[3] P. I. Bediz, *op. cit.*, p. 288.

[4] J. Brian Eby, "Salt Dome Interest Centers on Gulf Coast," *World Oil* (October, 1956), pp. 143 et seq.

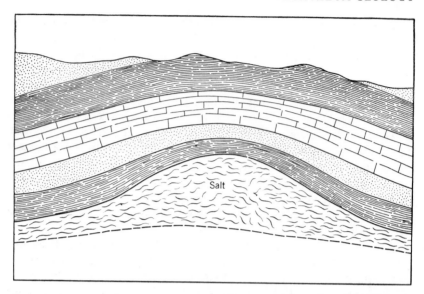

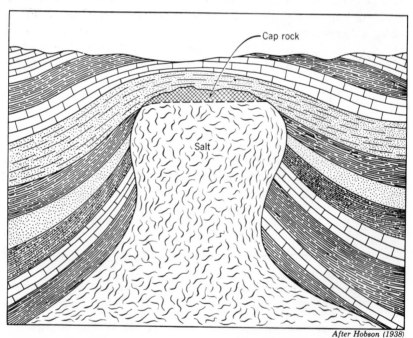

After Hobson (1938)

Fig. 12-1. Non-piercement and piercement salt domes. *Upper:* anticline produced by salt flowage. *Lower:* ruptured anticline or piercement salt dome. *From Science of Petroleum; permission to publish purchased from Oxford University Press.*

world. The prolific Gulf Coast salt domes of Texas and Louisiana are not the only North American occurrences of salt-cored structures. West and north of the "Four Corners," the common corner of New Mexico, Arizona, Utah, and Colorado, but mainly in Utah and Colorado, is a second province of this type.[5] It covers about 6000 square miles (Fig. 12–2). The gypsiferous rock which caps the salt plugs is exposed at several places, and drilling has shown that the salt is not far beneath. So far, no important hydrocarbon deposits have been found associated with these salt-cored structures. A third North American occurrence of salt structures is on the Isthmus of Tehuantepec in the state of Vera Cruz, Mexico (Fig. 12–3). Hanna[6] names eighteen domes which were known definitely at that time (1934) to have salt cores. This district has become important as a commercial source of salt-dome sulfur.

The German salt structure province is in the North German Zechstein Basin. All stages between undisturbed flat-lying sedimentary layers of salt and intrusive salt "stocks" are to be found there. The salt-cored structures are in a triangular area, with the three apices lying in the North Sea, the Baltic Sea, and the Harz Mountains. The Hanover, Bremen, and Hamburg districts lie within this triangular area.

Oil was produced by means of dug wells from salt-cored traps in Romania before the drilling of the Drake well in Pennsylvania. Romania has two salt-structure districts: (1) the outer edge of the Carpathians; and (2) the intermontane Transylvanian basin. Russia also has oil-producing salt-cored structures, in the Ural-Emba[7] district north of the Caspian. Many of these, like those in the Gulf Coast district of the United States, were discovered by geophysical methods.

The Middle East has three districts with salt-cored anticlines. One of these is near Qum, on the central Iranian plateau, where oil was discovered in 1956. The largest salt district underlies the Persian Gulf and the surrounding mainland. Some of the islands and shoal areas in the Gulf are thought to be topographic criteria of underlying salt plugs. The oil-producing Dammam dome (Fig. 12–14) in Saudi Arabia and the southeastern Kuwait anticline complex are possible salt-cored structures. The third Middle East salt-diapir district is in southwestern Arabia, beneath

---

[5] Thomas S. Harrison, "Colorado-Utah Salt Domes," *Bull. Am. Assoc. Petrol. Geol.*, Vol. 11 (February, 1927), pp. 111–133; H. W. C. Prommel and H. E. Crum, "Salt Domes of Permian and Pennsylvanian Age in Southeastern Utah and Their Influence on Oil Accumulation," *Bull. Am. Assoc. Petrol. Geol.*, Vol. 11 (April, 1927), pp. 373–393; John C. Benson and N. Wood Bass, "Eagle River Anticline, Eagle County, Colorado," *Bull. Am. Assoc. Petrol. Geol.*, Vol. 39 (January, 1955), pp. 103–106.

[6] Marcus A. Hanna, "Geology of Gulf Coast Salt Domes," *Problems of Petroleum Geology* (American Association of Petroleum Geologists, 1934), p. 633.

[7] C. W. Sanders, "Emba Salt Dome District, U.S.S.R., and Some Comparisons with Other Salt Dome Regions," *Bull. Am. Assoc. Petrol. Geol.*, Vol. 23 (April, 1939), pp. 492–516.

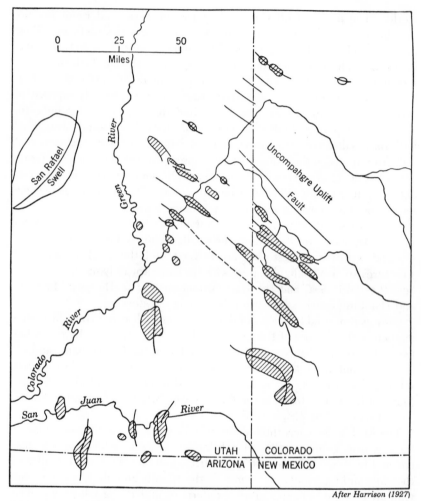

After Harrison (1927)

Fig. 12-2. Salt-cored structures in Utah and Colorado. *Courtesy American Association of Petroleum Geologists.*

and adjacent to the Red Sea. Salt-cored structures have also been reported in north Africa.

Distribution and Description of Gulf Coast Salt Domes. The salt-cored structure province of the Gulf Coast of North America extends along the coast line, both onshore and offshore, from Alabama to Mexico. The greatest concentration of salt intrusions occurs in a zone extending from about 200 miles east of the Louisiana-Texas line to a point about

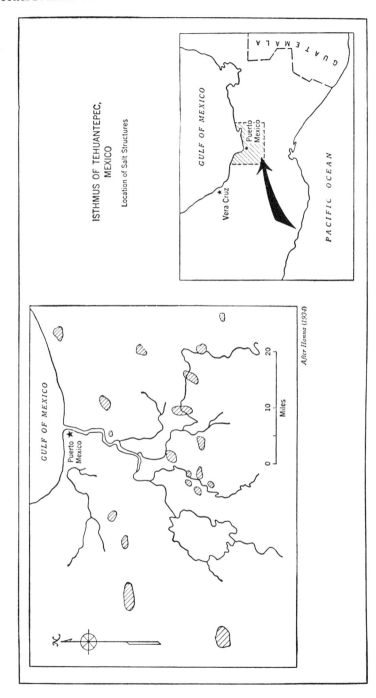

Fig. 12-3. Salt-cored structures on Isthmus of Tehuantepec. *Courtesy American Association of Petroleum Geologists.*

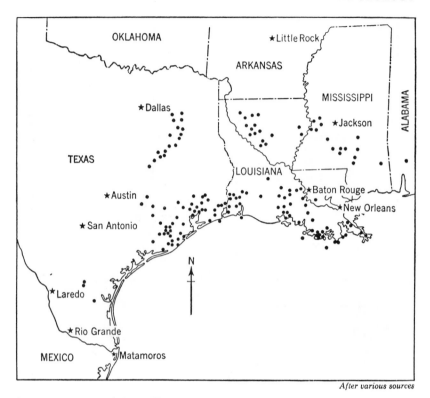

Fig. 12–4. Map of the Gulf Coast from Texas to Alabama, showing location of known salt domes.

200 miles west of this boundary (Fig. 12–4 and 12–5). There are also three inland groups of salt-cored structures. One group lies in the Tyler Basin of northeast Texas, one on the east flank of the Sabine uplift in northern Louisiana, and the third in eastern Louisiana, south central Mississippi, and western Alabama.

The Gulf Coast salt-cored structures are unique in their enormous per acre production of oil, in their size, and in the total absence, to the depths so far explored, of any appreciable diastrophic activity in the sediments not disturbed by the salt intrusion itself. These great bodies of salt are vertical cylinders, circular or elliptical in plan, with diameters varying from $\frac{1}{2}$ mile to 4 miles and with maximum vertical dimension perhaps as great as 30,000 feet.[8] The more shallow domes are invariably covered

---

[8] Marcus A. Hanna, "Geology of the Gulf Coast Salt Domes," *Problems of Petroleum Geology* (American Association of Petroleum Geologists, 1934), p. 646.

by cap rock, to be described and discussed subsequently. At least eleven of the Gulf Coast salt plugs are known to overhang at the top to a varying degree.[9] As much as 12,500 feet of salt has been penetrated between top and bottom of such overhangs (Fig. 12–6). A few plugs have bizarre shapes (Fig. 12–7).

The Gulf Coast salt cores average from 5 to 10 per cent disseminated anhydrite, with the rest halite.[10] The Tehuantepec cores are similar in degree of purity, but the Colorado-Utah salt is much more impure, and in Iran the injected material is a mixture of salt, gypsum, and red shale.[11]

The cap rock is a disc-like body at the top of the salt cylinder which may extend some distance down the flanks. Although cap rocks are best developed in the Gulf Coast, they are also found rather generally elsewhere, except in Romania. They are exposed at the surface in some places, and often have been found by drilling. However, in the salt-dome district of the Gulf Coast there are a number of probable salt-cored structures where drilling to depths in excess of 15,000 feet has failed to reach either the salt core or its overlying cap. The caps so far explored have varied in thickness from a few feet to over 1100 feet, with an average of about 350 feet. Although twenty-eight minerals[12] and varieties have been described in the Gulf Coast cap rocks, the only ones of any abundance are anhydrite, gypsum, and calcite, with sulfur present in important volume in a few places.

Anhydrite is invariably present immediately overlying the salt. It is massive and may be banded. Sand grains and other impurities may be present. The gypsum zone of the cap overlies the anhydrite and is gradational with it. Undoubtedly the gypsum is the result of hydration of anhydrite which originally occupied the gypsum zone level. The gypsum is usually coarsely crystalline selenite.[13] Above the gypsum zone, or immediately above the anhydrite where the gypsum zone is absent, is the so-called "limestone" cap rock. This rock is an aggregate of calcite crystals, but it is not a sedimentary stratum of limestone. It may be porous and even cavernous. Sulfur, where it is present, usually overlies the anhydrite in the calcite or gypsum zones.

[9] Sidney A. Judson and R. A. Stamey, "Overhanging Salt on Domes of Texas and Louisiana," *Gulf Coast Oil Fields* (American Association of Petroleum Geologists, 1936), pp. 141–169.

[10] Marcus A. Hanna, *op. cit.*, p. 637.

[11] P. I. Bediz, "Salt Core Structures and Their Importance in Petroleum Geology," *Mines Mag.*, Vol. 32 (May, 1942; June, 1942), p. 288.

[12] Marcus A. Hanna and Albert G. Wolf, "Texas and Louisiana Salt-Dome Cap Rock Minerals," *Gulf Coast Oil Fields* (American Association of Petroleum Geologists, 1936), pp. 119–132; F. W. Rolshausen, "Occurrence of Siderite in Cap Rock at Carlos Dome, Grimes County, Texas," *ibid.*, pp. 133–135.

[13] Marcus A. Hanna, *op. cit.*, p. 642.

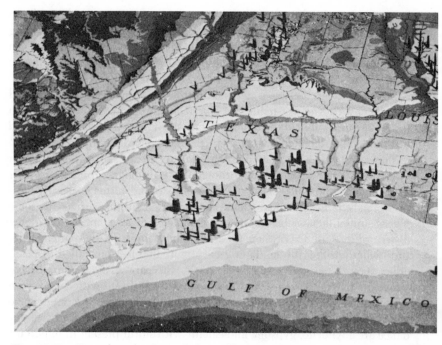

Fig. 12-5. How the salt plugs of the Gulf Coast of the United States would look if the top 20,000 feet of soil and rock were stripped off without disturbing the cylinders of salt. *Courtesy Humble Way, Humble Oil and Refining Company.*

Like all the piercement salt cores, those of the Gulf Coast have emplaced themselves by rupturing the overlying sediments. The overlying sediments, however, were flat-lying previous to the intrusion. As a result of the forcible injection of the salt mass, the initially horizontal strata have been dragged upward from a few hundred to several thousand feet around the flanks of the intrusion before being truncated. Dips have been increased from an original zero to as much as 60° to 90° and even more than 90° in the case of some of the overhangs. These strata also tend to thin toward the core, owing to attenuation by drag, to lesser original sedimentation because of concurrent uplift, or to erosion and overlap. Another result of rupture of the pre-existing sedimentary rocks is their brecciation. At depths great enough for the shales to be dry and brittle, the forcible emplacement of the salt plug produces a shale breccia.[14]

[14] Paul F. Kerr and Otto C. Kopp, "Salt-dome Breccia," *Bull. Am. Assoc. Petrol. Geol.*, Vol. 42 (March, 1958), pp. 548–560.

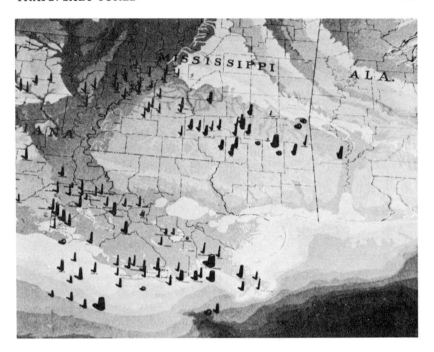

The strata overlying the salt and its cap have been arched upward, producing, as a general rule a quaquaversal dome. In many places, however, the overlying sediments are faulted as well as arched. Some Gulf Coast salt domes are bordered by peripheral faults;[15] in others the faults cross the dome.

A type of structure that is becoming increasingly common, owing to new discoveries, is the dome-with-graben, which overlies deep-seated salt cores. In this type, the arched sediments are crossed by a series of normal faults that create a keystone graben in the center of the domed area. These grabens are bordered by major faults which dip inward at angles ranging from 45° to 65°. The faults are normal, and the displacement varies from 100 feet to more than 900 feet. Subsidiary minor faults may parallel the major faults. Minor radial faults may also be present. Wallace[16] believes that the major faults bounding the grabens converge downward, meeting at the top of the salt core, and he suggests that the

[15] William R. Mais, "Peripheral Faulting at Bayou Blue Salt Dome, Iberville Parish, Louisiana," *Bull. Am. Assoc. Petrol. Geol.*, Vol. 41 (September, 1957), pp. 1915–1951.

[16] W. E. Wallace, Jr., "Structure of South Louisiana Deep-Seated Domes," *Bull. Am. Assoc. Petrol. Geol.*, Vol. 28 (September, 1944), pp. 1249–1312.

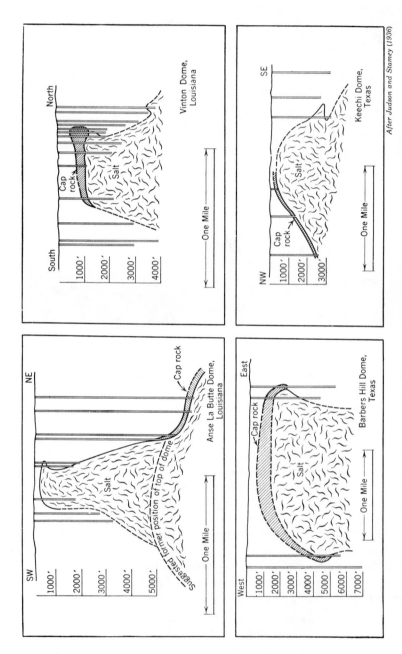

Fig. 12-6. Four examples of overhanging salt domes, Louisiana and Texas. *Courtesy American Association of Petroleum Geologists.*

After Judson and Stamey (1936)

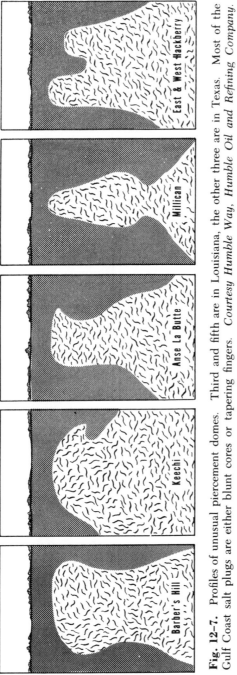

Fig. 12-7. Profiles of unusual piercement domes. Third and fifth are in Louisiana, the other three are in Texas. Most of the Gulf Coast salt plugs are either blunt cores or tapering fingers. *Courtesy Humble Way, Humble Oil and Refining Company.*

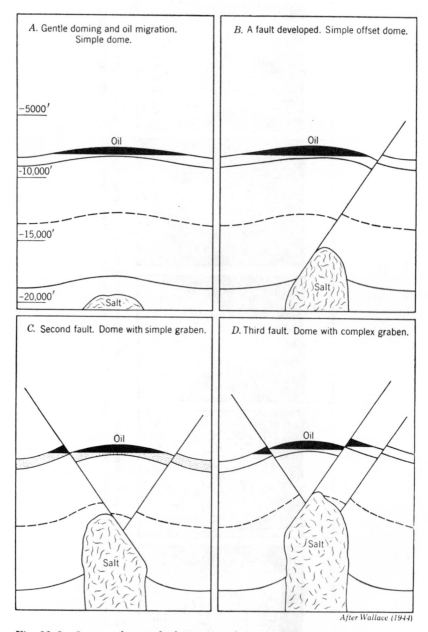

A. Gentle doming and oil migration. Simple dome.

B. A fault developed. Simple offset dome.

−5000′

Oil

−10,000′

−15,000′

−20,000′   Salt

C. Second fault. Dome with simple graben.

Oil

Salt

D. Third fault. Dome with complex graben.

Oil

Salt

After Wallace (1944)

Fig. 12–8. Suggested origin for keystone grabens in salt domes. *Courtesy American Association of Petroleum Geologists.*

grabens are actually undisturbed areas surrounded by blocks lifted up and tilted by movement along the diagonal faults caused by the intrusion of the salt stock (Fig. 12–8). Concurrent deposition in the grabens produces abnormal formation thickness in those areas.[17]

A possible explanation for the characteristic strike faults of the Gulf Coast area, in which the Gulf side has dropped, is that the faulting has accompanied the flow of salt into decidedly elongate anticlines ("salt ridges").[18]

Rim synclines may occur about the periphery of Gulf Coast salt domes. These are closed depressions with depths up to several hundred feet, and like the keystone grabens, they extend down to depths not yet explored by the drill. Possibly rim synclines are the result of the flowage of the adjacent bedded salt into the salt core, with subsequent sagging of the salt-roof rock.

Some of the shallow zones have sunken tops because of leaching of salt followed by collapse of the cap and overlying sediments.

Origin of The Gulf Coast Salt Domes. There can be little doubt that the salt in salt-cored structures everywhere was originally stratified sedimentary salt. The age of the mother salt beds ranges from the Cambrian (Iran) to the Tertiary (Romania). No well in the Gulf Coast area, exploring a salt dome, or drilling adjacent to a salt dome, has reached the source strata as yet. One borehole, testing the Wiggins anticline in southern Mississippi, recovered two feet of core consisting of one foot of anhydrite underlain by one foot of salt at approximately 20,445 to 20,447 feet. The total depth of this hole was 20,450 feet. but no samples were obtained in the basal three feet.[19] This evaporite section is pre-Smackover (Jurassic) in age, as is the thick salt bed penetrated in exploring beneath the Smackover, Arkansas, oil field. This salt has been assigned to the Jurassic by some and to the Permian by others.[20] The salt beneath the Texas and Louisiana Gulf shore area may or may not be of the same age as that in the interior. The source salt in the coastal belt is at least as old as Upper Cretaceous.[21]

[17]John B. Currie, "Role of Concurrent Deposition and Deformation of Sediments in Development of Salt-Dome Graben Structures," *Bull. Am. Assoc. Petrol. Geol.*, Vol. 40 (January, 1956), pp. 1–16.

[18]Miller Quarles, Jr., "Salt-Ridge Hypothesis on Origin of Texas Gulf Coast Type of Faulting," *Bull. Am. Assoc. Petrol. Geol.*, Vol. 37 (March, 1953), pp. 489–508.

[19]Paul L. Applin and Esther R. Applin, "The Cored Section in George Vasen's Fee Well 1 Stone County, Mississippi," *U. S. Geol. Survey Circular* 298 (1953), pp. 1–29.

[20]Roy T. Hazzard, W. C. Spooner, B. W. Blanpied, "Notes on the Stratigraphy of the Formations Which Underlie the Smackover Limestone in South Arkansas, Northeast Texas and North Louisiana," *Shreveport Geol. Soc., Rept.* II (1945), pp. 483–503.

[21]J. Ben Carsey, "Geology of Gulf Coastal Area and Continental Shelf," *Bull. Am. Assoc. Petrol. Geol.*, Vol. 34 (March, 1950), p. 367.

The origin of salt-cored structures has long been a subject of interest. Chief American investigators in recent years include Barton,[22] Hanna,[23] Nettleton,[24] Balk,[25] and Halbouty and Hardin.[26] Older theories, including volcanic intrusion and salt crystallization,[27] are untenable in the light of present knowledge of salt-cored structures. However, O'Brien[28] has recently revived volcanism as a possible *initiating* force in salt diapirism in south Persia. In that area are over 150 salt plugs, in which Cambrian salt has forced its way through more than 20,000 feet of sediments. Some of these plugs are mountains over 4000 feet high. The salt intrusions brought with them "exotic blocks" of both sedimentary and igneous rocks. The latter are unknown at the surface, and O'Brien postulates that they come from igneous intrusions which penetrated salt beds and did the initial breaching of the overlying sediments.

Only the plastic flow of the salt itself is adequate to explain all the following observational features: (1) the injection of bedded salt into the ruptured roofs of anticlines in Germany, Romania, Iraq, and Iran; (2) the uplift and truncation of sedimentary strata overlying the salt in the Gulf Coast district; (3) the presence of cap rocks; and (4) contorted banding within the salt masses. That salt can flow plastically has been demonstrated by numerous investigators in the laboratory and by witnessed natural flow of the salt "glaciers" in southern Iran.

Barton, Nettleton, and Balk have studied the mechanics of salt flowage, especially as applied to the genesis of the Gulf Coast domes. Parker and McDowell[29] have reproduced these salt domes with models in the labora-

---

[22] Donald C. Barton, "Mechanics of Formation of Salt Domes with Special Reference to Gulf Coast Salt Domes of Texas and Louisiana," *Gulf Coast Oil Fields* (American Association of Petroleum Geologists, 1936), pp. 20–78.

[23] Marcus A. Hanna, "Geology of Gulf Coast Salt Domes," *Problems of Petroleum Geology* (American Association of Petroleum Geologists, 1934), pp. 629–678.

[24] L. L. Nettleton, "Fluid Mechanics of Salt Domes," *Gulf Coast Oil Fields* (American Association of Petroleum Geologists, 1936), pp. 79–108; "Recent Experimental and Geophysical Evidence of Mechanics of Salt Dome Formation," *Bull. Am. Assoc. Petrol. Geol.*, Vol. 27 (January, 1943), pp. 51–63; "History of Concepts of Gulf Coast Salt-Dome Formation," *Bull. Am. Assoc. Petrol. Geol.*, Vol. 39 (December, 1955), pp. 2373–2383.

[25] Robert Balk, "Structure of Grand Saline Salt Dome, Van Zandt County, Texas," *Bull. Am. Assoc. Petrol. Geol.*, Vol. 33 (November, 1949), pp. 1791–1829; "Salt Structure of Jefferson Island Salt Dome, Iberia and Vermilion Parishes, Louisiana," *Bull. Am. Assoc. Petrol. Geol.*, Vol. 37 (November, 1953), pp. 2455–2474.

[26] Michel T. Halbouty and George C. Hardin, Jr., "Genesis of Salt Domes of Gulf Coastal Plain," *Bull. Am. Assoc. Petrol. Geol.*, Vol. 40 (April, 1956), pp. 737–746.

[27] Revived by Bailey Willis, "Artesian Salt Formations," *Bull. Am. Assoc. Petrol. Geol.*, Vol. 32 (July, 1948), pp. 1227–1264.

[28] C. A. E. O'Brien, "Salt Diapirism in South Persia," *Geologie en Mijnbouw* (September, 1957), pp. 357–376.

[29] Travis J. Parker and A. N. McDowell, "Model Studies of Salt-Dome Tectonics," *Bull. Am. Assoc. Petrol. Geol.*, Vol. 39 (December, 1955), pp. 2384–2470.

tory, using asphalt in place of salt. Unlike the Gulf Coast, in most if not all the other regions of salt-cored structures, the local sedimentary section has been subjected to lateral pressures during one or more periods of diastrophism. In Romania, especially, these tangential forces have been considerable, resulting in tight folding of the salt-bearing sedimentary rock series. Under such circumstances, the pressures needed to cause the relatively vulnerable salt to flow are obvious, and the salt moves in the direction of least resistance, which is through the fractured anticlinal crests. However, in the Gulf Coast district there is no visible evidence of the existence in the past of any appreciable tangential pressure. It is quite possible, if not probable, that there is an angular unconformity at depth beneath which the salt measures are folded and in which the original piercement of the salt roof rock took place,[30] but all the later injection has been through strata which had been hitherto undisturbed. The authorities ascribe this movement to (1) the great weight of the thick body of sediment overlying the salt-bearing strata, and (2) the lesser density of the salt compared with that of the "country rock." The static weight of the overburden pressing upon the relatively incompetent salt caused it to flow plastically, moving upward through whatever openings it could find in the roof. Once the salt body became a vertical feature, its lighter weight, compared with the weight of the laterally adjacent rocks, would create an isostatic inequilibrium which might cause further vertical movement. Barton points out that both the coastal and the interior salt domes of the Gulf area are in regions of pronounced downsinking. He believes, therefore, that much of the injection of the salt cores into the sediments has been actually residual accumulation of salt stocks by "downbuilding."[31] As the mother salt beds and the associated sediments sank, the plastic, lighter salt drained into the base of the nonsinking (or even up-rising) salt prism.

The value of interstitial water as a lubricant in aiding salt flow is stressed by some investigators and seriously questioned by others.[32]

Most students of Gulf Coast salt domes agree that the upward (relative) movement of the salt cores has been intermittent rather than continuous.[33]

[30] E. DeGolyer, "Origin of North American Salt Domes," *Bull. Am. Assoc. Petrol. Geol.,* Vol. 9 (August, 1925), pp. 831–874.

[31] Donald C. Barton, "Mechanics of Formation of Salt Domes with Special Reference to Gulf Coast Salt Domes of Texas and Louisiana," *Gulf Coast Oil Fields* (American Association of Petroleum Geologists, 1936), pp. 54 et seq.

[32] L. L. Nettleton, "Fluid Mechanics of Salt Domes," *Bull. Am. Assoc. Petrol. Geol.,* Vol. 18 (September, 1934), p. 1180.

[33] Gordon Atwater and McLain J. Forman, "Nature of Growth of Louisiana Salt Domes and Its Effect on Petroleum Accumulation," *Am. Assoc. Petrol. Geol., Program Annual Meeting* (1958), p. 45; F. M. Van Tuyl and Ben H. Parker, "The Time of Origin and Accumulation of Petroleum," *Quarterly Colo. School of Mines,* Vol. 36 (April, 1941), pp. 130–133.

This movement ceased at various times in the geologic past. Occasionally, post-Pleistocene movement has taken place, and some of these salt-core structures may be still growing. Various reasons that have been suggested for the cessation of upward movement of the salt include: (1) establishment of isostatic equilibrium; (2) reaching a balance between static pressure and frictional resistance; (3) cessation of regional downwarping; (4) exhaustion of salt supply; (5) removal of peripheral salt accompanied by a drawing together of the underlying and overlying strata.

The origin of the cap rocks and of the cap-rock minerals has also led to a lot of observation, cerebration, and publication.[34] The dominant and sole essential mineral in cap rocks is anhydrite. Its presence and position have been explained in several ways, but chiefly by either (1) the residual accumulation, through differential leaching, of anhydrite originally disseminated through the salt mass, or (2) by the punching out, followed by a ride upward on the top of the salt stock, of a slab of sedimentary anhydrite country rock. The best arguments for the second theory appear to be the sharpness and horizontality of the anhydrite-salt contact exposed in digging a shaft to the salt in the Texas Gulf Coast, and the presence of horizontal bands of sandstone at one level in the 900-foot thick anhydrite cap.[35] The arguments opposed to this concept are based mainly on improbability. So far as the writer knows, no 900-foot bed of sedimentary anhydrite has ever been noted. Furthermore, the question naturally rises as to why anhydrite, out of the many different types of rock passed through by the rising salt mass, should be the only one to be punched out and carried upward.

According to Hobson, "In Germany it has been widely accepted that if the salt rises into the zone of circulating subsurface waters the soluble materials at the top will be removed, leaving a more or less flat table of anhydrite, gypsum, and clay."[36] This theory also has become the more generally accepted one for the origin of the Gulf Coast cap rocks. The presence of 5 to 10 per cent of anhydrite in the Gulf salt cores has already been mentioned. Although both minerals are relatively soluble compared with most other minerals, salt is much the more soluble of the two. *Any*

[34] Marcus Goldman: "Gypsum-Anhydrite Cap Rock of Sulphur Salt Dome," *Geol. Soc. Am. Memoir* 50 (1952); "Origin of the Anhydrite Cap Rock of American Salt Domes," *U. S. Geol. Survey*, Prof. Paper 175 (1933), pp. 83–114; Marcus A. Hanna, "Geology of Gulf Coast Salt Domes," *Problems of Petroleum Geology* (American Association of Petroleum Geologists, 1934), pp. 648 et seq.; Ralph E. Taylor, "Origin of Cap Rock of Louisiana Salt Domes," *La. Geol. Survey*, Bull. 11 (1938); L. S. Brown, "Cap-Rock Petrography," *Bull. Am. Assoc. Petrol. Geol.*, Vol. 15 (May, 1931), pp. 509–529.

[35] L. P. Teas, "Hockley Salt Shaft, Harris County, Texas," *Gulf Coast Oil Fields* (American Association of Petroleum Geologists, 1936), pp. 136–140.

[36] G. D. Hobson, "Salt Structures: Their Form, Origin, and Relationship to Oil Accumulation," *Science of Petroleum* (Oxford University Press, 1938), Vol. 1, pp. 258–259.

contact of the surface of an anhydrite-containing salt core with ground
water would indubitably lead to an enrichment of that mineral in the
contact zone.

Opinions also vary as to the genesis of some of the associated minerals
in the cap rock. No doubt the gypsum that overlies the anhydrite is formed
by hydration of the anhydrite. The sulfur that is present in a few caps is
a probable alteration product of anhydrite as postulated for the commerical
sulfur deposits of Sicily. Experiments have tended to verify the theory
that anhydrite is reduced by sulfate-reducing bacteria to hydrogen sulfide
which subsequently oxidizes to native sulfur.[37] Since the calcite ("limestone") part of the cap always overlies the anhydrite-gypsum zone, it could
not be the result of residual accumulation of calcite crystals carried within
the salt. It is more likely the result of interaction between carbonate
waters and calcium sulfate which resulted in the precipitation of calcium
carbonate. The calcite zone is characteristically cavernous, and in many
places it consists of collapse breccia which contains intermingled fragments
of sedimentary limestone, sandstone, and other types of rock from the
overlying section. Some of the sandstone fragments found in the cap
rock are thought to be "xenoliths" carried upward by the salt along with
the anhydrite.[38]

Hanna[39] has divided the various stages in the evolution of the American
salt-cored structures into youth, maturity, and old age. Youth is the
stage of earliest salt flowage, when the bedded, flat-lying salt flows into
anticlines and arches the overlying sediments. In maturity the roof is
breached, the salt stock starts upward, and solution at the top of the stock
becomes active, which results in a concentration of anhydrite. During
old age the stock continues upward and the cap increases in thickness.
Solution activity spreads down the flanks of the salt core. Overhang of
both cap and salt may be produced. Rim synclines are developed in the
surrounding rocks. In extreme senility much salt is removed by circulating
water, and collapse of the cap results in depressed tops. The cycle ends
when upward movement and solution cease. There is some evidence that
a few Gulf Coast domes have been actively eroded.[40]

[37] Herbert W. Feely and J. Laurence Kulp, "Origin of Gulf Coast Salt-Dome Sulphur
Deposits," *Bull. Am. Assoc. Petrol. Geol.*, Vol. 41 (August, 1957), pp. 1802–1853; Galen
E. Jones, Robert L. Starkey, Herbert W. Feely, J. Laurence Kulp, "Biological Origin of
Native Sulfur in Salt Domes of Texas and Louisiana," *Science*, Vol. 123 (June 22, 1956),
pp. 1124–1125.

[38] Marcus A. Hanna, "Geology of Gulf Coast Salt Domes," *Problems of Petroleum Geology* (American Association of Petroleum Geologists, 1934), p. 643.

[39] Marcus A. Hanna, *op. cit.*, pp. 656 et seq.

[40] Marcus A. Hanna, "Evidence of Erosion of Salt Stock in Gulf Coast Salt Plug in Late
Oligocene," *Bull. Am. Assoc. Petrol. Geol.*, Vol. 27 (January, 1943), pp. 85–89, and Vol. 23
(April, 1939), pp. 604–607.

Oil Trapping by Gulf Coast Salt Domes. Salt-cored structures are important to the petroleum geologist because the intrusion of a salt stock creates local traps in which hydrocarbons can accumulate. There is no genetic relationship between salt and oil; the relationships are entirely physical. Oil and gas deposits may occur in three positions in respect to the salt core, namely: (1) in the cap rock, (2) in the sedimentary rocks on the flanks of the salt intrusion, and (3) in the sedimentary rocks arched over the top of the plug. No commercial oil or gas deposits have ever been found within the salt. Most fields in the Gulf Coast produce from only one of the three potential reservoirs, but some produce from two, and at least two fields (Spindletop, Texas, and Jennings, Louisiana) have produced from all three. By means of modern geophysical techniques, it is possible to discover and even "pin-point" salt-core structures. However, the area of accumulation may be so small, especially in the case of flank trapping, that the explorer may have to drill many failures before drilling the discovery. But the rewards are well worth the gamble. The Gulf Coast salt-cored fields are noted for their prolific yields—especially for their enormous production-per-acre figures.

Cap-rock production is almost entirely confined to the limestone cap. The limestone may be brecciated, because of either salt thrust or collapse. Solution porosity is common, and some caps have been found to be truly cavernous. Phenomenal yields have been obtained at Spindletop, Texas, which is illustrated in Figs. 12–9 and 12–11, and at Humble (Harris County, Texas) from cavernous caps. All production to date from cap-rock reservoirs has been from relatively shallow depths. Hanna[41] points out that thick limestone caps are found only near the surface.

Flank production is obtained from a wide variety of traps, but because their areas are small and their distance outward from the salt plug is variable, finding them presents the oil hunters with a real challenge. One salt dome, the Bay Junop on the Louisiana coast, was probed by 28 tests over a period of 24 years before oil was discovered a mile off the west side.

Flank traps are actually permeability pinchouts. Some were created by sedimentation and erosion activity prior to (or concurrent with) the salt intrusion, whereas others were created by the upward movement of the salt plug itself. In the former type, since the dragging upward of the permeability trap has increased many times both its closure and the size of its gathering area, the salt intrusion, although not creating the trap, has greatly improved its oil-storage potential. Examples of pre-existing permeability traps are sand lenses and angular unconformities. Intrusion-created traps include fault traps, pinched sands in the drag zone, and

[41]Marcus A. Hanna, "Geology of the Gulf Coast Salt Domes," *Problems of Petroleum Geology* (American Association of Petroleum Geologists, 1934), p. 669.

Fig. 12–9. Cross section through Spindletop diorama. Initial discovery was in cap rock; flank discovery came 25 years later. *Courtesy Humble Way, Humble Oil and Refining Company.*

abutment of dragged up and ruptured reservoirs against the salt overhang (Fig. 12–12) or against the shale or anhydrite sheath (gouge) on the sides of the salt plug. In the last-mentioned situation the trap is in near contact with the salt itself, but the fault, pinched-off sand, unconformity, and lenticular sand traps may be anywhere in the peripheral zone where the "country rock" has been dragged upward. The relatively small area of the flank traps is due to the steepness of reservoir rock dips, and to lateral closure by either tonguing of the sands or secondary dips at right angles to the main off-structure dip direction. As a general rule, flank production from any one reservoir is confined to but a small segment of the total circumference of the salt dome.

Fig. 12–10. Aerial view of Avery Island salt dome, Louisiana. Surface indications of dome structure are accentuated by drainage ditches. Black dots are producing oil wells. The field was discovered in 1942. *Courtesy Humble Way, Humble Oil and Refining Company.*

338

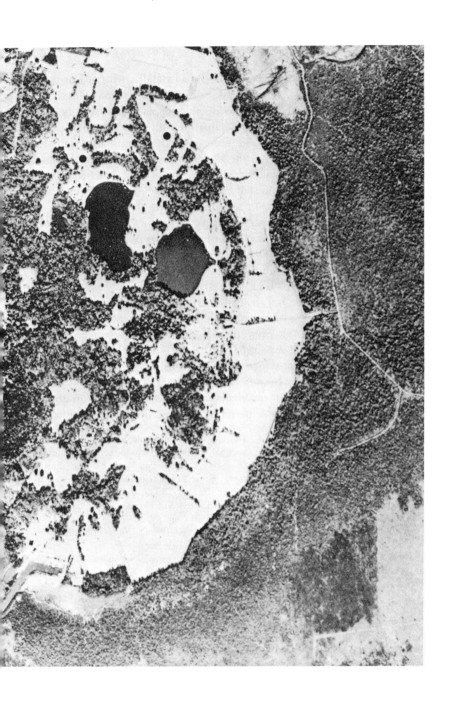

Salt-dome flank trapping has been described by Halbouty and Hardin,[42] Jackson,[43] and others. Atwater and Forman[44] call attention to the irregular and spasmodic character of salt-plug emplacement and its effect on oil accumulation as follows: "These intermittent shifts of upward movement within the salt stock in some cases have probably kept the petroleum accumulations around the dome in a game of musical chairs, constantly readjusting to positions of more favorable structural traps. The location of some of the present commercial accumulations represents only the result of the most recent readjustments."

Spindletop (Fig. 12-11) illustrates flank as well as cap accumulation. Other famous flank fields are Anse la Butte[45] and Jennings[46] in Louisiana and Barbers Hill,[47] Humble, Damon Mound, and Sour Lake in Texas.[48]

Super-cap accumulation is in normal reservoirs which have been arched upward into anticlinal traps by a salt intrusion. The usual laws of anticlinal accumulation apply, but the distribution of the hydrocarbons across the top of the structure may be complicated by the segmentation produced by faulting[49] or by lenticularity of the sandstone reservoirs. With a few relatively insignificant exceptions, all these salt stocks are so deeply buried that their actual existence has not been proved. The reservoirs are classified as super-cap because of the geographic location (in Gulf Coast structures), the quaquaversal shape of the dome, minimum gravity readings in some instances, and, perhaps, the presence of central grabens

[42] Michel T. Halbouty and George C. Hardin, Jr.: "New Geological Studies Result in Discoveries of Large Gas and Oil Reserves From Salt Dome Structures in the Texas-Louisiana Gulf Coast," Proc. Fourth World Petrol. Congress (1955), Section 1, pp. 83–101; "Factors Affecting Quantity of Oil Accumulation Around Some Texas Gulf Coast Piercement-Type Salt Domes," Bull. Am. Assoc. Petrol. Geol., Vol. 39 (May, 1955), pp. 697–711; "Types of Hydrocarbon Accumulation and Geology of South Liberty Salt Dome, Liberty Co., Texas," Bull. Am. Assoc. Petrol. Geol., Vol. 35 (September, 1951), pp. 1939–1977.

[43] J. Roy Jackson, "Unconformity Traps Around Piercement Salt Domes in South Louisiana," Am. Assoc. Petrol. Geol. Annual Meeting Program (1956), p. 37.

[44] Gordon Atwater and McLain J. Forman, "Nature of Growth of Louisiana Salt Domes and its Effect on Petroleum Accumulation," Am. Assoc. Petrol. Geol., Program Annual Meeting (1958), p. 45.

[45] F. W. Bates and Jay B. Wharton, Jr., "Anse la Butte Dome, St. Martin Parish, Louisiana," Bull. Am. Assoc. Petrol. Geol., Vol. 27 (August, 1943), pp. 1123–1156.

[46] M. T. Halbouty, "Geology and Geophysics of Southeast Flank of Jennings Dome, Acadia Parish, Louisiana, with Special Reference to Overhang," Bull. Am. Assoc. Petrol. Geol., Vol. 19 (September, 1935), pp. 1308–1329.

[47] M. T. Halbouty, "Geology and Economic Significance of Barbers Hill Salt Dome," World Petrol., Vol. 10 (January, 1939) pp. 40–55.

[48] Alexander Deussen, "Oil-Producing Horizons of Gulf Coast in Texas and Louisiana," Bull. Am. Assoc. Petrol. Geol., Vol. 18 (April, 1934), pp. 500–518.

[49] W. E. Wallace, Jr., "Structure of South Louisiana Deep-Seated Domes," Bull. Am. Assoc. Petrol. Geol., Vol. 28 (September, 1944), pp. 1249–1312.

or rim synclines. Super-cap accumulation is illustrated by the Conroe (Fig. 12-13) field, Texas, and by the Dammam (Fig. 12-14) field in Saudi Arabia.

Other oil-yielding structures that probably owe their genesis to the plastic flow of deeply buried salt are the domes of the Gulf interior, such as Hawkins, Texas, and the accumulations which have taken place along the Gulf Coast in anticlines lying at the edges of down-faulted blocks. The latter type is known as the Tepetate structure and is illustrated by the West Tepetate (Fig. 12-15) field, Louisiana. Other oil-producing anticlines are suspected to overlie deep-seated salt bulges. Burgan (Fig. 11-16) in Kuwait is an example. However these accumulations will not be classified as super-cap until more evidence of the presence of a salt core is available.

Searching for Oil and Gas in Salt-Cored Structures. The salt plugs themselves are relatively easily discovered by surface criteria (Fig. 12-10) or geophysical surveys, as described in Chapter 3. The trick is to find the oil accumulations, especially in flank traps. This calls for application of the highest type of three-dimensional geology, plus imaginative thinking.[50]

So far, at least, no salt dome on the Gulf Coast of Louisiana or Texas can be written off as barren, although many are as yet unproductive.[51] Several domes, hitherto condemned because of many dry holes, have become productive with further probing. Even the interior salt-dome district of east Texas, limited for many years to a single small field, has experienced a renewed and successful search for hydrocarbons.

The hydrocarbon reserve in the Gulf Coast salt-dome province, both coastal and inland, both onshore and offshore, must be considerable. Perhaps exploration of similar intensity in salt-dome (or diapir) areas elsewhere will have similar results.

[50] Gerry Bernatchez, "A Neglected Objective," *World Oil* (June, 1957), pp. 179 et seq.; Michel T. Halbouty and George C. Hardin, Jr., "New Exploration Possibilities on Piercement-Type Salt Domes Established by Thrust Fault at Boling Salt Dome, Wharton County, Texas," *Bull. Am. Assoc. Petrol. Geol.*, Vol. 38 (August, 1954), pp. 1725–1740; Paul Weaver, "Finding More Flank Production on Piercement-Type Salt Domes," *Oil and Gas Jour.* (June 23, 1952), pp. 90 et seq.

[51] Michel T. Halbouty and George C. Hardin, Jr., "64 Chances to Find New Gulf Coast Oil!" *Oil and Gas Jour.* (Oct. 10, 1955), pp. 321–324.

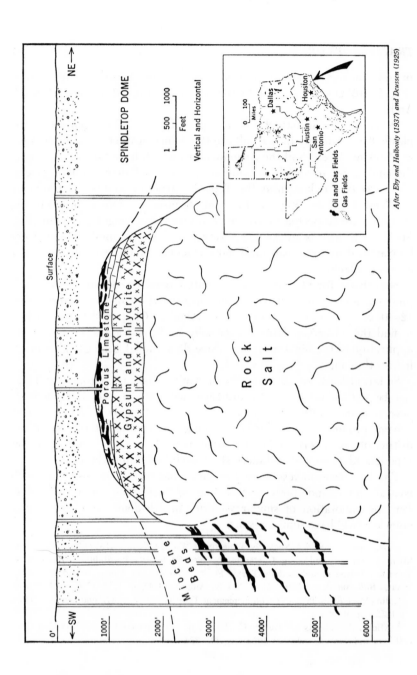

SPINDLETOP DOME

Surface

NE →
← SW

0'
1000'
2000'
3000'
4000'
5000'
6000'

Porous Limestone

× Gypsum and Anhydrite ×

Rock Salt

Miocene Beds

1   500   1000
Feet
Vertical and Horizontal

Dallas ★
Houston ★
Austin ★
San Antonio ★

0   100
Miles

Oil and Gas Fields
Gas Fields

After Eby and Halbouty (1937 and Deussen (1925)

TRAP CASE STUDIES   Cap and Flank Production

Fig. 12-11.   Spindletop, Texas.

The discovery of oil by Captain Lucas at Spindletop in January, 1901, marked the beginning of the Gulf Coast oil industry. The discovery well, drilled on the top of a topographic mound, was a gusher, estimated to flow 75,000 barrels daily. Sixty-four gushers of equal capacity were completed during the following 9 months.

The illustration is a composite from cross sections published by Deussen,° Barton and Paxson,† and Eby and Halbouty.‡ The initial discovery was in "limestone" cap rock which in reality is a highly cavernous dolomite. Coupled with the phenomenal early production rate was the rapid exhaustion of the reservoir; by the end of 5 years the field was down to 4000 barrels daily, relatively a mere trickle. The Spindletop cap-rock discovery was followed between 1901 and 1905 by similar discoveries at Sour Lake, Batson, and Humble, but there have been no important finds of this nature in the Gulf Coast area since.

Above the productive cap rock at Spindletop and Humble are several shallow sands belonging in the Lissie-Reynosa (Pleistocene) group in which some oil had accumulated, probably the result of leakage from the underlying cap.

Production from reservoirs flanking the salt cores was discovered first at Sour Lake in 1914. The southwest flank at Spindletop was found to be productive in 1926. A second Spindletop "boom" followed, and during the following decade the flank produced 50 per cent more oil than the prolific cap rock had produced since discovery. The flank reservoirs are sands in the Fleming formation of Miocene age.

The trapping at Spindletop has been due to various situations brought about by the intrusion of a salt stock. Over 50 million barrels of oil accumulated in the cavernous dolomitic cap rock which is overlain by relatively impervious sediment. Even greater in volume is the oil caught in the flank sands which were dragged up and pinched off by the salt intrusion. A little oil appears to have worked upward into arched supercap sands. Originally the oil must have traveled laterally into the flank and cap traps. *Courtesy American Association of Petroleum Geologists.*

° Alexander Deussen, "Oil-Producing Horizons of Gulf Coast in Texas and Louisiana," *Gulf Coast Oil Fields* (American Association Petroleum Geologists, 1936), p. 3.

† Donald C. Barton and Roland B. Paxson, "The Spindletop Salt Dome and Oil Field, Jefferson County, Texas," *Bull. Am. Assoc. Petrol. Geol.,* Vol. 9 (May-June, 1925), p. 601.

‡ J. Brian Eby and Michel T. Halbouty, "Spindletop Oil Field, Jefferson County, Texas," *Bull. Am. Assoc. Petrol. Geol.,* Vol. 21 (April, 1937), p. 486.

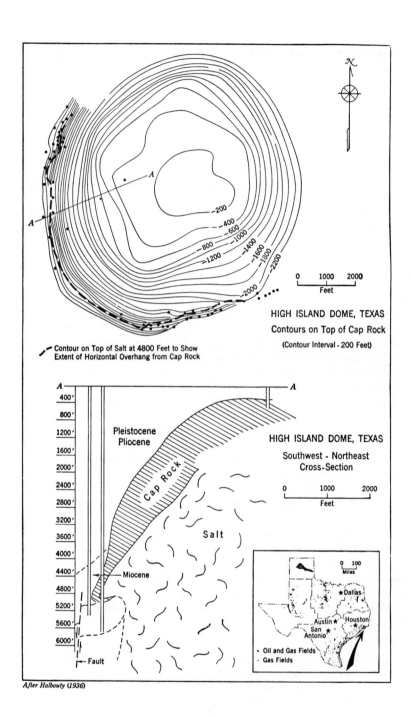

HIGH ISLAND DOME, TEXAS

Contours on Top of Cap Rock

(Contour Interval - 200 Feet)

Contour on Top of Salt at 4800 Feet to Show
Extent of Horizontal Overhang from Cap Rock

HIGH ISLAND DOME, TEXAS

Southwest - Northeast
Cross-Section

Pleistocene
Pliocene

Cap Rock

Salt

Miocene

Fault

0   100
Miles

Dallas

Austin
San
Antonio
Houston

- Oil and Gas Fields
· Gas Fields

After Halbouty (1936)

344

TRAP CASE STUDIES

Flank Production from beneath Overhang

Fig. 12-12. High Island Dome, Texas.

The High Island salt dome° lies about a mile inland from the Gulf of Mexico in Galveston County, Texas. It is marked at the surface by a mound that rises above the surrounding marshes for a height of about 20 feet. The presence of this hill, so similar in appearance to that at Spindletop, plus abundant seeps of sulfurous inflammable gas, led to an oil test in 1901. This test, and many more in the succeeding years, were unsuccessful. It was not until 1922 that High Island became a commercial producer. The first successful wells were drilled into porous cap rock. In 1931 the discovery well for the deep and prolific flank reservoirs was drilled. This well also proved the presence of an overhang of cap rock and salt with producing sands beneath the overhang as shown in the cross section on the opposite page.

Above the High Island salt core is the "true" cap consisting of anhydrite grading upward into gypsum with calcite at the top. The maximum thickness of this cap is 675 feet. Above the true cap is the "false" cap, which ranges from 200 to 1300 feet in thickness. It consists of hard sandstone and "lime rock", which is actually calcareous sandstone. The initial production at High Island was from the calcite zone in the true cap.

According to Halbouty the overhang had been proved for all flanks drilled at that time, which included the northwest, west, south, and southeast, and probably continued completely around the salt core. The horizontal extent of the overhang is as much as 800 feet. Both the cap rock and the salt itself occur in the overhanging section, as shown on the opposite page. The sedimentary formations below the overhang are faulted, broken, and distorted. On the western flank much gouge material is present in this zone. A peripheral fault is assumed, and a well-developed graben was discovered off the south flank.

The most prolific of the flank sands are in the Lower Miocene (MDv zone) and Middle Oligocene (*Discorbis* zone). These sands are dragged up so that in some places they are vertical. The intrusion of the salt core also shattered and faulted the reservoir rocks. Apparently the oil migrated through the Tertiary sandstones up-dip toward the salt stock and were trapped in the broken zone beneath the salt and cap overhang. *Courtesy American Association of Petroleum Geologists.*

° Michel T. Halbouty, "Geology and Geophysics Showing Cap Rock and Salt Overhang of High Island Dome, Galveston County, Texas," *Gulf Coast Oil Fields* (American Association Petroleum Geologists, 1936), pp. 909–960.

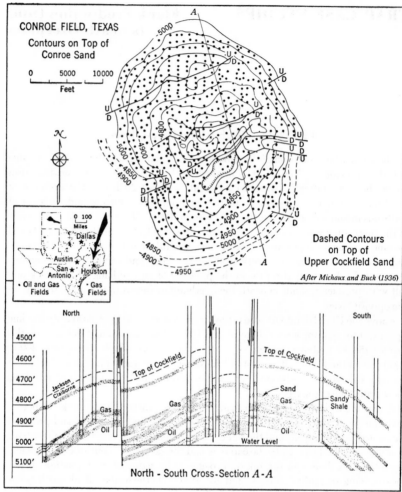

CONROE FIELD, TEXAS
Contours on Top of
Conroe Sand

0   5000   10000
Feet

Dashed Contours
on Top of
Upper Cockfield Sand

After Michaux and Buck (1936)

0   100
Miles

Dallas
Austin
San Antonio   Houston
• Oil and Gas Fields   • Gas Fields

North - South Cross-Section A-A

North   South

4500'
4600'
4700'
4800'
4900'
5000'
5100'

Jackson Claiborne
Top of Cockfield   Top of Cockfield
Gas   Gas   Sand
Oil   Gas   Sandy Shale
        Oil   Water Level   Oil

After Michaux and Buck (1936)

346

Fig. 12-13.   Conroe, Texas.

The Conroe oil field° is in Montgomery County, Texas, 40 miles north of Houston. Although preceded by Goose Creek and a few other fields, the Conroe discovery in 1931 ushered in a new era in Gulf Coast history. Since Conroe, most of the Gulf Coast oil has been produced from domes which are assumed to overlie a salt core, but the core is so deep-seated that it has not yet been penetrated by the drill. Today there are literally hundreds of fields of this type in Texas and Louisiana.

Although the arched sands above the cap produced some oil in the "piercement" type of structure at Spindletop and Jennings and perhaps other fields, the deep-seated supercap production at Conroe is analogous only in relative position and probable structural origin. At Conroe the assumed salt core is hundreds if not thousands of feet below the lowest "supercap" oil yet discovered, and the field has 17,200 producing acres which at the time of discovery exceeded the total aggregate productive acreage of the entire Gulf Coast salt-dome province of Texas and Louisiana.

The prediscovery surface indications at Conroe consisted of gas seeps and springs of hydrogen sulfide-bearing water. Some unusual clay outcrops were thought (but never proved) to be inliers.

The upper figure shows the geologic structure of the Conroe field mapped on the top of the main producing sand. Below is a northwest to southeast cross section. The Conroe structure is an unusually symmetrical dome with a structural relief in the Cockfield (Eocene) of over 800 feet and closure of 360 feet. Crossing the dome are three parallel normal faults. The fault pair to the south creates a keystone graben, $\frac{3}{4}$ mile wide, which occupies the exact center of the dome. The displacements range from zero at the ends to a maximum of 165 feet.

Accumulation has occurred in three sand members of the Cockfield, of which the lowest ("Conroe") is most important. However, the fault planes are not tight within the Cockfield, so there is free communication between sand members and they constitute one reservoir with a gas cap above and bottom water below. The gas cap is 170 feet thick, and the oil occupies an interval 130 feet in thickness. The elevation of the oil-water contact is rather uniformly 4990 feet below sea level. *Courtesy American Association of Petroleum Geologists.*

° Frank W. Michaux, Jr. and E. O. Buck, "Conroe Oil Field, Montgomery County, Texas," *Gulf Coast Oil Fields* (American Association Petroleum Geologists, 1936), pp. 789–832.

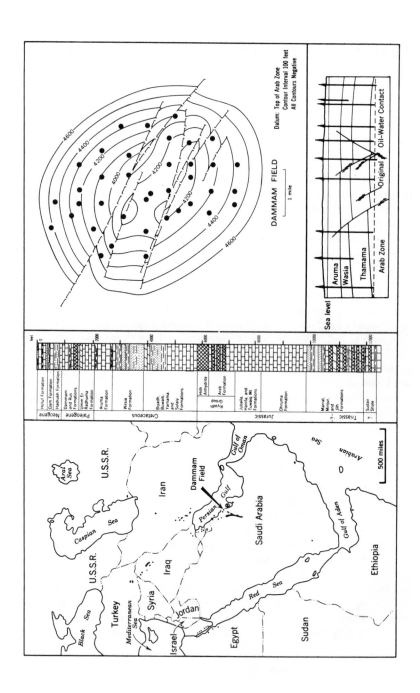

DAMMAM FIELD

1 mile

Datum: Top of Arab Zone
Contour Interval 100 feet
All Contours Negative

Sea level

Aruma
Wasia
Thamama
Arab Zone

Original Oil–Water Contact

Neogene — Holul Formation, Dam Formation, Hadrukh Formation
Paleogene — Dammam and Rus Formations, Umm Er Radhuma Formation
Cretaceous — Aruma Formation, Wasia Formation, Biyadh, Buwaib, Yamama, and Sulay Formations, Hith Anhydrite, Arab Formation, Riyadh Group, Jubaila, Hanifa, and Tuwaiq Mt. Formations, Dhruma Formation
Jurassic — Marrat, Minjur, and Jilh Formations
Triassic — Sudair Shale

feet
0
2000
4000
6000
8000
10000
12000

U.S.S.R.

Aral Sea

Iran

Dammam Field

Gulf of Oman

Arabian Sea

Persian Gulf

Saudi Arabia

Caspian Sea

U.S.S.R.

Iraq

Syria

Turkey

Black Sea

Mediterranean Sea

Israel

Jordan

Egypt

Red Sea

Gulf of Aden

Ethiopia

Sudan

500 miles

TRAP CASE STUDIES

Probable Supercap Production

Fig. 12–14. Dammam Oil Field

The Dammam° field was the discovery field of Saudi Arabia. Geologists working on the nearby island of Bahrain, where oil was discovered in 1932, suspected that the hill on the mainland which they could see from Bahrain Island was of structural origin, as was the hill on which they were standing. In 1933 a geological party landed on the shore of Saudi Arabia and confirmed by plane table mapping that this hill was actually the surface expression of a structural dome. First commercial production was obtained in 1938.

Unlike the other producing structures in Saudi Arabia, the Dammam dome is not associated with any major anticlinal axis. It is nearly circular, measuring 3¾ miles in one direction and 3 miles in the other. The closure at the top of the Arab zone is over 700 feet. As can be seen on the cross section on the opposite page, the central part of the dome contains a graben with displacements of up to 400 feet. The major faulting does not extend above the unconformity between the Wasia and the Aruma formations. The presence of a salt core at depth is assumed because of the presence of a gravity minimum centered between 1½ and 2 miles from the structurally highest point at the surface.

The reservoir rocks include all four members (A, B, C, and D) of the Arab formation of Upper Jurassic age. The reservoirs are oölitic and dolomitic limestones, and all four are sealed by beds of anhydrite.

The base of the Arab zone, which has a total thickness of about 400 feet, lies at a depth of about 4900 feet. During 1957 a deep test was drilled on the dome to a depth of 10,224 feet. It struck gas "in quantities sufficient to meet local requirements" in rocks of Permian age between 8500 and 8600 feet. This is the oldest formation so far found productive in Saudi Arabia. *Courtesy Arabian-American Oil Company.*

° Warren H. Thralls and R. C. Hasson, "Saudi Arabia's Oil," *Oil and Gas Jour.* (July 15 and July 22, 1957); W. H. Thralls and R. C. Hasson, "Geology and Oil Resources of Eastern Saudi Arabia," *Symposium on Deposits of Petroleum and Gas*, XX International Geol. Congress (1956), Vol. 2, pp. 9–32; R. C. Hasson, "Explorations in Saudi Arabia," *Mines Mag.* Vol. 45 (October, 1955), pp. 119–128; C. T. Barber, "Review of Middle East Oil," *Petroleum Times* (June 1948), pp. 75–76.

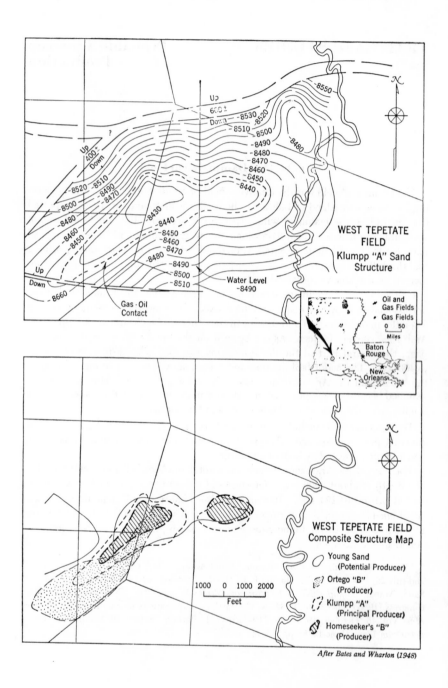

WEST TEPETATE
FIELD
Klumpp "A" Sand
Structure

Water Level
-8490

Gas - Oil
Contact

Oil and
Gas Fields
• Gas Fields
0   50
Miles

Baton
Rouge

New
Orleans

WEST TEPETATE FIELD
Composite Structure Map

⊘ Young Sand
   (Potential Producer)

▨ Ortego "B"
   (Producer)

⟨⟩ Klumpp "A"
   (Principal Producer)

◈ Homeseeker's "B"
   (Producer)

1000   0   1000 2000

Feet

After Bates and Wharton (1948)

TRAP CASE STUDIES

Gulf Coast Anticline
(Probably Salt Cored) in
Down-Faulted Block
("Tepetate-Type" Structure)

Fig. 12-15. West Tepetate, Louisiana.

Second only to the salt dome in importance as a trap for hydrocarbons along the Gulf Coast of Louisiana and Texas is the elongate anticline in which accumulation is in the downblock of a fault. Elsewhere than in these coastal districts accumulation is almost always in the upblock. The Gulf Coast faults parallel the shoreline and the strike of the Gulfward-dipping sediments. The shore-side block has dropped; on this block next to the fault plane is an anticline with long axis parallel to the fault. Oil moving up-dip from the direction of the Gulf has been trapped in this anticline, and either the fault on the north flank or the oil-water interface limits production in that direction.

Sometimes called the "Tepetate type" structure, because of its occurrence at Tepetate, Louisiana, downblock anticlinal accumulation is illustrated here by the adjacent West Tepetate° field. This field lies about 40 miles northeast of the city of Lake Charles in the northeastern corner of Jefferson Davis Parish.

The structure map contoured on the Klumpp A oil sand, which is the principal producer, is shown on the opposite page. The West Tepetate anticline is a low, gentle feature with only 60 to 70 feet of closure above the oil-water level. The dips range from about 100 to 200 feet per mile at the Klumpp level. The fault on the north side has a displacement of about 600 feet.

The lower figure is a composite structure map made by superimposing approximately equivalent contours from each of four structure maps. This composite shows that during the growth of the West Tepetate anticline the zone of greatest uplift shifted northward and lengthened in an east-west direction.

The producing section in the West Tepetate field lies below the Second Marginulina sand in a series of sandstones commonly considered to be Middle Miocene in age. The average depth of the Klumpp A reservoir is 8500 feet. Some oil and considerable gas distillate have been found in lower reservoirs.

The uplift is assumed to have been caused by the deep-seated intrusion of salt. After (or during) the arching of the strata above the salt core the hydrocarbons in the various permeable layers moved upward toward the structural crest or a fault where they became trapped. *Courtesy American Association of Petroleum Geologists.*

° Fred W. Bates and Jay B. Wharton, Jr., "Geology of West Tepetate Oil Field, Jefferson Davis Paris, Louisiana," *Bull. Am. Assoc. Petrol. Geol.*, Vol. 32 (September, 1948), pp. 1712–1727.

351

TRAPS: HYDRODYNAMIC AND FAULT

HYDRODYNAMIC TRAPS [1]

Tilted Oil-Water or Gas-Water Interfaces. In many oil and gas fields the contact surface between the oil (or the gas, where oil is not present) and the underlying water is not horizontal. This situation is not related to the topography, but is usually more pronounced in areas of marked relief, such as intermontane basins. In most of our known accumulations, the effect of the tilted interface is merely to extend production farther down one flank of an anticline than the other, but there are examples in which the entire accumulation is confined to one flank—all of it below the anticlinal axis (Fig. 13–1). Furthermore, when the tilt is greater than the dip of the beds, no hydrocarbons can remain in the anticline, and they are assumed to be "flushed" down-dip out of it.

Several theories have been developed to explain the tilted interface. [2] The least tenable theory is that the trapped oil or gas has failed to readjust to horizontal after a regional inclining of the reservoir bed (an

[1] M. King Hubbert: "Entrapment of Petroleum Under Hydrodynamic Conditions," *Bull. Am. Assoc. Petrol. Geol.*, Vol. 37 (August, 1953), pp. 1954–2026; "The Theory of Ground Water Motion," *Jour. Geol.*, Vol. 48 (November-December, 1940), pp. 785–944 (original publication of Hubbert's tilt equations and theory of interfaces between different fluids); Jack W. Knight, "Hydrodynamics, A Practical Exploration Tool," *Tulsa Geol. Soc. Digest*, Vol. 25 (1957), pp. 53–59.

[2] William L. Russell, "Tilted Fluid Contacts in Mid-Continent Region," *Bull. Am. Assoc. Petrol. Geol.*, Vol. 40 (November, 1956), pp. 2644–2668; S. T. Yuster, "Some Theoretical Considerations of Tilted Water Tables," *Jour. Petrol. Technol.*, Vol. 5 (May, 1953), pp. 149–156.

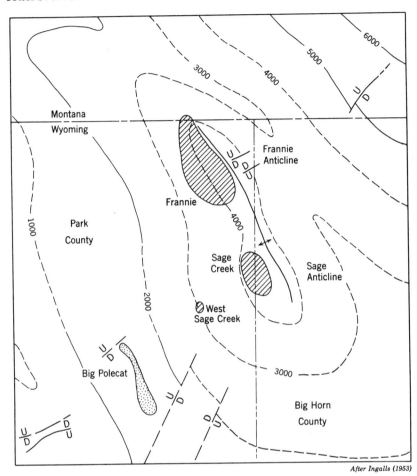

After Ingalls (1953)

Fig. 13-1. Tilted interfaces put both the Frannie and Sage Creek oil fields of northwestern Wyoming west of the anticlinal axis. Wells on the crest drill into water. The reservoir is the Tensleep sandstone (Pennsylvanian). *Courtesy the Oil and Gas Journal.*

event which occurs in a sagging basin). We have already seen in Chapter 7 that the viscosity of oil underground is low enough that, with adequate permeability of the reservoir rock, the oil can readjust to structural changes rapidly and perhaps can even keep pace with those changes. However, it is also probable that many accumulations are "frozen" into place by a plugging of the pores beneath the oil-water interface (Chapter 8), so that the interface tilts and stays tilted when the reservoir is inclined. This explanation appears to be the best for tilted interfaces in plains

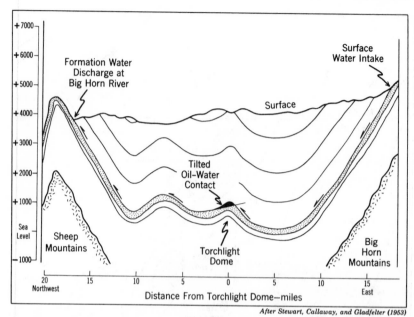

After Stewart, Callaway, and Gladfelter (1953)

Fig. 13–2. Relationship between reservoir water intake and outlet, and oil-water interface in Torchlight field, Wyoming. *Courtesy the Oil and Gas Journal.*

areas far from the mountains, where the potentiometric surfaces are practically horizontal.

In mountain-fringed basins the porous beds which crop out at high elevations may transport and discharge water at much lower elevations as springs or seepages. This system results in an inclined potentiometric (piezometric) surface. The potentiometric surface defines the heights to which artesian water would rise if wells were drilled into the carrier bed. An oil-containing aquifer with an inclined potentiometric surface also has an inclined oil-water contact surface. Such an accumulation is referred to as *hydrodynamic* (Fig. 13–2). If the oil-water interface is horizontal, the conditions are *hydrostatic*.

Tilted interfaces should not be confused with irregular interfaces due to variations in reservoir rock permeability.

Trapping by Hydrodynamics. The most obvious result of hydrodynamics is to give an oil or gas accumulation an eccentric position within an anticlinal trap. There is a definite relationship between the slope of gas-water and oil-water interfaces and the tilt of the potentiometric surface and this is shown in Fig. 13–3. Where oil is absent and gas directly overlies water, the interface slope parallels closely the potentiometric surface. But where oil and water are in contact, a greater oil density results in a steeper interface for the same potentiometric gradient. Knight[3] has calculated that for 35° oil the oil-water interface will have six times the gradient of the potentiometric surface. This relationship is constant only if the following two conditions are met: (1) the carrier bed must be homogeneous laterally (isotropic to oil and water); and (2) the oil must remain less fluid than the water.

Although pressure differentials may cause the top of the hydrocarbon accumulation to lie below the crest of an anticline (or monocline), lateral closure is still necessary to confine the oil or gas. Lateral closure is provided either by the warping of the reservoir bed around the flank of the anticline (an anticlinal flank by itself is a plunging anticline, or nose) so that the beds dip more steeply than the oil-water interface.

The Wheat field in west Texas is described under Trap Case Studies (Fig. 13–6) at the end of this chapter as an example of a hydrodynamic

[3] Jack W. Knight, *op. cit.*, p. 58.

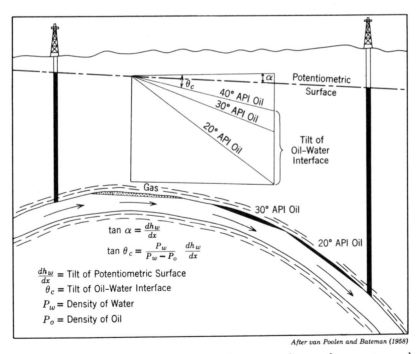

After van Poolen and Bateman (1958)

Fig. 13–3. Relationship between potentiometric surface and gas-water and oil-water interfaces, and formulas for calculating tilts. *Courtesy World Oil.*

trap. More examples are given by Hubbert[4] in his classic paper on hydrodynamic entrapment.

Searching for hydrodynamic traps. It has long been the practice in areas of notably inclined potentiometric surfaces to drill initial test wells on the basinward side of the anticlinal crest. To pinpoint properly the location for a test well it is necessary to (1) have an accurate picture of the structure of the reservoir to be tested, (2) estimate the approximate density of the oil expected in the reservoir, and (3) have a contour map of the potentiometric surface itself. To get this information with any degree of accuracy, earlier wells must have been drilled and tested for fluid pressures in the area. It is highly important that properly conducted and interpreted drill stem tests be made in order to obtain the accurate formation fluid pressure data vital to hydrodynamic studies.[5]

FAULT TRAPS

The earth forces that create folds also cause the rocks to break and fault. Normal faults result from vertical movements and horizontal tension, whereas reverse faults are the product of lateral compressional forces in the earth's crust. In some areas, notably the Gulf Coast of the United States, Romania, and the Middle East, the flowage of salt under pressure has caused fracture and displacement of associated more brittle rocks. These faults may be local, situated above and around a piercement dome, or they may be regional, in which case they are caused by large-scale deep-seated movement.[6] Salt-dome fault traps were discussed in the preceding chapter, but the regional type is considered here along with the more ordinary diastrophic faults.

Faulting plays various roles in the accumulation of oil. For example, graben faulting may produce a thick enough sedimentary section for oil occurrence in an area where the normal section is too thin. Overthrust faults may produce anticlinal traps in the thrust sheet, or they may mask possible traps in the underlying block. Without doubt, fault planes function in some areas as channels for vertical or transverse oil and water migration. Oil migration may be from deep-seated "carrier beds" upward into more shallow beds,[7] it may occur between various hydrocarbon-

[4] M. King Hubbert, *op. cit.*, pp. 2012–2023.

[5] H. K. van Poollen and Sam J. Bateman, "Application of DST to Hydrodynamic Studies," *World Oil* (July, 1958), pp. 90 et seq.

[6] Miller Quarles, Jr., "Salt-Ridge Hypothesis on Origin of Texas Gulf Coast Type of Faulting," *Bull. Am. Assoc. Petrol. Geol.*, Vol. 37 (March, 1953), pp. 489–508.

[7] John R. Fanshawe, "Faults as Sealing Factors and as Channels for Migration of Fluids in Intermontane Basins of Wyoming," *Am. Assoc. Petrol. Geol., Annual Meeting Program* (1956), pp. 27–28.

bearing strata, uniting them into one continuous reservoir, or the migration may be from a reservoir to the surface, producing seeps.

Whether a fault plane acts as a seal or as a fluid channelway depends upon several factors;[8] of special importance are the type of faulting and the lithology of the rocks cut by the fault. If the character of the faulting and the brittleness of the rocks cause brecciation along the fault plane, the result is a channel for fluids. If, however, the stage is set for pulverizing or plastic flowage of the rock along the fault contact, a seal may result. In sedimentary rock sections, especially where shales predominate, fault sealing is the general rule and fault channelways are exceptional. In crystalline rocks the reverse is true. Many hydrothermal vein deposits occur along faults cutting igneous and metamorphic rocks (and brittle sedimentary sections as well).

Where a sedimentary section consists of shale interbedded with brittle rocks it is possible for the fault to be impervious at some levels and pervious at others, depending upon the type of rock adjacent to the fault plane.

Trapping through Faulting. Normally, the following conditions are essential for a fault trap, if the presence of parallel seals (Chapter 8) above and below the reservoir rock is assumed: (1) the faulted reservoir bed must be either sealed with gouge or so faulted that impervious rock opposes it across the fault plane, (2) the fault zone must be impervious next to the reservoir bed; and (3) the fault must either cut across a plunging anticline so that edgewater semi-encircles the hydrocarbon accumulation from one point on the fault to another point on the fault, or the trap must be closed laterally by cross faults or permeability pinchouts. The usual fault trap is a plunging anticline cut normal to the plunge by a tight fault which has placed a shale or other impervious rock against the truncated reservoir bed. In some cases fault gouge alone traps the oil, with water-bearing permeable rock across the fault plane.

Oil-field faults, (those which cut oil or gas reservoirs), are of two structural types: (1) faults along the crest or high on the flanks of anticlines, and (2) strike faults which cut plunging anticlines on monoclines. The first faults are *not* traps, for the oil would have accumulated in the structure whether or not the faults were present. However, such faults may control the distribution of oil on the anticline by acting as a barrier across the path of the hydrocarbons migrating through the reservoir rock toward the top of the anticline. The reservoir bed beyond the fault plane may have dropped below the level of the oil-water interface so that here it contains only water. On some domes the faulting has no effect on ac-

[8]A. F. Woodward, "Factors Relating to Fault Seals in Some California Oil Fields," *Am. Assoc. Petrol. Geol., Annual Meeting Program* (1956), pp. 37–38.

Fig. 13–4. Fault trap. Diorama of Powell oil field, Texas. *Courtesy Humble Oil Company and Texas Memorial Museum, Austin.*

cumulation whatsoever, for the displacement has not been sufficient to offset completely the reservoir bed.

Faulted plunging anticlines on regional monoclines are true traps, for in these the oil would not have accumulated were it not for the dam created by the faulting across the pathway of the migrating hydrocarbons. Fault-trap pools tend to be elongate parallel with the fault line, and the accumulation is bounded by the fault on the up-dip side and by water on the down-dip side. Perhaps the greatest concentration of fault trap pools in the world occurs in Texas along the Mexia fault zone, where such pools as Mexia, Powell (Fig. 13–4), Richland, Wortham, and Talco have made this zone a leading oil-producing district.[9]   Oil was discovered in the Mexia district in the Woodbine sand in 1920.   The zone is dominated by a series of closely related parallel en echelon faults.   The displacement

[9]F. H. Lahee, "Oil and Gas Fields of the Mexia and the Tehuacana Fault Zone, Texas," *Structure of Typical American Oil Fields* (American Association of Petroleum Geologists, 1929), Vol. 1, pp. 304–388.

at the surface approaches 150 feet, but it increases severalfold with depth. The fault planes dip 35° to 60° west, steepening in the limestone and lessening in shale. Depending on the dip of the fault plane, production may be found up to a half mile west of the fault outcrop. The older pools are from ½ to 1½ miles wide and up to 7 miles long. The newer Talco field, which is considerably larger, is an excellent example of fault trapping (illustrated in Fig. 13–7).

Very similar in structural relationship to the fault-trap pools of the Mexia zone are the Pickens, Mississippi, field and the Gilbertown field in Choctaw County, Alabama.[10] These are illustrated in Fig. 13–8. At Gilbertown, the oil migrated up-dip through the sandy upper Eutaw formation and through a fracture system in the overlying Selma chalk until it reached the barrier formed by the fault. Fault trapping is also assumed for the Olinda end of the Brea Canyon-Olinda field, which lies along the Whittier fault in the Los Angeles basin. The greater part of the production at Olinda is obtained from inclined reservoirs that abut against this fault.[11] To the westward in the Brea Canyon part of the field is an elongate anticline the axis of which is parallel with the Whittier fault to the south. The accumulation there is controlled by this anticline. Still another example of fault trapping is in the Greater Oficina area in eastern Venezuela, where "accumulation is controlled largely by normal faults constituting barriers to migration of oil southward and eastward up the regional dip of the basin. Stratigraphic pinchouts are locally important adjuncts to accumulation."[12] Faulting has also been responsible for trapping in the Eocene rocks in the Mene Grande field in western Venezuela.[13]

Figure 13–5 is a drawing of a model of the Creole field[14] which lies 1¼ miles off the coast of Louisiana and was the first field discovered in the open waters of the Gulf of Mexico. This figure shows fault trapping in Miocene sandstone reservoirs. The Creole field was a very early example of development by drilling a series of slanted holes from one platform.

[10] A. N. Current, "Gilbertown Field, Choctaw County, Alabama," *Structure of Typical American Oil Fields* (American Association of Petroleum Geologists, 1948), Vol. 3, pp. 1–4.

[11] Walter A. English, "Geology and Oil Resources of the Puente Hills Region, Southern California," *U. S. Geol. Survey, Bull.* 768 (1926), pp. 80–81.

[12] H. D. Hedberg, L. C. Sass, H. J. Funkhouser, "Oil Fields of Greater Oficina Area, Central Anzoategui, Venezuela," *Bull. Am. Assoc. Petrol. Geol.*, Vol. 31 (December, 1947), p. 2090.

[13] Staff of Caribbean Petroleum Company, "Oil Fields of Royal Dutch-Shell Group in Western Venezuela," *Bull. Am. Assoc. Petrol. Geol.*, Vol. 32 (April, 1948), p. 576.

[14] Theron Wasson, "Creole Field, Gulf of Mexico, Coast of Louisiana," *Structure of Typical American Oil Fields* (American Association of Petroleum Geologists, 1948), Vol. 3, pp. 281–298.

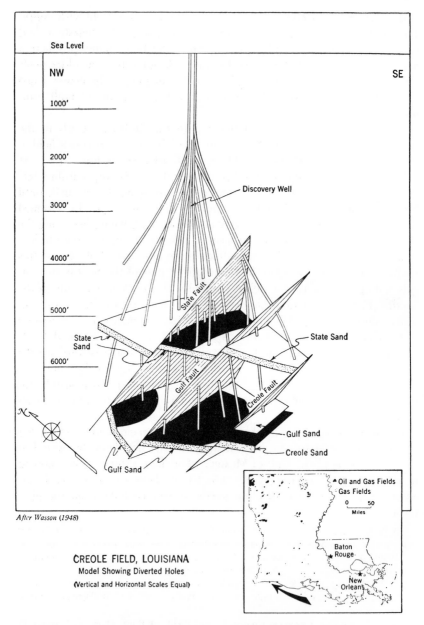

Fig. 13-5. Multiple-fault trap. Model of Creole field, offshore Louisiana, showing both fault-controlled accumulation and slanted boreholes drilled from one platform. *Courtesy American Association of Petroleum Geologists.*

Searching for Fault Traps.[15]   Searching for fault traps is in the same general category as looking for the proverbial needle.   To locate a well at the surface so that it will penetrate reservoir beds (1) dipping away from the fault, (2) in a plunging anticline or other closed trap, and (3) above the edgewater boundary, calls for three-dimensional geology of the highest quality.   If the reservoir rock has a steep dip, the producing area will be very narrow and this situation lessens the chances for successful probing.

In order to locate the test well properly the geologist has to know the strike, dip, and position of the fault and the detailed structure of the sedimentary beds.   Unfortunately, faults have the reprehensible habit of changing dip with depth (they usually lessen), and the strata between the reservoir bed and the structure datum may have enough convergence to shift the axis of the plunging anticline.   Other complications include unconformities in the section and intersecting faults.[16]   Finally, even if all the structure data prove to be correct and the descriptive geometry perfect, the test still may be a dry hole because of the usual hazards of accumulation timing, erratic reservoir permeability, or lack of hydrocarbons to accumulate.

Unlike anticline exploring, fault-trap tests have to be located separately for each potential reservoir, because unless the fault is very steep, each test penetrates only one prospective reservoir on the proper side of the fault near the top of the trap.   For all these reasons the success ratio in fault-trap hunting is less than for anticline testing.   Perhaps this is the reason why only 1.2 per cent of the known oil in the major fields of the free world is fault-trapped.[17]   In all probability the actual percentage of oil so trapped is considerably higher.

Most successful fault-trap searches have been the result of subsurface geology.   Under ideal conditions seismic surveys[18] obtain the fault and structure data necessary to locate the exploratory tests.

[15] Robert M. Knebel, "The Fashing Fault Find," *Petrol. Eng.* (February, 1957), pp. B 21; G. D. Hobson, "Faulting and Oil Accumulation," *Inst. Petrol.*, Vol. 42 (1956), pp. 23–26.

[16] George Dickinson, "Subsurface Interpretation of Intersecting Faults and Their Effects Upon Stratigraphic Horizons," *Bull. Am. Assoc. Petrol. Geol.*, Vol. 38 (May, 1954), pp. 854–877.

[17] G. M. Knebel and Guillermo Rodriguez-Eraso, "Habitat of Some Oil," *Bull. Am. Assoc. Petrol. Geol.*, Vol. 40 (April, 1956), p. 554.

[18] Stanley M. Leventhal, "Buried Fault Indicated by Seismic Record Data," *World Oil* (Aug. 1, 1958), pp. 75–76.

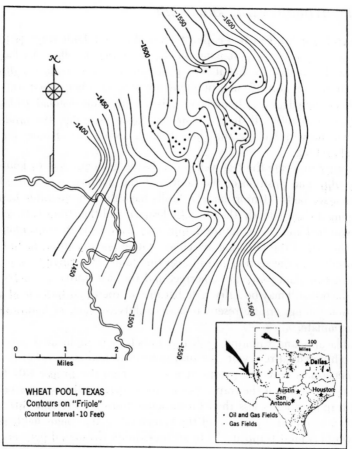

WHEAT POOL, TEXAS
Contours on "Frijole"
(Contour Interval - 10 Feet)

0   1   2
Miles

• Oil and Gas Fields
• Gas Fields

0   100
Miles

Dallas

Austin
San Antonio

Houston

After Adams (1936)

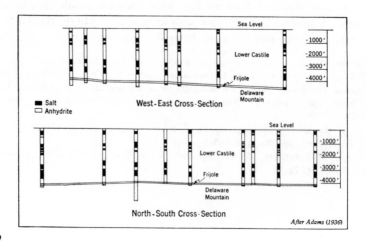

Sea Level

Lower Castile

Frijole

Delaware Mountain

-1000'
-2000'
-3000'
-4000'

■ Salt
☐ Anhydrite

West - East Cross - Section

Sea Level

Lower Castile

Frijole

Delaware Mountain

-1000'
-2000'
-3000'
-4000'

North - South Cross - Section

After Adams (1936)

362

TRAP CASE STUDIES

Possible
Hydrodynamic
Accumulation

Fig. 13–6. Wheat Pool, Texas.

The Wheat pool° lies near the center of the Delaware basin on the east side of Pecos River in Loving County, Texas. It began to produce commercially in 1925. There is no surface evidence of structure of any sort at Wheat, for the bed rock is buried beneath a thick veneer of caliche, alluvium, and sand.

The stratigraphic section so far explored at Wheat extends from the Cenozoic down into the Upper Permian. About 1500 feet of Upper Castile (Permian) evaporites, including anhydrite, salt, and polyhalite (potash salt), are present in the wells drilled at Wheat. Below is about 2000 feet of Lower Castile, which likewise consists of anhydrite and salt, but no potash. The evaporites are underlain by the thin Frijole shale at the top of the Delaware Mountain group, also Permian in age. Below the Frijole is the reservoir sandstone, which actually consists of about 70 per cent very fine sand and 30 per cent silt.

The figure on the left page shows the subsurface structure as contoured on the top of the Frijole formation which overlies the reservoir rock. The west-east cross section shown above, and the contour map itself, indicate a gentle dip to the east and a noticeable flattening in the middle, productive part of the area. The north-south cross section, in conjunction with the structure contour map, shows a fairly consistent tendency for structural nosing in those parts of the area where the producing wells are indicated. The oil occurs in the zone of 50-feet-per-mile dip; to the west and east, above and below the oil accumulation, the dip is in the neighborhood of 100 feet-per-mile and the sandstone is water-bearing. "There is no evidence of faulting, and dry holes on the west, northwest, and southwest indicated good permeability. These facts preclude the interpretation of the field being a fault or stratigraphic trap, and in the light of present information it appears to be a hydrodynamic trap produced by water flowing eastward from the Delaware Mountains." †

It should be noted that so far as the north-south cross section is concerned, the trapping at Wheat is anticlinal. In other words the lateral closure is supplied by structure. *Courtesy American Association of Petroleum Geologists.*

° John Emery Adams, "Oil Pool of Open Reservoir Type," *Bull. Am. Assoc. Petrol. Geol.,* Vol. 20 (June, 1936), pp. 780–796.

† M. King Hubbert, "Entrapment of Petroleum Under Hydrodynamic Conditions," *Bull. Am. Assoc. Petrol. Geol.,* Vol. 37 (August, 1953), p. 2016.

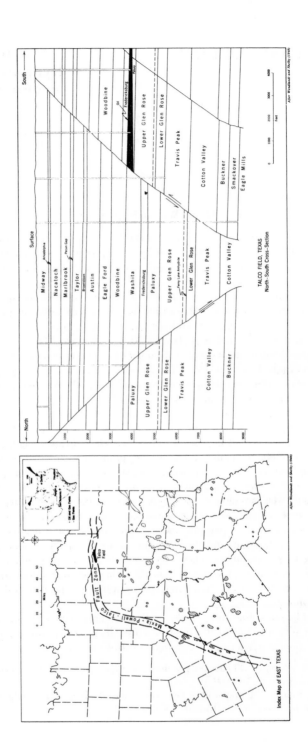

TALCO FIELD, TEXAS
North-South Cross-Section

After Woodland and Shelby (1948)

Index Map of EAST TEXAS

After Woodland and Shelby (1948)

364

Fig. 13-7. Talco Field, Texas.

The Talco field° is in northeastern Texas, about 55 miles north of the East Texas field. Its geographic position in respect to other northeastern Texas fields is shown on the index map. The Talco field lies in the Mexia-Powell-Talco fault zone which parallels the great Balcones fault as it swings around the southeastern side of the Llano uplift and turns north through Limestone and Navarro counties, where the older fault pools such as Mexia, Powell, and Wortham, have been producing for many years, to Hunt County where the fault zone turns eastward south of the Ouachita uplift across northern Franklin and Titus counties where lies the Talco field.

The Talco fault was mapped by surface geology in 1924. Failure to discover oil in the Woodbine sand north of the Powell field, where the Woodbine is the reservoir rock, discouraged exploration of the Talco fault, and it was not until 1933 (after the discovery of the East Texas field, a prolific Woodbine producer) that wells were drilled at Talco to test the Woodbine, and in 1936 the discovery well of the Talco field was drilled into the deeper Paluxy reservoir.

The right figure is a cross section of the Talco field and faults, showing the type of the trapping. As the oil migrated up the Paluxy sands from down the dip to the south it met a dam formed by the impervious Washita-Fredricksburg shales and limestone on the opposite side of the fault plane. The Talco field when first discovered extended 13¼ miles along the fault and averaged about a mile in width, larger in area than any of the other fields so far discovered along the Mexia-Powell-Talco fault zone. The oil in the reservoir is underlain everywhere by salt water, and as the field has been exploited it has decreased in area by salt-water encroachment.

The Paluxy reservoir consists of a series of porous and permeable sands which are individually lenticular, but with some degree of intercommunication, so that as a general rule these variable and erratic sands function as a single reservoir. About one-fourth of the Paluxy section above the bottom water level in the Talco field consists of saturated oil sand. *Courtesy American Association of Petroleum Geologists.*

° E. A. Wendlandt and T. H. Shelby, Jr., "Talco Oil Field, Franklin and Titus Counties, Texas," *Structure of Typical American Oil Fields* (American Association of Petroleum Geologists, 1948), Vol. 3, pp. 432–451.

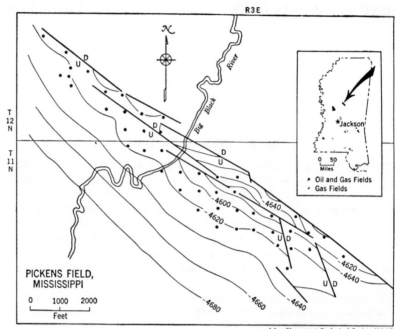

PICKENS FIELD,
MISSISSIPPI

0   1000   2000

Feet

After Shreveport Geological Society (1946)

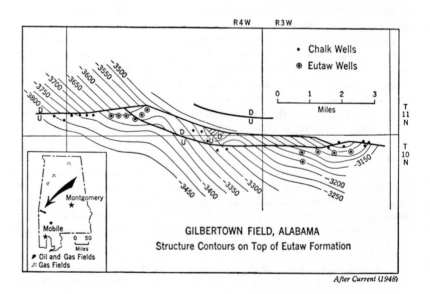

GILBERTOWN FIELD, ALABAMA

Structure Contours on Top of Eutaw Formation

After Current (1948)

Fig. 13-8. Pickens, Mississippi, and Gilbertown, Alabama.

Both figures on the opposite page illustrate fault trapping in the "interior" Gulf Coast district of Mississippi and Alabama. This type of accumulation is fairly common both here and to the westward in northern Louisiana and southern Arkansas. The upper figure is the Pickens field° in central Mississippi, and the lower figure is Alabama's first oil field, the Gilbertown,† on the west edge of southern Alabama.

In both fields the fault line is parallel to the strike. At Gilbertown it is also parallel to the shoreline, but the swing of the strike into the Mississippi embayment gives the Pickens fault a northwest trend. At both Pickens and Gilbertown and in the other interior fault fields the accumulation is on the "up" side of the fault which is on the side down the regional dip. Accumulation is on the same side in the Gulf Coast (Tepetate) type (Fig. 12-15) of fault structure, but there the down-dip side of the fault is also the downblock.

Both the fields on the opposite page illustrate accumulation of oil in plunging anticlines dammed up the regional dip by faults. At Pickens a single plunging anticline nearly 6 miles long carries oil in its highest parts, adjacent to the fault zone. The trap is closed laterally by the turning of the flanks of the fold into the faults. The Gilbertown structure shows three plunging anticlines, with two intervening synclines, abutting the fault zone. Dry holes have been drilled in the synclines, and the producers are found only above the oil-water interface level on the plunging anticlines.

The discovery at Pickens was the result of surface mapping followed by a detailed geophysical survey. The reservoir rock is a sandstone (Wilburn) member of the Austin formation of Upper Cretaceous age. The average producing depth is about 4800 feet.

The Gilbertown field was discovered in 1944 as a result of surface geological studies. Oil is produced from two reservoir rocks. The upper one is a fractured zone in the Selma chalk (Upper Cretaceous) which lies at depths ranging from 2300 to 2800 feet. The other is a sandstone lying about 900 feet deeper in the Upper Cretaceous Eutaw formation. So far attempts to obtain Tuscaloosa production have not been successful. *Courtesy Shreveport Geological Society and American Association of Petroleum Geologists.*

° Shreveport Geological Society, "Pickens Field, Madison and Yazoo Counties, Mississippi," *1945 Reference Rept.*, Vol. 1 (1946), pp. 295-298.

† A. M. Current, "Gilbertown Field, Choctaw County, Alabama," *Structure of Typical American Oil Fields* (American Association of Petroleum Geologists, 1948), Vol. 3, pp. 1-4.

*T*RAPS:
VARYING PERMEABILITY

In a reservoir rock of permeability adequate for the movement of oil or gas, an upward disappearance of permeability creates just as effective a trap for the accumulation of hydrocarbons as that created by an anticline or a fault. This upward termination of adequate permeability may be abrupt, where, for instance, an inclined reservoir rock has been truncated and subsequently sealed, or it may be gradual, as in a facies change. Since the trapping is more closely connected with stratigraphy than with diastrophism, accumulations caused by varying permeability are generally referred to as "stratigraphic" traps. However, in most places, tilting of the sedimentary layers has been an essential event in causing accumulation in traps of this type.

A large number of oil and gas pools, described in the literature as stratigraphic trap accumulations, actually do not merit this classification. These are the pools in which varying permeability merely limits the *distribution* of hydrocarbons across the top of an anticline or other types of structural traps. If an affirmative answer can be given to the question "Would hydrocarbons have accumulated here had the reservoir rock been consistently permeable?" then the pool should not be classified as a varying-permeability or stratigraphic trap.

Although known accumulations of hydrocarbons in varying permeability traps are as old as the oil industry, it was the discovery of the East Texas field in 1930 that really focused attention on this type of trap. This field (Fig. 14–14) owes its existence to the wedging out of a sandstone reservoir beneath an unconformity. At the close of 1957 it was still the world's largest oil field in terms of yield to date, with a cumulative pro-

368

duction of 3.3 billion barrels. Naturally its discovery created a great awareness of the possibilities of varying permeability traps.

Subsequent discoveries of prolific oil accumulation in reefs in Alberta and Texas, further kindled interest in stratigraphic traps. Today in every oil-producing area, as soon as all of the obvious (or easily discoverable) structural traps have been explored, more and more attention is paid to permeability trap possibilities. [1]

In the following discussion, varying-permeability traps are classified genetically into three types: (1) those in which the permeability differences are acquired initially, during the depositional or sedimentation stage; (2) those in which circulating ground waters have been responsible for locally increasing or decreasing the porosity; and (3) those in which an inclined permeable stratum has been truncated and the eroded edge sealed by a covering of relatively impervious material. The last of these categories is most important in terms of oil-production figures.

Varying Permeability Caused by Sedimentation. Sedimentation traps are of three types: (1) reefs, (2) lenticular sandstone bodies, and (3) reservoir rocks with change in facies.

Reefs.[2] Reefs have been defined by Cloud as follows: "Organic reefs are or were actually or potentially wave resistant mounds, platforms, or linear or irregular masses that were constructed under organic influence and they rise or rose significantly above the sea floor."[3] He advocates retaining the term "bioherm" for ancient reef-like organic masses which may or may not be wave-resistant. Further quoting Cloud, "Although coral reefs appear to be the rule among existing organic reefs, their rule is by no means universal to-day, and many important reef structures of the past were built with little or no help from corals. It is also well to keep in mind that the frame-building organisms do not necessarily or even characteristically comprise the bulk of the reef. They merely serve to hold it together and the frame they build is a trap for clastic or chemically precipitated sediments . . . Important organic components of reefs, according to locality and time, are corals, calcareous algae, stromatoporoids, sponges, bryozoans, the aggregated tubes of sessile gastropods,

---

[1] Various authors, "Stratigraphic Type Oil Accumulations in Rocky Mountains," (Symposium), *Bull. Am. Assoc. Petrol. Geol.*, Vol. 41 (May, 1957).

[2] Preston E. Cloud, Jr., "Facies Relationships of Organic Reefs," *Bull. Am. Assoc. Petrol. Geol.*, Vol. 36 (November, 1952), pp. 2125–2149 (70 references); Seismograph Service Corp., *Bibliography of Organic Reefs, Bioherms, and Biostromes* (Tulsa, Oklahoma, 1950); Philip B. King, "Geology of the Southern Guadalupe Mountains, Texas," *U. S. Geol. Survey*, Prof. Paper 215, (1948); Norman D. Newell et al., "The Permian Reef Complex of the Guadalupe Mountains Region, Texas and New Mexico," W. H. Freeman (San Francisco, 1953). There is also a large literature on modern reefs.

[3] Preston E. Cloud, Jr., *op. cit.*, p. 2126.

calcareous worm tubes, foraminifera, and, of course, the whole gamut of organic detritus.[4] Many ancient reefs, originally composed of calcite, have been completely dolomitized.

Reefs are very erratic as to shape and size. They have been classified by shape as: (1) patch reefs, which are irregular organic structures; (2) table reefs, flat-topped, isolated, open ocean features; and (3) linear reefs, including fringing reefs, barrier reefs, and the encircling outer reefs of atolls.[5] Some are barely noticeable nodes on the upper surface of a limestone stratum, but others are abrupt vertical features rising hundreds of feet into younger sediments. Reefs vary in area from protuberances no greater than a washtub in diameter to massive features underlying many square miles. The rigid framework that constitutes the core of the reef does not compact under the weight of overburden. It "enables a reef margin to grow upward and outward at much steeper angles (even vertical) than is the case with sedimentary clastic rocks. Reefs are commonly characterized by lack of well-developed stratification. Differential settling in rocks adjacent to them usually causes draping of strata over reefs. The extra weight of reefs may cause downward bending of strata under them."[6]

Link[7] believes that a series of interconnecting reefs may be deposited by a transgressing sea, with each succeeding reef a little higher and offset landward until the point of maximum submergence is reached. Then regression starts, and a second continuous series of reefs, extending outward from the now rising land mass, is formed, which results in a recumbent < with the apex at the point of maximum submergence (Fig. 14–1). Evaporites might be deposited on the shoreward side of the regressive reefs. If the reef building in the transgressing and regressing sea took place on the same vertical plane, a well could be drilled through both legs of the <, penetrating first the regressive bioherm with associated evaporites and at a lower level reaching the zone of transgressive bioherms with associated clastics.

Reefs are widely distributed both in space and time. Modern coral banks have been described in the north Atlantic Ocean and along the coast of Norway at depths below 300 feet and in water temperatures as

[4] *Op. cit.*, p. 2128.

[5] Preston E. Cloud, Jr., *op. cit.*, p. 2125.

[6] W. B. Wilson, "Reef Definition," *Bull. Am. Assoc. Petrol. Geol.*, Vol. 34 (February, 1950), p. 181.

[7] T. A. Link, "Leduc Oil Field, Alberta, Canada," *Bull. Geol. Soc. Am.*, Vol. 60 (March, 1949), pp. 318–402; "Theory of Transgressive and Regressive Reef (Bioherm) Development and Origin of Oil," *Bull. Am. Assoc. Petrol. Geol.*, Vol. 34 (February, 1950), pp. 263–294.

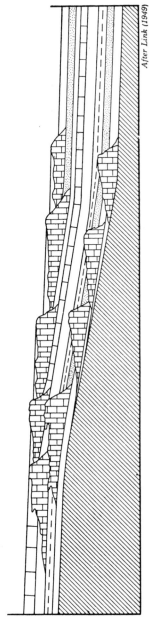

Fig. 14–1. Ideal cross section showing series of reefs formed in transgressing (lower) and regressing (upper) seas. *Courtesy Geological Society of America.*

After Link (1949)

low as 4° C.[8]   Apparently, although water temperatures may control the abundance of reefs, they do not restrict reef-making organisms to the benign latitudes.

Twenhofel[9] describes reefs of algae in the Precambrian, reefs built of algae and sponges in the Cambrian, with coral-containing reefs appearing in the Ordovician. Corals and algae were both very important in building Silurian and Devonian reefs. Crinoids also aided during these periods and were mainly responsible for large reefs occurring in the Mississippian rocks. Enormous reefs were built during the Permian, especially in western Texas and southeastern Russia. Reefs are relatively scarce in the Mesozoic and Tertiary rocks of North America, but the Middle East contains reefs of both Cretaceous and Tertiary age.[10]

The principal reason for the prolific oil production of some reefs lies in the extreme porosity and permeability of the reservoir rock. This porosity may be initial or it may be induced. As a matter of fact, it is highly probable that the great reef reservoirs had considerable original porosity which expedited the development of secondary openings. Therefore, the extreme porosity is actually a combination of both types. Initial porosity is due to both the presence of the abandoned chambers in which the animals lived and the inevitable empty spaces between the outer walls of the shells in a motley assemblage of organisms. The porosity produced by random shell arrangement is combined from the start with permeability. The living chambers, on the other hand, may be sealed off from the outside and be unavailable for the storage of oil unless boring organisms have riddled the framework of shell material and created connections from room to room and from room to intershell void.

Induced porosity is that which results from: (1) the leaching of a reef mass by circulating waters when the reef lies close to the surface; (2) an excess of solution over precipitation during dolomitization; [11] and (3) the fracturing of the rigid reef rock due to subsequent earth movements. Some oil-producing reefs have been at or near the surface in the geologic past and have developed a honeycomb porosity due to the dissolving activity of circulating waters, and have also been fractured to such an

[8] Curt Teichert, "Cold-and Deep-Water Coral Banks," *Bull. Am. Assoc. Petrol. Geol.*, Vol. 42 (May, 1958), pp. 1064–1082.

[9] W. H. Twenhofel, "Characteristics and Geologic Distribution of Coral and Other Organic Reefs," *World Oil*, Vol. 129 (July 1, 1949), pp. 61–64; "Coral and Other Organic Reefs in Geologic Column," *Bull. Am. Assoc. Petrol. Geol.*, Vol. 34 (February, 1950), pp. 182–202.

[10] F. R. S. Henson, "Cretaceous and Tertiary Reef Formation and Associated Sediments in Middle East," *Bull. Am. Assoc. Petrol. Geol.*, Vol. 34 (February, 1950), pp. 215–238.

[11] Kenneth K. Landes, "Porosity through Dolomitization," *Bull. Am. Assoc. Petrol. Geol.*, Vol. 30 (March, 1946), pp. 305–318.

extent that a considerable volume of oil can be stored in the fracture fissures alone. The completely dolomitized Devonian reefs of Leduc, Alberta, have extensive porosity and permeability which have been described by Link[12] as primary. The pores appear from the published pictures[13] to have undergone considerable enlargement by leaching. Many of the vugs are lined with dolomite crystals, which suggests that the leaching may have accompanied dolomitization.

In addition to the porosity within the reef rock itself, some reefs have been elevated above the sea and have been eroded and blanketed with a veneer of clastic reef material. In the Marine, Illinois, oil pool such reef "sand" is the principal oil reservoir rather than the underlying solid reef rock.[14]

The origin of the oil within the reefs is a matter of controversy. Some, like Link,[15] believe that the oil is indigenous to the reef. Others suspect that the reef was merely a convenient body of porous rock which afforded storage space to the oil in the same manner as did the serpentine rock reservoirs in the Gulf Coastal Plain of Texas, which are described in the next section. It is obviously true that the reefs were once scenes of teeming life. It is not so obvious that any appreciable percentage of the soft body parts of the reef-building organism were trapped and preserved within the reefs, later to become oil which accumulated beneath the reef cover.

The trapping of oil and gas in reefs is the most easily understood part of the natural history of a bioherm oil deposit. All oil-producing reefs are highly porous and permeable bodies of rock, completely surrounded, at least above the water level, by fine sediment impervious to the passage of hydrocarbons in any volume. Ordinarily the pores in a bioherm are filled with water just as are the pores in any other type of buried reservoir rock. Oil in the reef rock, whether indigenous or migrant, rises through the water-filled pores until the overseal is reached. Many reefs have the same succession below the overseal of gas, oil, and water as found in anticlinal accumulations.

Oil has been produced for many years from reefs in several regions,

[12]T. A. Link, "Leduc Oil Field, Alberta, Canada," *Bull. Geol. Soc. Am.*, Vol. 60 (March, 1949), pp. 389–390.

[13]D. B. Layer et al., "Leduc Oil Field, Alberta, A Devonian Coral Reef Discovery," *Bull. Am. Assoc. Petrol. Geol.*, Vol. 33 (April, 1949), p. 586 (Plate 2, middle picture). Also W. W. Waring and D. B. Layer, "Devonian Dolomitized Reef, D-3 Reservoir, Leduc Field, Alberta, Canada," *Bull. Am. Assoc. Petrol. Geol.*, Vol. 34 (February, 1950), pp. 301–307.

[14]Heinz A. Lowenstam, "Marine Pool, Madison County, Illinois, Silurian Reef Producer," *Structure of Typical American Oil Fields* (American Association of Petroleum Geologists, 1948), Vol. 3, pp. 153–188.

[15]T. A. Link, *op. cit.*

especially the Southern Field of Mexico, and in western and northern Texas.[16] The Norman Wells reef limestone field in Northwest Territory, Canada, was discovered in 1920.[17] This field lies less than 100 miles south of the Arctic Circle. It was not until the advent of war created demand in 1942 that Norman Wells was fully developed. However, it was the discovery of the Leduc field of Alberta, a prolific reef producer, in 1947 that really started the intensive search for the bioherm type of accumulation. This field has been chosen to illustrate reef traps (Fig. 14–9). The Leduc discovery was followed shortly by Redwater, Golden Spike, and other new strikes in the Edmonton district, and within a few months the estimate of the reserve in reef-oil accumulation in Alberta had risen to 1 billion barrels.

At the same time, limestone reef fields were being discovered in Scurry County in west Texas. The first strike was made in July, 1948, and within the year five areas in the western part of the county had been found to be underlain by oil pools trapped in reefs. Subsequent drilling may connect some of these separate fields. Several wells have penetrated nearly 700 feet of oil-producing reef rock.[18] Another Texas producing reef is a crinoidal limestone of Pennsylvanian age in the Todd field of Crockett County.[19]

Lenticular sandstone bodies. A considerable volume of oil is produced each year from lenticular sands. Classified in this group are those sandstones which, owing to local conditions of sedimentation, were deposited as separate and distinct bodies of sand completely surrounded by finer sediment which subsequently became shale. Not included in this classification are the sandstone reservoirs with lenticular porosity due to lateral facies change during sedimentation. These reservoirs are discussed in a subsequent paragraph.

Lenticular sands vary in shape from extremely elongate "shoestring" sands to highly irregular-shaped bodies. They are more likely, however, to have a definite "trend" than to be equidimensional in plan. It should also be noted that even the shoestring sands are not continuous but are broken every few miles by a shale-filled gap.

[16] DeWitt C. Van Sicklen, "Organic Reefs of Pennsylvanian Age in Haskell County, Texas," *Geophysics*, Vol. 22 (July, 1957), pp. 610–629.

[17] J. S. Stewart, "Norman Wells Oil Field, Northwest Territory, Canada," *Structure of Typical American Oil Fields* (American Association of Petroleum Geologists, 1948), Vol. 3, pp. 86–109.

[18] D. H. Stormont, "Scurry County, West Texas, Limestone Reef Development," *Oil and Gas Jour.*, Vol. 48 (July 7, 1949), pp. 54 et seq.

[19] Robert F. Imbt and S. V. McCollum, "Todd Deep Field, Crockett County, Texas," *Bull. Am. Assoc. Petrol. Geol.*, Vol. 34 (February, 1950), pp. 239–262.

The origin of the lenticular sands, especially those of decided elongation, has been a subject of considerable investigation.[20] Among the various environments of deposition which have been postulated as causing the development of lenticular sand bodies are: stream channel fillings; channel fillings in the distributaries of deltas; beaches, especially beach ridges; dune sands which have drifted into and remained in the water,[21] spits, hooks, and bars; offshore bars separated from the mainland by shallow brackish-water lagoons which may be replete with plant and animal life; and sea-bottom sand accumulations which have the appearance of giant ripples on the present sea floor and which may be the result of currents or storm waves or a combination of both.[22]

Typical examples of lenticular sands are the famous Pennsylvanian shoestring sands of Greenwood and adjacent counties, Kansas (Fig. 14-10), and the Devonian sand "trends" in the Venango district, northwestern Pennsylvania. Other occurrences of shoestring sands include Pennsylvanian sands considerably younger than those in Greenwood County which produce at shallow depths in Anderson County in northeastern Kansas;[23] gas-producing Mississippian "stray" sands in Michigan;[24] the Gay-Spencer-Richardson trend in West Virginia;[25] and the Music Mountain pool in Pennsylvania.[26]

[20] N. Wood Bass, Constance Leatherock, W. Reese Dillard, and Luther E. Kennedy, "Origin and Distribution of Bartlesville and Burbank Shoestring Oil Sands in parts of Oklahoma and Kansas," *Bull. Am. Assoc. Petrol. Geol.*, Vol. 21 (January, 1937), pp. 30–36; N. Wood Bass, "Origin of Shoestring Sands of Greenwood and Butler Counties, Kansas," *State Geol. Survey Kans.*, Bull. 23 (1936); Homer H. Charles, "Oil and Gas Resources of Kansas, Anderson County," *State Geol. Survey Kans.*, Bull. 6, part 7 (1927); John L. Rich, "Shorelines and Lenticular Sands as Factors in Oil Accumulation," *Science of Petroleum* (Oxford University Press, 1938), Vol. 1, pp. 230–239; John L. Rich, "Submarine Sedimentary Features on Bahama Banks and Their Bearing on Distribution Patterns of Lenticular Oil Sands," *Bull. Am. Assoc. Petrol. Geol.*, Vol. 32 (May, 1948), pp. 767–779; R. E. Sherrill, P. A. Dickey, and L. S. Matteson, "Types of Stratigraphic Oil Pools in Venango Sands of Northwestern Pennsylvania," *Stratigraphic Type Oil Fields* (American Association of Petroleum Geologists, 1941), pp. 507–538; Oren F. Evans, "Internal Structure of Shoestring Sands," *World Oil*, Vol. 131 (July 1, 1950), pp. 66–70.

[21] Kenneth K. Landes, "Detroit River Group in the Michigan Basin," *U. S. Geol. Survey* Circular 133 (September, 1951).

[22] John L. Rich, *op. cit.*

[23] Homer H. Charles, *op. cit.*

[24] Max W. Ball, T. J. Weaver, H. D. Crider, Douglas S. Ball, "Shoestring Sand Gas Fields of Michigan," *Stratigraphic Type Oil Fields* (American Association of Petroleum Geologists, 1941), pp. 237–266.

[25] E. T. Heck, "Gay-Spencer-Richardson Oil and Gas Trend, Jackson, Roane, and Calhoun Counties, West Virginia," *Stratigraphic Type Oil Fields* (American Association of Petroleum Geologists, 1941), pp. 806–829.

[26] Charles R. Fettke, "Music Mountain Oil Pool, McKean County, Pennsylvania," *Stratigraphic Type Oil Fields* (American Association of Petroleum Geologists, 1941), pp. 492–506.

Less elongate in outline, but nevertheless lenticular, is the Devonian Bradford sand, the reservoir rock for the Bradford pool of Pennsylvania and New York.[27] Lenticular sands are oil reservoirs in many other parts of the Appalachian province also, but some accumulations so credited may be due to facies change rather than depositional lenticularity.[28] Furthermore, much of the Appalachian accumulation is actually structurally controlled, but with varying permeability responsible for the distribution of hydrocarbons across the tops of the anticlines.

Accumulation along old strand lines where the sands pinch out up-dip in Jim Hogg County, Texas,[29] has been described. The Bisti field in northwestern New Mexico also produces from Upper Creatceous sands interpreted to be shoreline in origin.[30] Lower Cretaceous shoestring sands have been described in the Denver basin.[31] Lenticular sands of Eocene age are productive in the Uinta Basin, Utah.[32] The Osage field in eastern Wyoming produces mainly from "locally thick discontinuous sandy bodies enclosed by marine shale" which constitute the Newcastle sandstone member of the Graneros (Upper Cretaceous) shale along the east flank of the Powder River basin.[33] The Sunburst and Moulton production at Cut Bank, Montana, is from Lower Cretaceous lenticular sandstone beds, but the main production is from the Cut Bank sand, in which trapping is due to facies change.[34] Sand lenticularity and facies changes in Miocene sediments are also responsible for much of the accumulation in the Maracaibo basin oil fields of western Venezuela.[35]

The trapping of hydrocarbons in lenticular sands is due to the presence of a porous and permeable reservoir completely surrounded by impervious material. The best guess as to the source of the oil is that it has been

[27] Wallace W. Wilson, "Practical Application of Geology to Reservoir Analysis," *Petrol. Engineer*, Vol. 17 (September, 1946), pp. 152 et seq.

[28] A. H. McClain, "Stratigraphic Accumulation in Jackson-Kanawha Counties Area of West Virginia," *Bull. Am. Assoc. Petrol. Geol.*, Vol. 33 (March, 1949), pp. 336–345.

[29] James E. Freeman, "Strand Line Accumulation of Petroleum, Jim Hogg County, Texas," *Bull. Am. Assoc. Petrol. Geol.*, Vol. 33 (July, 1949), pp. 1260–1270.

[30] J. Q. Tomkins, "Bisti Oil Field, New Mexico," *Bull. Am. Assoc. Petrol. Geol.*, Vol. 41 (May, 1957), pp. 906–922.

[31] H. F. Murray, "Stratigraphic Traps, Denver Basin," *Bull. Am. Assoc. Petrol. Geol.*, Vol. 41 (May, 1957), pp. 839–847.

[32] M. Dane Picard, "Red Wash-Walker Hollow Field, Eastern Uinta Basin, Utah," *Bull. Am. Assoc. Petrol. Geol.*, Vol. 41 (May, 1957), pp. 923–936.

[33] C. E. Dobbin, "Exceptional Oil Fields in Rocky Mountain Region of the United States," *Am. Assoc. Petrol. Geol.*, Vol. 31 (May, 1947), p. 801.

[34] John E. Blixt, "Cut Bank Oil and Gas Field, Glacier County, Montana," *Stratigraphic Type Oil Fields* (American Association of Petroleum Geologists, 1941), pp. 327–381.

[35] Staff of Caribbean Petroleum Company, "Oil Fields of Royal Dutch-Shell Group in Western Venezuela," *Bull. Am. Assoc. Petrol. Geol.*, Vol. 32 (April, 1948), pp. 517–628.

squeezed out of the surrounding shale into the relatively incompactible
and permeable sandstone.   A very local (vertically) source for the oil in
the shoestring sands of southeastern Kansas and northeastern Oklahoma
is indicated by the fact that the crude oils from 33 pools in the Burbank
sand distributed through an area 150 miles long are alike, but quite
dissimilar from that found in other sands separated vertically from the
Burbank by but a few feet of shale.[36]

Reservoir rocks with change in facies.   Trapping due to facies
change is usually brought about by up-dip "shaling" of sandstone or
limestone.   More rarely it is created by the merging of sandstone into
limestone.   Facies-change traps are due to environmental differences at
the time of deposition.   They occur (1) where coarser sediment merges
with finer sediment which was deposited in quieter water or at a greater
distance from the source of supply, or (2) where porous carbonate rock
merges with fine clastic material in what was the landward direction at
time of deposition.   In order for the facies change to become a trap (1)
the sedimentary beds must be tilted in such a direction as to place the
impervious facies up-dip from the porous rock, and (2) the tilted rocks
must be creased by a longitudinal (plunging) anticline so as to close the
trap on the sides.   However, condition (2) can be avoided if the zone of
disappearing permeability is curved, with its concave-side down-dip
thereby sealing the trap on the sides as well as across the top.

An outstanding example of the up-dip shaling of carbonate rock reservoirs is illustrated by the Hugoton field of southwestern Kansas (Fig.
14–11).   Gas has accumulated below the facies barrier in this area for
many miles down-dip.   More recently the Cottonwood Creek field in
Wyoming has been discovered and developed (Fig. 14–2).   Here dolomite
of the Phosphoria (Permian) formation changes up-dip to red shale.
Anhydrite present in the shale assists in the sealing of the trap.   Lateral
closure is partly by structural nosing but mainly by the concave shape
of the permeability boundary.   An additional distinction of Cottonwood
Creek is that it is an important stratigraphic trap accumulation sought for
through surface and subsurface studies which was discovered by design
rather than by accident.[37]

Trapping by the up-dip shaling of a sandstone is the reason for the
accumulation at Pembina, one of Canada's great oil fields which has a

[36] Tulsa Geological Society Research Committee, "Relationship of Crude Oils and
Stratigraphy in Parts of Oklahoma and Kansas," *Bull. Am. Assoc. Petrol. Geol.*, Vol. 30
(May, 1946), pp. 747–748; L. M. Neumann, et al., "Relationship of Crude Oil and Stratigraphy in Parts of Oklahoma and Kansas," *Bull. Am. Assoc. Petrol. Geol.*, Vol. 25 (September, 1941), pp. 1801–1809.

[37] John D. Pedry, "Cottonwood Creek Field," *Bull Am. Assoc. Petrol. Geol.*, Vol. 41
(May, 1957), pp. 823–838.

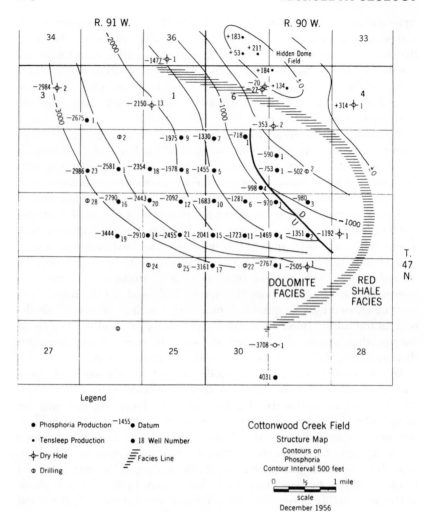

Fig. 14-2. Cottonwood Creek Field, Washakie County, Wyoming. A facies change trap. *Courtesy American Association of Petroleum Geologists.*

daily production (1957) of over 100,000 barrels and an estimated recoverable primary reserve of over one billion barrels.[38] Here the Cardium formation, of Upper Cretaceous age, changes up-dip along a hinge line many miles in length from a sand to a shale facies (Fig. 14-3). At least four separate oil-producing sandstones and a conglomerate pinch out in

[38] A. M. Patterson and A. A. Arneson, "Geology of Pembina Field, Alberta," *Bull. Am. Assoc. Petrol. Geol.*, Vol. 41 (May, 1957), pp. 937–949.

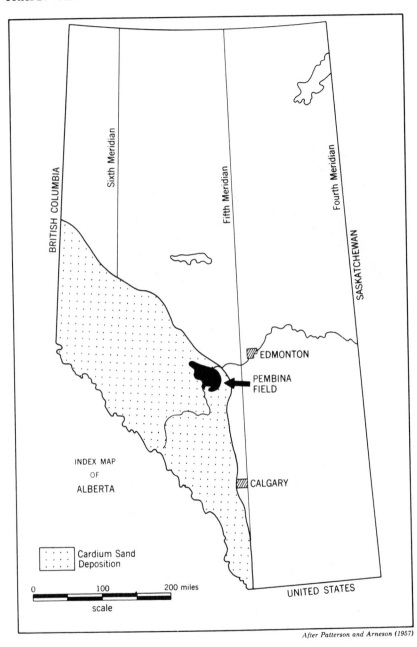

After Patterson and Arneson (1957)

Fig. 14-3. Location of Pembina Field in Alberta, and its position in respect to the edge of the Cardium sand. *Courtesy American Association of Petroleum Geologists.*

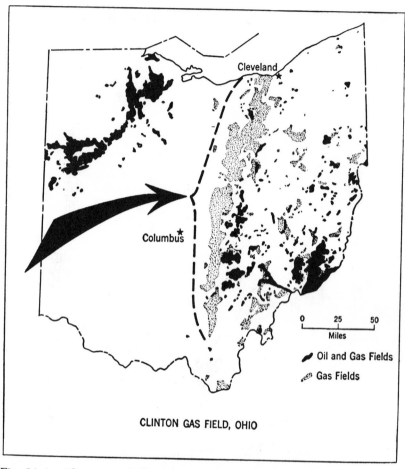

Fig. 14–4. Clinton gas field, Ohio. Trapping due to up-dip termination of porosity in reservoir rock. *United States Geological Survey.*

this zone. The Pembina trap is closed on the sides, as well as at the top, by decreasing permeability.

Another possible illustration of trapping by up-dip shaling is the Clinton gas field of central Ohio. Accumulation here has been along the western edge of the reservoir rock on the east flank of the Cincinnati arch. The pools, mostly gas but some oil, occur in a belt extending from Lake Erie southward to the Ohio River (Fig. 14–4). The "Clinton" sandstone reservoir is in reality a white sandstone belonging to the Medina

(Silurian) rather than the Clinton. This sandstone thins rapidly westward and is absent or "very indefinite" before reaching the central part of eastern Ohio.[39] There is some doubt whether the western boundary of the Clinton sand is actually a change in facies from sand to shale. The westward thinning of the sandstone, with the source of supply to the east, makes the concept of a change to finer sediment, deposited contemporaneously, appear logical. However, some have interpreted the meager subsurface data available to indicate the deposition of discontinuous lenses of sand as offshore bars in a regressive Medina Sea.

The sandstone reservoir rocks in the Bryson field, Texas, grade up-dip into shale and silstone with the oil trapped down-dip below the zone of facies change.[40] The Cut Bank sand (Lower Cretaceous), previously mentioned as the chief oil-reservoir rock in the Cut Bank field, Montana, is a cherty, conglomeratic, pervious sand where it contains oil, but it merges up-dip and laterally into tight, chert-free, sandstone and siltstone. This facies change may be due to irregular sedimentation in a floodplain and delta environment.[41] The possible role of facies change in oil accumulation in the Jackson-Kanawha district of West Virginia[42] has also been mentioned previously, as well as the impounding of hydrocarbons below facies barriers in the Maracaibo basin, Venezuela.[43] Although the principal cause for trapping in the Benton field of Illinois is an anticline, a facies change in the upper Mississippian from sandstone into shale in the northeast part of the field completes the closure needed for oil accumulation.[44]

It is not always easy to distinguish traps produced by a sandstone-to-shale up-dip facies change from lenticular sands. The erratic New Castle sandstone is classified in the preceding section as a lenticular sand in the Osage field of eastern Wyoming, but in the near-by Marsh Creek-Skull

[39] W. Stout, R. E. Lamborn, D. T. Ring, J. S. Gillespie, and J. R. Lockett, "Natural Gas in Central and Eastern Ohio," *Geology of Natural Gas* (American Association of Petroleum Geologists, 1935), p. 908; Richard H. Denman, "The Clinton Gas Field of Ohio," *The Compass*, Vol. 22 (March, 1942), pp. 164–170.

[40] T. C. Hiestand, "Bryson Oil Field, Jack County, Texas," *Stratigraphic Type Oil Fields* (American Association of Petroleum Geologists, 1941), pp. 539–547.

[41] John E. Blixt, "Cut Bank Oil and Gas Field, Glacier County, Montana," *Stratigraphic Type Oil Fields* (American Association of Petroleum Geologists, 1941).

[42] A. H. McClain, "Stratigraphic Accumulation in Jackson-Kanawha Counties Area of West Virginia," *Bull. Am. Assoc. Petrol. Geol.*, Vol. 33 (March, 1949), pp. 336–345.

[43] Staff of Caribbean Petroleum Company, "Oil Fields of Royal Dutch-Shell Group in Western Venezuela," *Bull. Am. Assoc. Petrol. Geol.*, Vol. 32 (April, 1948), p. 543.

[44] J. V. Howell, "Geology of Benton Field, Franklin County, Illinois," *Bull. Am. Assoc. Petrol. Geol.*, Vol. 32 (May, 1948), pp. 745–766.

Creek area the trapping is perhaps better described as due to facies changes from sandstone to shale.[45]

Varying Permeability Due to Ground-Water Activity. Circulating underground waters have the ability to increase rock porosity by solution and to diminish porosity by precipitation of mineral matter in the pre-existing voids. Fortunately for mankind, more reservoirs have been created by solution than have been destroyed by precipitation. Although solution may occur in any type of rock, even sandstone (Chapter 6), the reservoirs most commonly aided by solution activity are carbonate rocks. In many carbonate rock reservoirs the oil is stored in cavities created or at least enlarged by the dissolving action of circulating ground waters. This leaching is selective, being most pronounced in the more susceptible layers and perhaps absent altogether in strata less vulnerable to this type of attack. Within a given stratigraphic zone, however, the extent of the permeability created by ground-water solution has a fair degree of lateral consistency, even though some zones are notoriously erratic in the degree of permeability across short distances. Because of the usual widespread character of solution porosity it is difficult to find examples of pools in which the trapping has been solely due to local solution porosity in carbonate rock. An exception is dolomitization porosity, which is discussed in a subsequent paragraph. Accumulation in solution-porosity reservoirs is either anticlinal or is due to faulting, truncation and overlap, or some other conventional type of trap. Locally erratic porosity may result in the drilling of many dry holes inside the periphery of oil pools of this sort, however. Variations in permeability therefore may control the *distribution* of oil within a trap.

A quite different situation is the porosity which is produced by and during dolomitization and which may be local in extent and without any structural relationship. An example of this type of accumulation is the Deep River pool of Michigan, in which the oil has accumulated without regard to rock structure in a long, narrow porous zone where the Devonian limestone has been locally dolomitized. Accumulation at Deep River is illustrated in Fig. 14–12. Similar in the manner of trapping are the many oil and gas pools which form the Lima-Indiana district.[46] Although these fields lie either on the flanks or across the top of the broad Findlay arch, their localization at any one point is not a result of the structural situation but a result of the fact that there and there only the Trenton (Ordovician)

---

[45] H. E. Summerford, E. E. Schieck, T. C. Hiestand, "Oil and Gas Accumulation Controlled by Sedimentary Facies in Upper Cretaceous New Castle Sandstone, Wyoming," *Bull. Am. Assoc. Petrol. Geol.*, Vol. 34 (September, 1950), pp. 1850–1865.

[46] J. Ernest Carman and Wilber Stout, "Relationship of Accumulation of Oil to Structure and Porosity in the Lima-Indiana Field," *Problems of Petroleum Geology* (American Association of Petroleum Geologists, 1934), pp. 521–529.

limestone has been dolomitized and has sufficient porosity to function as a reservoir rock. The oil has accumulated in the dolomitized zone because that is the only place in a considerable stratigraphic section where there is enough porosity and permeability for the rock to be a reservoir. As in the lenticular sands, the trapping is due to the presence of a reservoir of limited extent completely surrounded by impermeable limestone and shales.

An unusual condition exists in the Gulf Coastal Plain of Texas and elsewhere where igneous rock bodies of limited extent have been altered, leached, weathered, and fractured into porous and permeable serpentines, as at Chapman, shown in Fig. 14–13. These porous rock masses are likewise surrounded by material impervious to the movement of hydrocarbons and have afforded a haven to migrating hydrocarbons. They function as reservoirs in exactly the same way as the previously described dolomitized zones, lenticular sandstones, and isolated reefs.

Thorough cementation of the reservoir rock can create an up-dip barrier to further migration by trapping the oil below, but examples of this situation are scarce. Perhaps the best illustration is in the Rose Hill field of southwest Virginia where the oil occurs in fractured Trenton (Ordovician) limestone beneath a folded regional overthrust fault. Miller[47] ascribes the actual trapping to cementation of the fractured limestone and dolomite up-dip from the oil pool by circulating carbonate-depositing ground waters. In the Tri-County oil field of southwestern Indiana, differential cementation has been effective in limiting the distribution of oil, but accumulation at that particular spot has been primarily due to structure.[48] The same is true in the New Harmony field[49] in Illinois and Indiana. Accumulation of oil has been influenced by eccentric cementation of the reservoir rock in other fields, including East Tuskegee,[50] Oklahoma, Greasewood,[51] Colorado, and Stephens,[52] Arkansas.

Many have noted the confinement of oil to zones of coarse, loosely

[47]Ralph L. Miller, "Carbonate Cementation as a Sealing Factor in Oil Accumulation, Southwest Virginia," *Am. Assoc. Petrol. Geol., Annual Meeting Program* (1956), pp. 38–39.

[48]Joseph M. Wanenmacher and Wendell B. Gealy, "Surface and Subsurface Structure of the Tri-County Oil Field of Southwestern Indiana," *Bull. Am. Assoc. Petrol. Geol.,* Vol. 14 (April, 1930), pp. 423–431.

[49]George V. Cohee, "Lateral Variation in Chester Sandstones Producing Oil and Gas in Lower Wabash River Area," *Bull. Am. Assoc. Petrol. Geol.,* Vol. 26 (October, 1942), p. 1606.

[50]J. L. Borden and R. A. Brant, "East Tuskegee Pool, Creek County, Oklahoma," *Stratigraphic Type Oil Fields* (American Association of Petroleum Geologists, 1941), pp. 436–455.

[51]C. S. Lavington, "Greasewood Oil Field, Weld County, Colorado," *ibid.,* pp. 19–42.

[52]William C. Spooner, "Stephens Oil Field, Colombia and Ouachita Counties, Arkansas," *Structure of Typical American Oil Fields* (American Association of Petroleum Geologists, 1929), Vol. 2, pp. 1–17.

cemented sandstone surrounded by fine and tightly cemented sandstone. The general tendency is to ascribe this situation to cementation *after* oil accumulation has taken place.[53] The oil enters the coarser sands initially because of the greater room in the interstices between the water-coated grains of sand. Then comes the cementation, which is confined largely to the non-oil-bearing rocks, because where oil is present in the voids it prevents cementation. Evidence of precementation oil in a limestone reservoir has been described by Wegemann: "Fragments of limestone thrown out of wild wells in the South Fields of Mexico and kept for years will, when struck with a hammer, break with explosive force due to confined gases cemented off in the pores of the limestone, proving cementation after accumulation."[54] Sherrill, Dickey, and Matteson note the presence of oil in Pennsylvania in comparatively unconsolidated gravel beds surrounded by fine, thoroughly cemented sands, and they conclude likewise that the lack of cementation in the coarser clastics was due to the early entrance of oil into them.[55]

If the cementation is postaccumulation, then the trap must be precementation in order for the accumulation to have taken place, and unless there has been subsequent tilting, the tightly cemented rock would not be a factor in the capturing of the oil.

The "freezing" of already trapped oil or gas by reservoir cementation around the periphery of the hydrocarbon accumulation was described in Chapter 8. It was also cited as a possible cause for tilted interfaces in Chapter 13. If the "postfreezing" tilt is of enough magnitude to wipe out the pre-existing structural closure, the entrapped oil has all the appearance of either a hydrodynamic or an up-dip facies-change accumulation, but of course it is not either one.

Varying Permeability Due to Truncation and Sealing. This type of trap has been responsible for the accumulation of enormous deposits of oil. The first step in the natural history of such accumulations is the emergence, folding and tilting, and beveling by erosion of a sedimentary section including reservoir rocks. The second step is the sealing of the edges of the reservoir layers, where they are folded into a plunging anticline, by impervious material. In rare instances this impervious material has been asphalt formed from the seeping oil itself, but most accumulations of this type have been beneath beds of shale or dense limestone which were deposits on top of the truncated, beveled strata during a subsequent submergence. This type of trap is referred to as an overlap seal.

[53] F. M. Van Tuyl and Ben H. Parker, "The Time of Origin and Accumulation of Petroleum," *Quarterly Colo. School of Mines*, Vol. 36 (April, 1941), Chapter 8, pp. 62–66.
[54] Carrol H. Wegemann, in F. M. Van Tuyl and Ben H. Parker, *op. cit.*, p. 63.
[55] *Op. cit.*, p. 538.

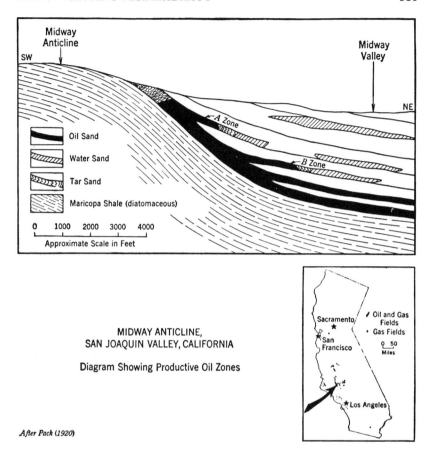

MIDWAY ANTICLINE,
SAN JOAQUIN VALLEY, CALIFORNIA

Diagram Showing Productive Oil Zones

After Pack (1920)

Fig. 14–5. Solid hydrocarbon (asphalt) seal, Midway anticline, California. *United States Geological Survey.*

Solid hydrocarbon seal. One of the relatively few examples of sealing by solid hydrocarbons is along the southwest edge of the Midway field in California[56] (Fig. 14–5). Oil moving to the southwest up the flank of the Midway syncline has been trapped just below the reservoir-rock outcrop by the accumulation of asphalt there. It has been suggested that this tar was formed by the interaction between mineralized ground waters and petroleum hydrocarbons. Solid hydrocarbon seals also play a role in the trapping of oil in other California fields, especially East Coalinga and McKittrick. Although some of the oil in the Whittier field in California has been trapped by the termination of the reservoir rock

[56] R. W. Pack, "The Sunset-Midway Oil Field, California," *U. S. Geol. Survey*, Prof. Paper 116 (1920), Fig. 13, p. 115.

against the Whittier fault, there has also been trapping due to the presence of asphaltic residue at the outcrop of the reservoir.[57] A solid hydrocarbon seal is present also in the Cacheuta field in Argentina.[58] Five to fifteen feet of asphalt-impregnated sandstone prevents the oil from reaching the top of buried sand bars in Ohio County, Kentucky.[59]

Overlap and offlap seals. Overlap seals can be divided into two types: (1) the strata overlap a truncated plunging anticline on the flank of a monocline; (2) the overlapping sediments overlie a truncated closed anticline, which has been aptly referred to as a "baldheaded structure."

Angular unconformities are a well-known type of trap and the most prolific of those that fall within the varying-permeability classification.[60] The outstanding example of sealing by overlap of monoclinal sediments is the East Texas field, in which the beveled Woodbine sandstone has been effectively sealed by overlapping Austin chalk. Accumulation in this field is illustrated in Fig. 14–14. Eastern Venezuela has many examples of this type of hydrocarbon trap.[61]

If the pre-overlap tilted sediments are of unequal hardness, differential erosion may scallop the surface into hogback ridges and strike valleys. In the Apco field in Pecos County, Texas, a hogback of permeable Ellenburger dolomite was submerged and buried beneath impervious Permian sediments, where it formed an ideal trap for oil accumulation (Fig. 14–15). Very similar in its structural pattern to East Texas is the West Edmond oil field of central Oklahoma which was discovered in 1943.[62] At West Edmond the tilted, truncated, and overlapped reservoir rock is the Bois d'Arc limestone of Hunton (Siluro-Devonian age), sealed by unconformably overlying Lower Pennsylvanian shale. At both West

[57] W. H. Holman, "Whittier Oil Field," *Oil and Gas Fields of California* (*Calif. Div. Mines, Bull.* 118, 1943), p. 290.

[58] "YPF Holds Leading Place in Argentine Oil Development," *World Petrol.*, Vol. 12 (September, 1941), pp. 72–79.

[59] Frank H. Walker, "Sealing Factors in the Buford or Barnett Creek Pool in Ohio County, Kentucky," *Am. Assoc. Petrol. Geol., Annual Meeting Program* (1956), pp. 32–33.

[60] Frank J. Gardner, "Relationship of Unconformities to Oil and Gas Accumulation," *Bull. Am. Assoc. Petrol. Geol.*, Vol. 24 (November, 1940), pp. 2022–2031; A. I. Levorsen, "Relation of Oil and Gas Pools to Unconformities in the Mid-Continent Region," *Problems of Petroleum Geology* (American Association of Petroleum Geologists, 1934), pp. 761–784.

[61] H. H. Renz, H. Alberding, K. F. Dallmus, J. M. Patterson, R. H. Robie, N. E. Weisbord, and José MasVall, "The Eastern Venezuela Basin," *Habitat of Oil* (American Association of Petroleum Geologists, Special Vol. 1958), pp. 551–598.

[62] D. A. McGee and H. D. Jenkins, "West Edmond Oil Field, Central Oklahoma," *Bull. Am. Assoc. Petrol. Geol.*, Vol. 30 (November, 1946), pp. 1797–1829; Robert M. Swesnik, "Geology of West Edmond Oil Field, Oklahoma, Logan, Canadian and Kingfisher Counties, Oklahoma," *Structure of Typical American Oil Fields* (American Association of Petroleum Geologists, 1948), Vol. 3, pp. 359–398.

Edmond and East Texas the reservoir rock extends down-dip westward into a large basin which supplies gathering area of considerable magnitude. To the south of West Edmond, along the eastern margin of the Anadarko basin, are additional unconformity accumulations including Oklahoma City, a baldheaded structure described in a subsequent paragraph. Similar traps are found in various other parts of the world. Examples include many fields in California, especially in the San Joaquin Valley.[63]

Overlap across bald-headed structures may also be complicated by pre-overlap topography. If the strata in the anticlinal fold are similar in their degree of resistance to erosion, the plane of the unconformity will be flat and the picture the same as for an overlapped beveled monocline except that the reservoirs may be repeated, in reverse order, on the opposite side of the buried anticline. Many times, however, the rocks in the anticlinal fold prove to be of unequal resistance, and if the core is harder than the younger flanking rocks, the buried eroded anticline will also be a hill. Sometimes later folding along the same line of weakness sharpens the relief of the hill by further upward arching of the plane of unconformity. It is also possible for a hill containing flat-lying permeable strata to be submerged and covered by impervious overlapping rock, so that it becomes a trap for oil accumulation. Such situations, however, are uncommon.

Sealing on a bald-headed structure is usually brought about by the blanketing effect of the overlapping younger sediments as illustrated by the Kraft-Prusa field (Fig. 14–16). Another example of this type of trap is the Oklahoma City field,[64] one of the great oil fields of the world. There the core ("baldhead") of the truncated dome is Arbuckle limestone of Cambro-Ordovician age. Arranged in a semicircular pattern (Fig. 14–6), on the west are successively younger Ordovician strata, including various sandstones, whereas to the east, cutting across the side of the dome, is a major fault which has dropped the Mississippian rocks down to the level of the Arbuckle, and buried the Ordovician reservoirs to great depth (Fig. 14–7). This eroded faulted dome is overlain unconformably by Pennsylvanian formations. The principal reservoir rocks are the Arbuckle limestone at the top of the dome beneath the uncon-

[63] George M. Cunningham and W. D. Kleinpell, "Importance of Unconformities to Oil Production in the San Joaquin Valley, California," *Problems of Petroleum Geology* (American Association of Petroleum Geologists, 1934), pp. 785–805.

[64] D. A. McGee and W. W. Clawson, Jr., "Geology and Development of Oklahoma City Field, Oklahoma County, Oklahoma," *Bull. Am. Assoc. Petrol. Geol.*, Vol. 16 (October, 1932), pp. 957–1020.

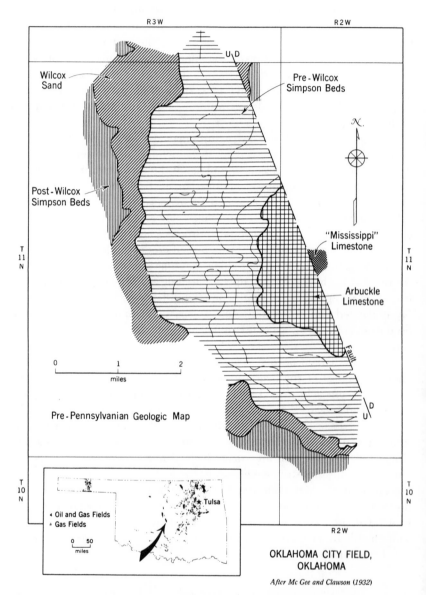

Fig. 14–6. The Oklahoma City oil field, an example of sealing by overlap across a truncated (bald-headed) faulted dome. *Courtesy American Association of Petroleum Geologists.*

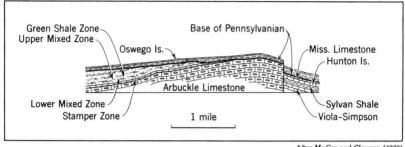

After Mc Gee and Clawson (1932)

Fig. 14-7. Transverse cross section through Oklahoma City Field, Oklahoma. *Courtesy American Association of Petroleum Geologists.*

formity, and various truncated sandstone formations belonging to the Simpson group of Ordovician age.

Baldheaded structures buried beneath unconformities have been important producers of oil in many other localities.[65] At El Dorado,[66] Kansas, the core of the truncated anticline is the completely buried Nemaha granite ridge. In the Texas Panhandle the conditions are somewhat similar, but here the core rock is the east-west trending Amarillo Mountain range. At Healdton, Oklahoma, complexly folded sediments, instead of granite, underlie the unconformity. The Golden Lane pool of Mexico, south of Tampico, produces oil from a great buried anticlinal ridge of porous limestone. Other fields which illustrate this situation are Cushing and Seminole, Oklahoma, Gorham and Fairport (Fig. 3–4), Kansas, and Yates, Texas.[67]

Sealing at unconformities by the abutting of the reservoir rock against impervious older rock is also possible. This occurrence is an offlap rather than an overlap relationship, and it results from regression of the ancient sea in which reservoir-rock deposition took place. The trapping of hydrocarbons in some of the producing sands in the Amarillo district of the Texas Panhandle has been a result of the deposition of granite wash material on the flanks of the now-buried Amarillo Mountains. These

---

[65] John L. Ferguson and Jess Vernon, "The Relationship of Buried Hills to Petroleum Accumulation," *Science of Petroleum* (Oxford University Press, 1938), Vol. 1, pp. 240–243; Joseph A. Kornfeld, "Stratigraphic Traps, Source of Major Production over Central Kansas Uplift," *Oil Weekly*, Vol. 100 (Jan. 15, 1941), pp. 13–19; A. I. Levorsen, "Stratigraphic versus Structural Accumulation," *Oil and Gas Jour.*, Vol. 34 (March 26, 1936), pp. 41 et seq.

[66] John R. Reeves, "El Dorado Oil Field, Butler County, Kansas," *Structure of Typical American Oil Fields* (American Association of Petroleum Geologists, 1929), Vol. 2, pp. 160–167.

[67] John L. Ferguson and Jess Vernon, *op. cit.*

arkosic sediments wedge out against the granite surface and have been covered by younger impervious rock.[68]   At the southeast edge of the Oklahoma Anadarko basin in Garvin and McClain Counties, northwest of the Arbuckle Mountains, the Late Pennsylvania sea lapped up on the truncated edges of various older Pennsylvanian formations after the Paul's Valley uplift[69] had tilted the plane of the eroded surface to the west. During a transgressive phase of the Late Pennsylvanian sea, the Deese sands were deposited in sheet form extending out into the basin in one direction and abutting against the plane of angular unconformity in the other (Fig. 14–8). In both the Texas Panhandle and Deese sand fields, accumulation is the result of the impounding of oil moving up-dip and reaching an impervious barrier where the reservoir rocks terminate against the steeper surface of uplifted impervious older rocks.

Only a beginning has been made in searching for the offlap type traps. Krueger[70] has pointed out several places in the Rockies where Paleozoic sediments, some of them prolific reservoirs not far away, abut against "old positive elements." Similarly in southern Israel, Paleozoic sediments terminate against Precambrian granite hills somewhere below the surface.

Stratigraphic Trap Finding. As a rule, stratigraphic traps, unlike structural traps, give no surface indications of their presence. There are, however, two possible exceptions to this generalization. Differential compaction of shale above a shoestring sand may result in a slight arching of the strata.[71] Obviously it is necessary in order to determine the presence of buried sandstones of this type to employ extreme accuracy and detail in mapping the structure in the surface bed-rock formations. The other possible exception is that resulting from the draping and differential compaction of the sediments overlying limestone reefs. If the reef is of sufficient magnitude, its presence may be reflected to the surface by the configuration of the younger strata. That is the situation in the Marine pool of Illinois, where a Pennsylvanian limestone only 250 feet below the surface has 60 feet of closure. The much deeper reef has double that closure.[72]

[68] Henry Rogatz, "Geology of Texas Panhandle Oil and Gas Field," *Bull. Am. Assoc. Petrol. Geol.*, Vol. 19 (August, 1935), pp. 1089–1109; Vol. 23 (July, 1939), pp. 983–1053.

[69] "Anadarko Basin Discoveries," *The Link*, Vol. 12 (May, 1947), pp. 1–5.

[70] Max L. Krueger, "Strat Oil—Rockies' Biggest Challenge," *Oil and Gas Jour.* (April 19, 1954), pp. 156 et seq.

[71] John L. Rich, "Application of Principle of Differential Settling to Tracing of Lenticular Sand Bodies," *Bull. Am. Assoc. Petrol. Geol.*, Vol. 22 (July, 1938), pp. 823–833.

[72] Heinz A. Lowenstam, "Marine Pool, Madison County, Illinois, Silurian Reef Producer," *Structure of Typical American Oil Fields* (American Association of Petroleum Geologists, 1948), Vol. 3, pp. 153–188.

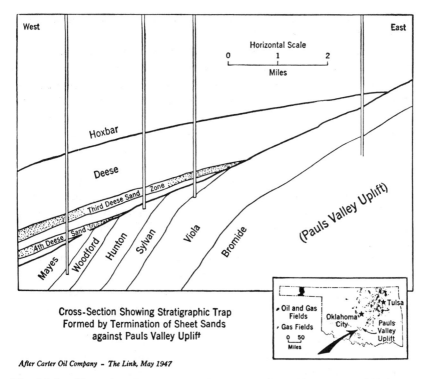

Cross-Section Showing Stratigraphic Trap
Formed by Termination of Sheet Sands
against Pauls Valley Uplift

After Carter Oil Company - The Link, May 1947

Fig. 14-8. Unconformity seal caused by onlap. Pauls Valley area, Oklahoma
Courtesy The Link and the Carter Oil Company.

The techniques of geophysical exploration are of greatest value in mapping folds. Where the sediments are arched over a reef, geophysical instruments may determine that arching as in the case of other anticlines. The Leduc field in Alberta was discovered in this way.[73] Theoretically the reef itself, if of any magnitude, should be discoverable by geophysical methods, but geophysical techniques have not yet progressed to the point at which their application has had any spectacular success. Gravity surveys tend to be complicated by unevenness on the crystalline rock basement floor, and seismic surveys tend to draw a blank above reef-containing areas. The latter, a negative criterion, may prove to be useful in reef-finding, however. Refinements in both instrumentation and interpretation are steadily increasing the value of both the seismograph and

[73]T. A. Link, "Leduc Oil Field, Alberta, Canada," *Bull. Geol. Soc. Am.,* Vol. 60 (March, 1949), pp. 381–402.

the gravity meter as subsurface tools, and better identification of buried reefs is a definite future possibility.[74]

There has been no notable success as yet in locating sand lenses, or zones of up-dip shaling of sandstones and carbonate rocks, by geophysical instruments, but endeavors continue.[75] New methods of presentation of reflection seismic data make better interpretation of the geology possible. Isochron (equal time) seismic maps, like isopach maps, have possibilities for locating stratigraphic wedge-outs and draping over sand bars.[76]

With these exceptions and possible exceptions, the sole technique for finding stratigraphic traps is the use of subsurface geology.[77] Wells already drilled have to be squeezed dry of every bit of available information. When the subsurface information is integrated with the surface geology, a three-dimensional picture develops of the regional paleogeography and paleogeology. The general areas of (1) overlap and offlap, (2) wedge-outs beneath unconformities, (3) river beds, shorelines and deltas,[78] (4) facies changes, and (5) favorable reef environment are depicted. It may be necessary to drill stratigraphic-structural tests in order to locate more closely the varying permeability traps to be tested with exploratory wells.

To develop the necessary three-dimensional picture from subsurface and other data the geologists must sharpen his old tools and create new tools. Of great value are facies studies,[79] studies of both recent[80] and

[74] Richard A. Pohly, "New Gravity Approach Aids Reef Interpretation," *World Oil* (May, 1953), pp. 116 et seq.; F. McQueen Rozelle, "Seismic Exploration for Reefs," *World Oil* (June, 1952), pp. 83 et seq.; Fred J. Agnich and William P. Harvey, "Primary Seismic Evidence of Limestone Reefs," *World Oil*, Vol. 131 (September, 1950), pp. 83–86.

[75] Gerald H. Westby, "Discovery of Stratigraphic Traps by the Reflection Seismograph," *Oil and Gas Jour.* (March 17, 1958), pp. 144–156; Neal Clayton, Richard A. Pohly, "New Techniques Being Used to Find Stratigraphic Traps," *World Oil* (April, 1955), pp. 116 et seq.

[76] Gerald H. Westby, *lecture*, University of Michigan, Ann Arbor, Michigan (Nov. 14, 1958).

[77] V. E. Monnett, "Stratigraphic Exploration and Future Discoveries," *Oil Weekly*, Vol. 101 (March 31, 1941), pp. 26 et seq.

[78] Daniel A. Busch, "Deltas Significant in Subsurface Exploration," *World Oil*, (January, 1955), pp. 82 et seq.

[79] Parke A. Dickey and Richard E. Rohn, "Facies Control of Oil Occurrence," *Habitat of Oil* (American Association of Petroleum Geologists, Special Vol., 1958), pp. 721–734; L. L. Sloss, "Location of Petroleum Accumulation by Facies Studies," *Proc. Fourth World Petrol. Congress* (1955), Section I, pp. 315–335; J. Dufour, "Facies Shift and Isochronous Correlation," *Proc. Third World Petrol. Congress* (1951), Section I, pp. 428–438.

[80] Francis P. Shepard, "Stratigraphic Research That May Pay Off," *Oil and Gas Jour.* (Feb. 21, 1955), pp. 184 et seq.; W. C. Krumbein, "Recent Sedimentation in the Search

ancient[81] sediments, paleoecology,[82] petrology,[83] and geochemistry. The last can be used in seeking zones of dolomitization porosity by means of iso-magnesia or magnesia-lime ratio lines.[84]

It is obvious that the search for new stratigraphic traps is more expensive than the search for structural traps. But as the undiscovered traps become fewer and fewer, the emphasis on the search for stratigraphic or varying-permeability traps will increase. It is equally apparent that this search is possible only by the employment of the finest type of scientific geology.

---

for Petroleum," *Bull. Am. Assoc. Petrol. Geol.*, Vol. 29 (September, 1945), pp. 1233–2161; "Principles of Sedimentation and the Search for Stratigraphic Traps," *Econ. Geol.*, Vol. 36 (December, 1941), pp. 786–810.

[81] Earl H. Linn, "Sedimentation . . ." *Oil and Gas Jour.* (April 30, 1956), pp. 165–170.

[82] Samuel P. Ellison, Jr., "Economic Applications of Paleoecology," *Econ. Geol.—50th Anniversary*, Part II (1905–1955), pp. 867–884.

[83] R. Passega, "Texture as Characteristic of Clastic Deposition," *Bull. Am. Assoc. Petrol. Geol.*, Vol. 41 (September, 1957), pp. 1952–1984; Morton K. Blaustein, "Relation of Petrology and Structure to Productivity in a Stratigraphic Trap, Lindman Oil Field, Rennels County, Texas," *Am. Assoc. Petrol. Geol., Annual Meeting Program* (1956), pp. 73–74; André Vatan, "Les Progres recents de la sedimentologie et ses applications a la recherche du petrole," *Proc. Fourth World Petrol. Congress* (1955), Section I, pp. 5–17; D. J. Doeglas, "Sedimentology and Petroleum Geology," *Proc. Third World Petrol. Congress* (1951), Section I, pp. 439–445.

[84] Kenneth K. Landes, "Porosity Through Dolomitization," *Bull. Am. Assoc. Petrol. Geol.*, Vol. 30 (March, 1946), pp. 305–318; Richard L. Jodry, "Rapid Method for Determining Magnesium-Calcium Ratios of Well Samples and its Use in Predicting Structure and Secondary Porosity in Calcareous Formations," *Bull. Am. Assoc. Petrol. Geol.*, Vol. 39 (April, 1955), pp. 493–511.

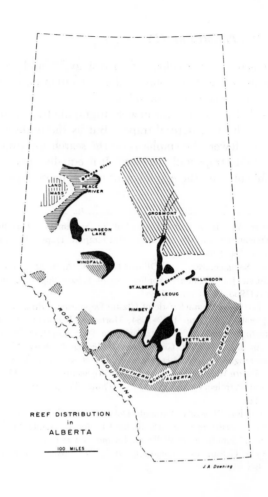

REEF DISTRIBUTION
in
ALBERTA

100 MILES

J.A. Downing

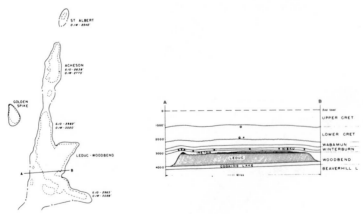

TRAP CASE STUDIES   Reef Trap

Fig. 14–9. Leduc Field, Alberta.

The Leduc field ° is 15 miles southwest of Edmonton, Alberta. It was discovered in February, 1947, and the first well was completed with an initial potential of 1000 barrels per day from Upper Devonian carbonate rocks at a depth of approximately 5000 feet. Development of the field has led to more than 1200 wells with recoverable oil estimated in excess of 200 million barrels.

The Leduc oil occurs in two reefoid reservoirs. The lower Leduc formation is a reef bioherm, the D–3 zone, while the upper is the Nisku formation, a biostromal reef deposit known as the D–2 † zone. The two reservoirs are separated by a hundred feet or more of impervious shale. The location of the Leduc field to the overall reef (D–3) distribution in Alberta is shown in A. It lies midway along a chain of biohermal reefs which extend north-northeast for more than 200 miles. The cross section accompanying B shows the Leduc reef and the overlying strata; the vertical scale is exaggerated approximately 5 to 1. The D–3 oil column is shown in solid black with overlying gas cap in fine stipple. The cross-hatched portion of the reef is the water laden section. The structure on the Nisku (D–2) results from "drape" over the underlying bioherm. The overlying formations also reflect this structure, but to a lessening degree. The average thickness of the oil column is 35 feet in the Leduc (D–3) reef, and 63 feet in the Nisku (D–2).

The Nisku (D–2) formation, which was the reservoir rock for the Leduc discovery well, is a porous dolomite containing abundant corals and bryozoa. Below is a greenish grey, calcareous shale overlying the Leduc (D–3) bioherm. The reef rock is almost entirely a crystalline dolomite with residual structures of bryozoans, stromatoporoids, corals and algal organisms. Porosity, averaging about 8 per cent, is mainly vug and intercrystalline with near-vertical fractures common in some wells. In the Leduc field the bioherm is some 600 feet thick.

Accumulation in the Leduc (D–3) zone is entirely stratigraphic, controlled by the presence of a highly porous reef dolomite completely surrounded by impervious shale. The accumulation of oil in the D–2 zone, however, is both anticlinal and stratigraphic. *Courtesy Theodore A. Link.*

° D. B. Layer et al., "Leduc Oil Field, Alberta, A Devonian Coral Reef Discovery," *Bull. Am. Assoc. Petrol. Geol.*, Vol. 33 (April, 1949), pp. 572–602; T. A. Link, "Leduc Oil Field, Alberta Canada," *Bull. Geol. Soc. Am.*, Vol. 60 (March, 1949), pp. 381–402.

† Geological Staff of Imperial Oil Ltd., "Devonian Nomenclature in the Edmonton Area, Alberta, Canada." *Bull. Am. Assoc. Petrol. Geol.*, Vol. 34 (September, 1950), pp. 1807–1825.

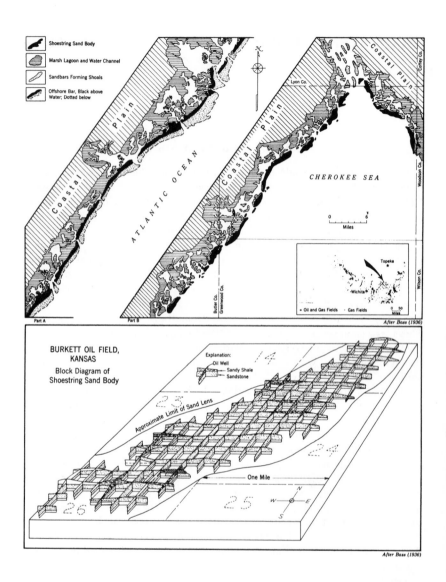

Shoestring Sand Body

Marsh Lagoon and Water Channel

Sandbars Forming Shoals

Offshore Bar, Black above Water; Dotted below

Lyon Co.

CHEROKEE SEA

Coastal Plain

ATLANTIC OCEAN

Coastal Plain

Coastal Plain

Miles

Butler Co.

Greenwood Co.

Woodson Co.

Wilson Co.

Topeka

Wichita

● Oil and Gas Fields · Gas Fields

Miles

Part A

Part B

After Bass (1936)

BURKETT OIL FIELD, KANSAS

Block Diagram of Shoestring Sand Body

Explanation:
Oil Well
Sandy Shale
Sandstone

Approximate Limit of Sand Lens

One Mile

N
W — ⊙ — E
S

After Bass (1936)

TRAP CASE STUDIES   "Shoestring Sand" Pools

Fig. 14-10. Greenwood County, Kansas.

Probably the best development of the shoestring-sand type of accumulation is to be found in Greenwood and adjacent counties in southeastern Kansas.° The elongate lenses of quartz sand which carry the oil lie within the lower part of the Cherokee shale of Pennsylvanian age. They are not all of exactly the same age; in northern Greenwood County a northwestward-trending sand body is crossed by a northeastward sand trend of slightly younger age.

Shown in black in part *B* of the upper illustration are the shoestring-sand oil fields of the Sallyards and Lamont trends. Each one of the areas shown in black is a separate oil pool and a separate and isolated body of sand. These elongate lenses are completely surrounded by shale. The shoestring-sand bodies range in thickness from 50 to more than 100 feet, in width from ½ to 1½ miles, and in length from 1 to 7 miles. The trends, composed of individual sand lenses arranged end to end, vary in length from 25 to over 45 miles.

The lower figure is a block diagram of the Burkett shoestring-sand body which lies toward the northeast end of the Sallyards trend. The sand and associated rock is shown by means of a series of intersecting fences. Each intersection is an oil well; the wells are spaced 660 feet apart. The sand body is shown by the stippled pattern, and the diagram demonstrates its lenticularity in every direction. These sand bodies contain little or no water but were filled with oil at time of discovery, regardless of structural position. The surrounding shale, especially the shale at the same stratigraphic level to the northwest, is abnormally rich in organic matter, and it is a logical conclusion that the oil moved into the sand from this shale, probably during compaction.

The origin of the shoestring-sand bodies has excited considerable interest. Bass believes the Greenwood-Butler County shoestring sands to be the remnants of offshore bars in the Cherokee Sea. The sketch at the upper left is a part of the New Jersey Coast taken from United States Coast and Geodetic Survey charts. The offshore bars are shown in black. The parallelism between these bars and the shoestring sands of Greenwood County is most marked. *Courtesy State Geological Survey of Kansas.*

° N. Wood Bass, "Origin of the Shoestring Sands of Greenwood and Butler Counties, Kansas," *State Geol. Survey of Kans., Bull.* 23 (1936).

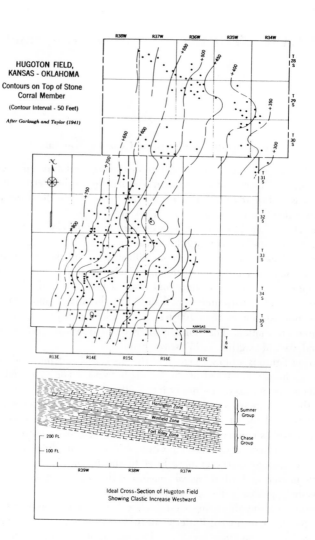

HUGOTON FIELD,
KANSAS - OKLAHOMA

Contours on Top of Stone
Corral Member

(Contour Interval - 50 Feet)

After Garlough and Taylor (1941)

Ideal Cross-Section of Hugoton Field
Showing Clastic Increase Westward

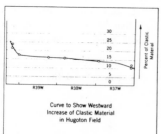

Curve to Show Westward
Increase of Clastic Material
in Hugoton Field

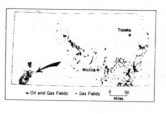

Oil and Gas Fields   Gas Fields

398

Fig. 14-11. Hugoton Gas Field, Kansas.

An up-dip change in facies is responsible for the accumulation of gas in the great Hugoton field° of southwestern Kansas. The proved gas-producing area now covers parts of eight counties near the southwestern corner of Kansas and the adjoining part of Texas County in the Oklahoma Panhandle. The Hugoton gas field lies almost due north of the northwest end of the Amarillo gas field in the Texas Panhandle, and the two fields connect through Texas County, Oklahoma. The birth date for the Hugoton field is generally given as 1927, when a 6 million cubic foot gas well was completed near Hugoton, the county seat of Stevens County.

The Hugoton gas occurs in three reservoir zones which lie within the Big Blue series of the Permian. The total thickness of the three zones is about 250 feet, but this is by no means all reservoir rock. All three zones are composed of carbonate rock, either drusy dolomite or oölitic limestone. The greater part of the porosity is ascribed to solution processes.

The upper figure is a subsurface structure map of the Hugoton gas field showing the dip of the top of the middle of the three reservoir zones. The strata have a fairly consistent dip from 800 feet above sea level on the west side of the field to 300 feet above sea level on the east side. There is obviously no relationship between the structural picture and the accumulation of gas.

The lower left drawing shows the cause of the trapping of the gas in the Hugoton field. The carbonate-rock zones become more and more shaly up-dip until the porosity diminishes beyond the point where commercial quantities of gas can be obtained. The lower right figure shows the percentage of clastic material as against the calcite and dolomite percentage along a line crossing the field from east to west. The sharp increase in percentage in clastic material which takes place in R 40 W explains the virtual absence of gas wells in that row of townships. The great volume of gas in the Hugoton field has been due to impounding relatively high on the flanks of a great basin of sedimentation up which the gas migrated. *Courtesy American Association of Petroleum Geologists.*

° John L. Garlough and Garvin L. Taylor, "Hugoton Gas Field, Grant, Haskell, Morton, Stevens and Seward Counties, Kansas and Texas County, Oklahoma," *Stratigraphic Type Oil Fields* (American Association of Petroleum Geologists, 1941), pp. 78–104.

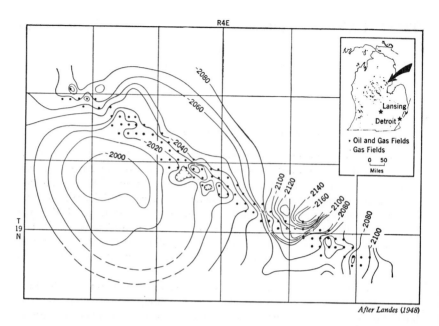

After Landes (1948)

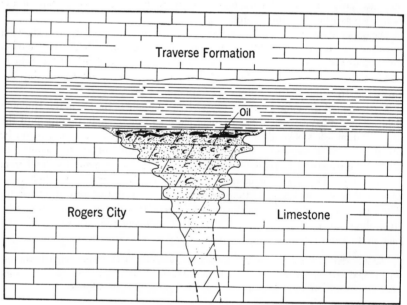

Hypothetical Cross-Section

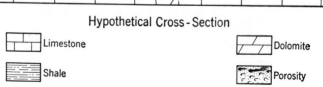

TRAP CASE STUDIES   Localized Dolomite Porosity

Fig. 14-12.   Deep River, Michigan.

A relatively new field illustrating the accumulation of oil in a patch of locally dolomitized limestone is the Deep River pool in Arenac County, Michigan.° This pool was discovered on Dec. 30, 1943. Subsequently about 150 wells were drilled, of which over one-third were failures. This abnormal percentage of dry holes after the discovery of a pool was due to the extremely narrow width of the producing zone, as can be seen in the upper figure on the opposite page.

The upper figure is a subsurface structure map contoured on the base of the Devonian Traverse formation which immediately overlies the reservoir rock. The structure is that of a dome, but the Deep River oil field, where on the dome at all, is far down on the northeast flank, and there is absolutely no relationship between accumulation and structure. The Deep River gas field, producing from the Mississippian Berea sandstone, occupies the highest part of the dome.

The oil reservoir rock at Deep River is the Rogers City formation of Devonian age. Normally the Rogers City formation is limestone, but at Deep River, *where productive of oil*, it is a porous dolomite. The unsuccessful wells failed to find oil, porosity, or dolomite. Conversely, where the limestone has been dolomitized it is both porous and oil-bearing in this general vicinity.

The lower figure is a hypothetical cross section of the Deep River producing zone. The areal distribution of the dolomitized zone beneath the Traverse can be accurately mapped, owing to the presence of some 57 dry holes surrounding the Deep River oil pool. However, the third dimension is entirely unknown; it is assumed that the dolomitized zone is vertical or nearly so, because none of the dry holes, regardless of depth drilled, have passed through this zone. It has been suggested that the localized porosity has been the result of an excess of solution over precipitation by dolomitizing solutions passing along a vertical fissure through the Rogers City formation. Apparently this dolomitized porous zone afforded sanctuary for oil which moved in after dolomitization had taken place. Presumably, but yet to be proved, the oil is underlain by water and has been impounded at the top of the porous zone by the presence of impervious shale at the base of the overlying Traverse formation. *Courtesy American Association of Petroleum Geologists.*

Subsequently an oil field trend, the Albion-Pulaski-Scipio, in southermost Michigan was discovered and became Michigan's most prolific source of oil. The trapping is similar to that at Deep River, except for the reservoir rock, which is dolomitized Trenton (Ordovician) limestone.

° Kenneth K. Landes, "Deep River Oil Field, Arenac County, Michigan," *Structure of Typical American Oil Fields* (American Association Petroleum Geologists, 1948), Vol. 3, pp. 299–304.

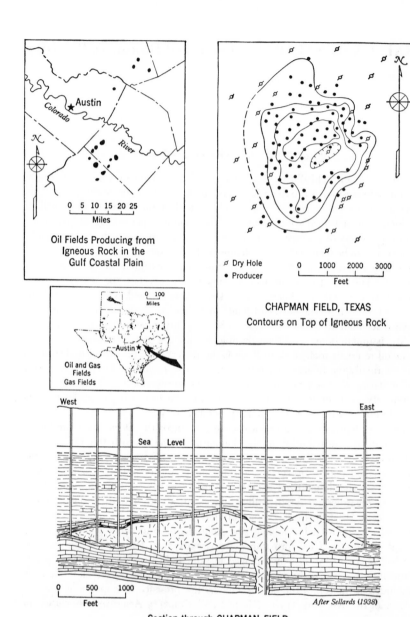

Oil Fields Producing from
Igneous Rock in the
Gulf Coastal Plain

Oil and Gas
Fields
Gas Fields

Dry Hole
Producer

CHAPMAN FIELD, TEXAS
Contours on Top of Igneous Rock

Section through CHAPMAN FIELD

After Sellards (1938)

402

TRAP CASE STUDIES

Altered Igneous
Rock Reservoir

Fig. 14-13. Chapman Field, Texas.

The picture on the upper left on the opposite page shows the location of oil fields producing from altered igneous rock in the Gulf Coastal Plain of Texas. Both intrusion and extrusion of igneous rock occurred over a large area during Upper Cretaceous time in the Gulf province. The igneous activity continued through a considerable span of time, but the igneous rocks which subsequently became oil reservoirs were formed immediately after Austin time and were overlain and surrounded by clays and marls belonging to the Cretaceous Taylor formation. These igneous rocks, which may have had some initial porosity due to vesicularity, were altered to serpentine and allied rocks with considerable increase in permeability in pre-Taylor time. Oil is stored both in the igneous rock and in reworked igneous material deposited on the flanks of the serpentine masses. The depths to the igneous reservoir depend upon the position in respect to the monocline.

In all the serpentine fields the oil is believed to have moved into the porous igneous rock long after alteration and weathering took place and after the igneous rock had become submerged beneath the Taylor sea and covered with a thick blanket of sediment.

The figure on the upper right ° is a map of the Chapman oil field in Williamson County contoured on top of the igneous rock with contour interval of 100 feet. This field produced over 3.7 million barrels of oil between the time of its discovery in 1930 and the end of 1934. The total producing area is 476 acres. Initial productions were highly variable, running as high as over 5000 barrels. The large lower figure ° is a hypothetical cross section through the Chapman oil field. The igneous rock body is somewhat bulbous upward with a probable feeder vent at the point indicated in the cross section, where a well penetrated 954 feet of igneous rock without passing through this material. The rock beneath the igneous rock is Austin chalk, and the overlying rock is Taylor. Most, but not all, of the oil is produced from wells drilled varying distances into the igneous rock itself. However, some oil has been obtained from the limestone shown immediately overlying the west flank of the serpentine body. Apparently the oil was squeezed out of the compacting Taylor clay and obtained sanctuary in any porous rock it could find. *Courtesy Texas Bureau of Economic Geology and the American Association of Petroleum Geologists. Upper left figure after Lonsdale* (1927).

°E. H. Sellards, "Oil Accumulation in Igneous Rocks," *Science of Petroleum* (Oxford University Press, 1938), pp. 261–265.

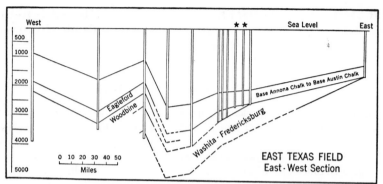

West   ★ ★   Sea Level   East

500
1000
2000
3000
4000
5000

Eagleford
Woodbine

Base Annona Chalk to Base Austin Chalk

Washita - Fredericksburg

0 10 20 30 40 50
Miles

EAST TEXAS FIELD
East - West Section

After Minor and Hanna (1941)

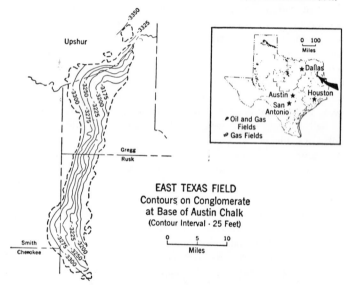

Upshur

Gregg
Rusk

Smith
Cherokee

0   100
Miles

★ Dallas

Austin ★   Houston ★
San ★
Antonio
⚲ Oil and Gas
    Fields
⚲ Gas Fields

EAST TEXAS FIELD
Contours on Conglomerate
at Base of Austin Chalk
(Contour Interval · 25 Feet)

0   5   10
Miles

404

TRAP CASE STUDIES   Overlapped Sand Wedge

Fig. 14-14.   East Texas Field.

The East Texas field° is in northeastern Texas about 50 miles southwest of the northwestern corner of Louisiana and the southwestern corner of Arkansas. The field is over 40 miles long and is nearly 5 miles in average width. The discovery well was completed on Oct. 3, 1930. Within a few years after discovery the East Texas field had produced more oil than the total production of any other field in the United States. It has been estimated that the amount of oil originally present which can be successfully recovered from drilled wells is in excess of 4 billion barrels, or 30,000 barrels per acre.

Geologically the East Texas field lies along the wedgeout zone of the westward-dipping Woodbine sand of Upper Cretaceous age. To the west is the Tyler basin, and to the east is the Sabine uplift with its highest point in northwestern Louisiana. The uppermost figure on the opposite page is a cross section drawn from Powell, a fault-trap field in Navarro County, northeastward and eastward across the northern end of the East Texas pool to the Louisiana line. The two wells in the cross section marked by stars are producing wells in the East Texas field. Accumulation is due to the pinching out of the Woodbine sand in a plunging anticline on the flank of the Sabine uplift. Wells drilled north, west, and south of the field boundary enter the Woodbine below the level of the oil-water interface.

The middle figure is a structure contour map of the base of the Austin chalk which blankets the truncated edges of the Woodbine and Eagle Ford formations. It is, therefore, a contour map showing the present configuration of an old erosion surface. No doubt a contour map of the structure of the Woodbine sandstone would show even greater dip to the west.

The Woodbine sand in the East Texas field consists of a series of thin, lenticular, but nevertheless interconnecting, sands interstratified with shales and silts. The average reservoir thickness is about 30 feet, and the maximum amount of sand which has been observed in cores is 75 feet.

Oil entering the Woodbine sand on the east flank of the Tyler basin has migrated up-dip until impounded below the overlapping Austin chalk.

° H. E. Minor and Marcus A. Hanna, "East Texas Oil Field, Rusk, Cherokee, Smith, Gregg, and Upshur Counties, Texas," *Stratigraphic Type Oil Fields* (American Association Petroleum Geologists, 1941), pp. 600–640.

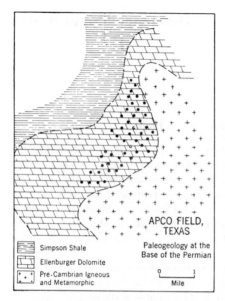

APCO FIELD, TEXAS

Paleogeology at the Base of the Permian

Simpson Shale

Ellenburger Dolomite

Pre-Cambrian Igneous and Metamorphic

0   1
Mile

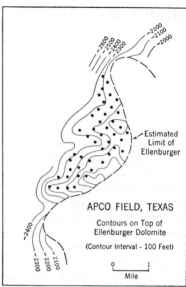

APCO FIELD, TEXAS

Contours on Top of Ellenburger Dolomite

(Contour Interval - 100 Feet)

0   1
Mile

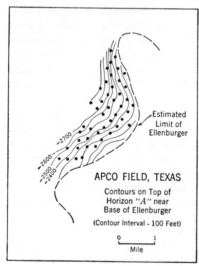

Estimated Limit of Ellenburger

APCO FIELD, TEXAS

Contours on Top of Horizon "A" near Base of Ellenburger

(Contour Interval - 100 Feet)

0   1
Mile

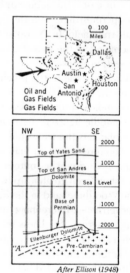

After Ellison (1948)

TRAP CASE STUDIES   Overlapped Dolomite Wedge

Fig. 14–15.   Apco Field, Texas.

The Apco field ° is south of the Pecos River in southwestern Texas. Although shallow oil had been produced in this area since 1929 the deep pay was not discovered until 1939.

The surface rocks in the vicinity of the Apco field are Quaternary and Recent alluvial sediments, ancient deposits of the Pecos River. The subsurface section includes Cretaceous, Triassic, Permian, Ordovician, and Precambrian rocks. The greater part of the stratigraphic section between the thin veneer of surficial deposits and the oil reservoir consists of Permian dolomites.

The main producing zone of the Apco field lies within the Ellenburger formation of Lower Ordovician age. The Ellenburger is about 800 feet thick at Apco and consists primarily of gray to buff siliceous dolomite. It apparently underwent considerable weathering and leaching during pre-Permian emergence, with the result that vuggy porosity has been developed in three persistent zones lying from 150 to 300 feet above the base. The interstitial porosity has been augmented by fractures. Exposure to weathering also produced an enrichment of siliceous material at the Ellenburger surface by differential leaching.

Accumulation in the Apco field is in a buried hogback of Ellenburger dolomite which laps up on the flank of a hill of Precambrian rock. Trapping is due to the blanket of Permian shale which overlies not only the dip slope of the hogback but also the scarp face and the Precambrian hill. Oil traveling up through porous zones in the tilted Ellenburger dolomite was impounded by this seal.

The upper left figure on the opposite page shows the paleogeology at the base of the Permian. The regional dip is to the northwest, and pre-Permian erosion beveled the Ordovician Simpson shale, the underlying Ellenburger dolomite, and the Precambrian complex. The figure to the right shows, by means of contour lines, the topography of the top of the Ellenburger where it is in contact with the unconformable overlying Permian shale. Fairly consistent dips to the west show the dip slope of the hogback. Also shown is the rim of the hogback and the eroded scarp face. The drawing at the lower left shows the actual rock structure, a fairly consistent westward-dipping monocline. The cross section from northwest to southeast through the Apco field shows the Ellenburger hogback. *Courtesy American Association of Petroleum Geologists.*

° Samuel P. Ellison, Jr., "Apco Field, Pecos County, Texas," *Structure of Typical American Oil Fields* (American Association of Petroleum Geologists, 1948), Vol. 3, pp. 399–418.

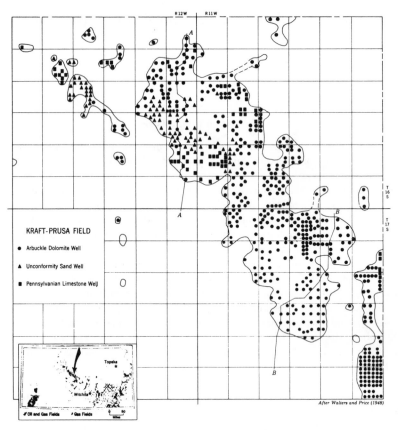

KRAFT-PRUSA FIELD

● Arbuckle Dolomite Well

▲ Unconformity Sand Well

■ Pennsylvanian Limestone Well

After Walters and Price (1948)

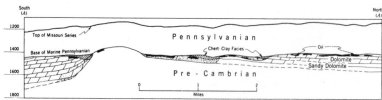

Cross Sections through KRAFT - PRUSA FIELD

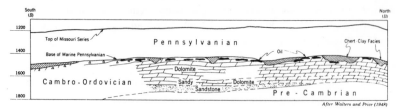

After Walters and Price (1948)

408

Fig. 14–16.   Kraft-Prusa Field, Kansas

The Kraft-Prusa oil field ° on the central Kansas uplift is notable for the fact that in spite of the presence of four reservoirs at different stratigraphic levels, one out of four wells drilled has been a failure.

The upper picture shows the main Kraft-Prusa field and its many satellite pools, some of which consist of only a single producing well.   The cross sections below give an account of the geological history of the area as well as demonstrating how oil is trapped.   At the north end of the field pre-Pennsylvanian erosion not only uncovered the Precambrian surface ("bald head") but also carved the ancient crystalline rocks into a low hill rising above the surrounding truncated and domed layers of the Cambro-Ordovician Arbuckle dolomite formation.   During this emergence sink holes were formed in the Arbuckle topographic surface which subsequently became filled with debris.   Sheet wash, off the granite hill, spread a layer of sand over nearby parts of this erosion surface.   Subsequently this area was submerged beneath the Pennsylvanian sea and hundreds of feet of sediment was deposited upon the pre-Pennsylvanian unconformity.   Renewed movements along the line of uplift which domed the sediments in pre-Pennsylvanian time caused an arching of the Pennsylvanian sediments above the bald head.

The youngest reservoirs are Pennsylvanian limestones which overlie "hills" on the Precambrian surface and in which the trapping is anticlinal.   The next older reservoir rock is the unconformity sand.   Here the trapping is entirely due to varying permeability, as the porous and permeable sand on the flanks of the uplift is overlain by relatively impervious Pennsylvanian sediment.   The greatest production at Kraft-Prusa has come from the arched truncated Arbuckle dolomite layers immediately beneath the angular unconformity, but even here the porosity is quite erratic.   In some places only a little solution of the dolomite has taken place, and so the wells are not commercial.   In other places so much solution took place that collapse occurred and the sinks became filled with impervious material.   Therefore the optimum situation for the occurrence of oil in the Arbuckle is a moderate amount of leaching producing adequate, but not too much, porosity.   The stratigraphically lowest reservoir is the fractured Precambrian quartzite which is productive in one well.   *Courtesy American Association of Petroleum Geologists.*

° Robert F. Walters and Arthur S. Price, "Kraft-Prusa Oil Field, Barton County, Kansas," *Structure of Typical American Oil Fields* (American Association of Petroleum Geologists, 1948), Vol. 3, pp. 239–280.

15

REGIONAL ASPECTS
OF ACCUMULATION

In the preceding chapters on oil and gas traps we reviewed the local conditions necessary for the concentration of petroleum hydrocarbons in a commercial deposit. Now we shall see that oil and gas accumulations have regional as well as local controls. In addition to learning how these deposits are pinpointed, we also want to know what broad areas are most desirable for driving pins.

We can first eliminate from further consideration the ocean basins, on account of inaccessibility and other reasons. That leaves the continental platforms—great rock-floored plateaus rising above the ocean basins, with 85 to 90 per cent of their total area also above present sea level. Upon the continents some areas are far more hospitable to oil and gas accumulations than others; the controlling factor is the *tectonic history*.

Tectonic Controls on Oil and Gas Field Distribution.[1] If a world map of oil and gas fields is superimposed upon geologic, tectonic, and relief maps, certain relationships immediately become apparent. As a general but not invariable rule, the oil and gas fields shun the uplift areas and concentrate in the downwarped segments of the earth's crust. Some of the greatest concentrations of oil and gas fields occur along the continental margins, especially on the coastal plains. Inland belts that are also topographically low are preferred oil-field areas. Even on plateaus of considerable elevation, the topographic position of the oil and

[1] L. G. Weeks, editor, "Habitat of Oil," *Am. Assoc. Petrol. Geol., Special Vol.* (1958); K. F. Dallmus, "Mechanics of Basin Evolution and its Relation to the Habitat of Oil in the Basin," *ibid.*, pp. 883–931; John W. Harrington, "Some Criteria by Which Basin-forming Mechanisms May Be Recognized," *ibid.*, pp. 932–947.

gas fields tends to be low in comparison with the surrounding areas. Locally, however, especially where anticlines are reflected at the surface as hills, oil fields may be at higher-than-average elevations.

Most of the oil and gas fields of the world occur in sedimentary basins. This relationship is strikingly displayed in maps compiled by Weeks[2] showing the sedimentary basins and oil fields of each continent. Fig. 15-1 shows the distribution of the sedimentary basins only, in the western and eastern hemispheres. Sedimentary basins are downwarped areas; most of them were submergent through considerable periods of geologic time. Furthermore, these basins have been subject to intermittent sinkings, and the sediments deposited during such times of depression increase markedly in thickness between the edge and center of the basin.

The reasons for oil in downwarps are obvious.[3] Not only do such areas contain an adequate thickness of sedimentary rock for the occurrence and retention of large deposits of oil and gas, but also the geological history has virtually insured that included within the thick stratigraphic section will be sediments rich in organic material (source rock) and rocks containing adequate porosity and permeability to function as reservoirs. Furthermore, as a general rule, such downwarped areas have been relatively quiet diastrophically, so that the oil contained within has not been destroyed. The presence of traps to capture the oil on the flanks of sedimentary basins can be assumed. It is quite unlikely that sedimentary rock areas of any size exist which do not contain potential oil and gas traps. There have been enough movements of the earth's crust to insure the presence of both structural and varying-permeability traps. The relative importance of one type over another, and the variety of the traps present, are dependent upon the local diastrophic and sedimentation history.

Low-relief regional uplifts may also play a role in accumulation. Oil and gas deposits occur on some uplifts that lie between downwarps. Many of these positive features are probably not actual areas of uplift, but are structurally residual segments that stood still while the surrounding crust sank. Although erosion has stripped off most, if not all, of the sedimentary cover across many uplift areas, in such places as the Central

[2] L. G. Weeks, "Highlights on 1947 Developments in Foreign Petroleum Fields," *Bull. Am. Assoc. Petrol. Geol.*, Vol. 32 (June, 1948), pp. 1093–1160; "Highlights on 1948 Developments in Foreign Petroleum Fields," *ibid.*, Vol. 33 (June, 1949), pp. 1029–1124.

[3] Wallace E. Pratt, "Distribution of Petroleum in the Earth's Crust," *Bull. Am. Assoc. Petrol. Geol.*, Vol. 28 (October, 1944), pp. 1506–1509; H. de Cizancourt, "Location of Oil Fields in Tectonic Belts," *Science of Petroleum* (Oxford University Press, 1938), Vol. 1, pp. 244–246; William Wyman Mallory, "Continental Framework and Petroleum Exploration in Western United States," *Bull. Am. Assoc. Petrol. Geol.*, Vol. 37 (November, 1953), pp. 2490–2497.

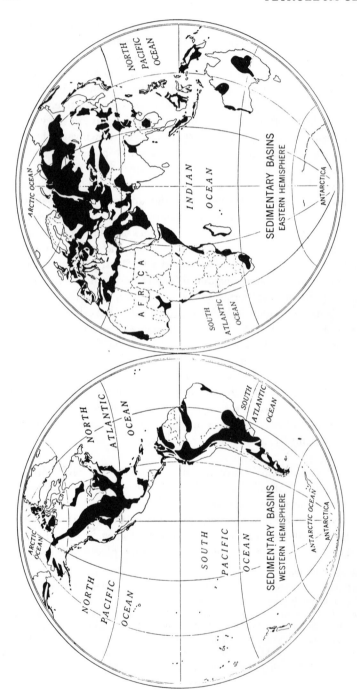

Fig. 15-1. Sedimentary basins of the world. Basins in black. *Courtesy L. G. Weeks and American Association of Petroleum Geologists.*

Kansas uplift and the Cincinnati arch enough of the sedimentary section has remained to permit accumulation of oil and gas.

Origin and Types of Basins. At one time it was rather generally believed that sediment-filled basins subsided because of the weight of that sediment. We now have an increasing body of evidence to the effect that sinking can and does take place without the addition of any load. Subsidence merely makes deposition possible by providing space beneath water for sediment accumulation. One line of evidence supporting this conclusion is the fact that sea-floor basins have sunk in recent times off the California coast. Those near shore have become partly filled with sediment, whereas those farther offshore have received almost no sediment.

Other evidence is the presence on the continents of "starved" basins. "A starved basin is defined as a basin which received a thinner section of deposits than adjoining areas, because the rate of subsidence was materially greater than the rate of deposition. In contrast to ocean basins which have remained starved over long periods of time, intracratonic basins, that show these characteristics, were starved for only portions of their history." [4]

With our present state of knowledge, about all that we can say regarding the motivation for basin subsidence is that the earth's crust appears to get a "sinking feeling," as so aptly stated by Jeffreys. [5] We do know that this feeling is (1) deep-seated, extending far into the crust, and (2) periodic, rather than continuous. Isopach maps have shown that some basins have sagged a half-dozen times or more. Some of these sags are slight, but others have been of such magnitude as to permit the deposition of thousands of feet more sediment in the center than around the edges.

Most sedimentary basins are approximately round, like a wash basin. Special types of basins are *geosynclines*, which are elongate troughs of sedimentation, and *embayments*. Embayments are to be found along the continental margins, where the oceans in the geologic past have lapped up on the continental borders to a greater extent than at the present. This overlap by marine waters has been accompanied by a downwarping of the inundated areas with a resultant seaward thickening of the sediment deposited, just as has happened in the more conventional basins.

Because of the periodic subsidences, the regional dip in all basins is toward the center. Obviously the older, deeper layers, which experienced the most sagging, have the greater dips. There may be spread across the top one or more young formations which were deposited since the last

[4] Hugh N. Frenzel, "Starved Basins," *Am. Assoc. Petrol. Geol., Convention Program* (1955).
[5] Harold Jeffreys, quoted by O. T. Jones, "Gravity as a Geological Factor," *The Observatory* (London, 1950), Vol. 70, p. 110.

subsidence. These formations are not only unconformable with the underlying sediments but also do not reflect the basin structure in any way.

Sedimentary basin studies have also shown that the deepest downwarping did not always take place at exactly the same spot. Therefore, the basin bottom is at different places for different stratigraphic intervals, but the wandering is never very far away from the average center in terms of basin dimensions.

During the active history of a basin the floor is at times below water, so that it is receiving sediment, and at other times it is above water, with erosion removing sediment. The environmental conditions shift in place and time so that sand, clay, reef-making organisms, and evaporites appear and disappear from the scene. As a result of this varied history, including diastrophic activity, the sedimentary section contains shoreline deposits, reefs, salt beds, thick shales, and all types of unconformities.

Folding from one cause or another is a common feature of basins. Many of the folds are features sharp enough to more than overcome the regional dip, thereby producing a *reversal* of the regional dip—that is, an anticline. Consequently, basin flanks (and basin bottoms) are crenulated features, fortunately for the oil hunter. Geosynclines are much more susceptible to "pleating" by compressive forces in the earth's crust than are embayments and quaquaversal basins. Therefore the sides of geosynclines may be furrowed with great anticlines and synclines, many miles in length, plus overthrust blocks. However, the general trough-like shape, with regional dips downward toward the geosynclinal axis, is still maintained, regardless of the anticlinal, synclinal, and fault interruptions.

As a result of the sedimentary and tectonic history of all basins, regardless of type, traps for oil accumulation are always present. No basin has ever been condemned on account of lack of traps.

Basin Architecture.

Oil occurrence . . . is more directly controlled by the environment of deposition than by any other factor or combination of factors. The environment that exists on the deposition bottom and in the sediments directly below is to a large degree dependent on the basin-bottom form or configuration and the supply of sediments. These, in turn, are mainly determined by the fundamental basin architecture or physical framework, which varies with the manner in which basins of different genetic types form and develop.[6]

---

[6] L. G. Weeks, "Factors of Sedimentary Basin Development that Control Oil Occurrence," *Bull. Am. Assoc. Petrol. Geol.*, Vol. 36 (November, 1952), p. 2124.

The architecture of a basin depends upon its environment and tectonic history. The simplest basin is the sort in which sagging takes place in a stable area. These basins tend to be circular in plan and symmetrical in cross section. They consist of an outermost *shelf* completely surrounding the basin. Some formations that are confined to the basin thin to extinction on the sloping shelf. The central part of the basin is the *deep.* Here the original floor lies at maximum depth, perhaps many thousands of feet, and the sedimentary section is the thickest. However, at no one time during the geologic history of the basin has the sea floor actually been deep in terms of ocean deeps. It was merely a few tens of feet deeper in the center of the basin than it was on the submerged shelf. Between the deep and the shelf is the *hinge belt.* Here is the zone of demarcation between the more intense downwarping of the basin center and the modest inward tilting of the shelf.

In regions of more active diastrophism the basins are troughs or geosynclines, and one side of the elongate downwarp is less stable than the other side. The active side is referred to as the *mobile rim.* Geosynclinal troughs tend to be assymetric, with a much steeper flank plunging basinward from the mobile rim. This steep flank shifts the bottom of the basin toward the active side. The hinge belt and shelf lie far up the gentle flank on the stable side of the trough.

The shelf is the optimum environment for the deposition of reservoir rocks. Here, nearshore, are currents of sufficient force to transport and deposit sand. Slightly farther out the waters are more quiet, and organic limestones can accumulate. Slight variation in sea level elevate these deposits above sea level, permitting reworking and winnowing of the sands and the development of solution porosity in the limestones.

Limestone deposition continues out to and beyond the hinge belt. Here is the preferred, but by no means exclusive[7] environment of the reef. Shales become increasingly abundant between the shelf and the deep.

On the flank below the hinge line, and across the bottom of the deep, conditions are excellent for source-bed (and seal) deposition. Clays and calcareous oozes, enriched with bottom-seeking planktonic remains, accumulate in the quiet, relatively deep waters where an anaerobic condition makes hydrocarbon preservation more likely.

However, nothing is stable very long in geological history, and the shelf and deep situations just described are liable to interruptions. A temporary rise in the ocean level permits the deposition of source materials and potential seals between layers of reservoir rock, and the geological column of nearly every basin shows such alternations of sea-floor conditions.

---

[7] Curt Teichert, "Cold- and Deep-Water Coral Banks," *Bull. Am. Assoc. Petrol. Geol.,* Vol. 42 (May, 1958), pp. 1064–1082.

Wierich[8] presents evidence to the effect that all of the genesis, migration, and accumulation of oil in the fields of the McAlester basin of eastern Kansas and eastern Oklahoma took place only over the shelf.

In the same manner the temporary shallowing of a basin permits the deposition of reservoir rocks in the central area. The production of oil from close to the centers of both the Illinois and the Michigan basins are proof of this statement.

The mobile rims, when they are present, may be zones of intense folding and even overthrusting. In some areas the diastrophism has been of such magnitude as to destroy any hydrocarbons that may have been there. This disturbance does not prevent, however, subsequent migrations of oil or gas from entering this zone.

A special type of basin is the *silled* basin in which a restriction at shallow depth separates the downwarping area from the open ocean. Here the conditions are optimum for large-scale evaporite deposition. Where the climate is at least semi-arid, the greater water evaporation in the relatively shallow basin area creates a constant inflow of normal salt water across the sill. However, the heavier bottom-seeking brines which result from the greater evaporation within the basin cannot escape because of the sill. In time, when the water becomes saturated with calcium sulfate and eventually with sodium chloride, precipitation takes place. Evaporite beds make the best seals, as described in Chapter 8. The cycle can be interrupted at any time by diastrophism, or infilling to sill top, so that normal marine sediments may overlie evaporites.

Preferred Habitats of Known Oil Accumulations.[9] Knebel and Rodriguez-Eraso[10] made a statistical study of the locale of accumulation of major oil deposits. A total of 236 fields, each with an ultimate recovery estimated at over 100 million barrels, was classified as to basin position, type of trap, and other data. These fields represent 82.5 per cent of the ultimate recovery from all of the oil fields in the free world.

It was found that, of the *known* oil in the free world, over 50 per cent occurs in the hinge zone and 13½ per cent is on the shelf—a total of nearly 64 per cent on the stable side of basins. The mobile rim is second with 21 per cent and the deep basin last with 11½ per cent. Occurrences outside of these 4 basin locales account for 3½ per cent of the oil. In terms of *numbers* of oil fields rather than barrels of oil, the shelf leads, with the

---

[8]Thomas Eugene Wierich, "Shelf Principle of Oil Origin, Migration, and Accumulation," *Bull. Am. Assoc. Petrol. Geol.*, Vol. 37 (August, 1953), pp. 2027–2045.

[9]L. G. Weeks, "Habitat of Oil and Some Factors That Control It," *Habitat of Oil* (American Association of Petroleum Geologists Special Vol., 1958), pp. 1–61.

[10]G. M. Knebel, and Guillermo Rodriguez-Eraso, "Habitat of Some Oil," *Bull. Am. Assoc. Petrol. Geol.*, Vol. 40 (April, 1956), pp. 547–561.

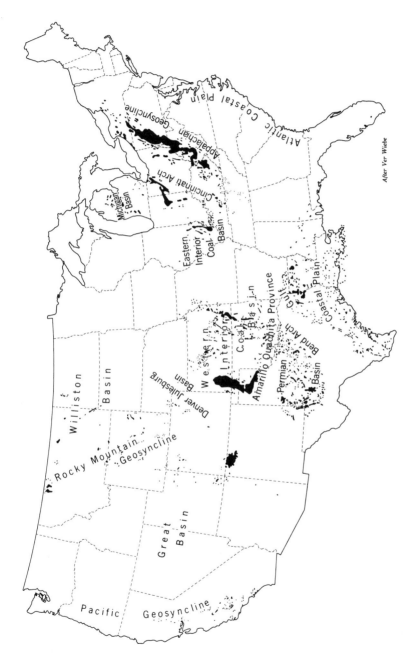

Fig. 15-2. Petroliferous provinces of the United States.

After Ver Wiebe

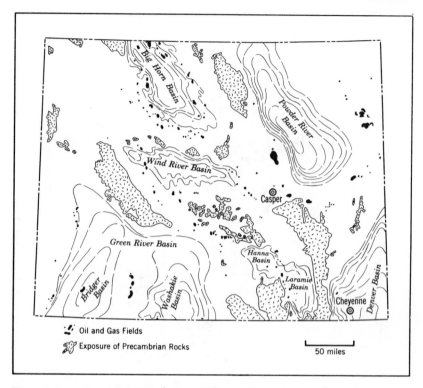

Fig. 15-3. Map of Wyoming, showing the relationships between oil fields and sedimentary basins; each basin constitutes a petroleum district within the much larger Rocky Mountain petroliferous province.

deep basin, hinge, other locales, and mobile rim following. This great discrepancy is due largely to the fact that a relatively few fields in the Middle East, Venezuela, California, and the Rocky Mountains occurring on mobile rims, have enormous untimate recoveries.

Dallmus [11] explains the concentration in the hinge belt as a result of the maximum distention and therefore minimum pressure in this zone.

Tectonic Classifications of Oil and Gas Field Distribution. Ver Wiebe [12] proposed a tectonic classification for the oil fields of the United States in 1929. In this classification the United States is divided into several "petroliferous provinces," each of which is named after the

[11] K. F. Dallmus, *lecture* before Michigan Basin Geological Society (October, 1958).

[12] W. A. Ver Wiebe, "Tectonic Classification of Oil Fields in the United States," *Bull. Am. Assoc. Petrol. Geol.*, Vol. 13 (May, 1929), pp. 409–440; "Oil and Gas in the United States," *Science of Petroleum* (Oxford University Press, 1938), Vol. 1, pp. 66–95.

dominating tectonic element. A modernized version of Ver Wiebe's provinces is shown in Fig. 15-2. Where desirable, the petroliferous provinces can be subdivided into "districts" and named after either a secondary tectonic feature or the geographic position. The manner in which the Rocky Mountain geosyncline province has been divided into districts in Wyoming is included as an example (Fig. 15-3).

Pratt [13] emphasizes the great concentration of oil and gas that is known to occur in three intercontinental depressions in the earth's crust and calls attention to a fourth untested depression similar in tectonic setting to the other three. One of these intercontinental depressions is that lying within, and adjacent to, the Gulf of Mexico and the Caribbean Sea, between the continents of North and South America. Here are found the greatest oil deposits of the western hemisphere, including those in Colombia, Venezuela, Trinidad, and Mexico, as well as the fields of the Gulf Coast of Louisiana and Texas. The second tectonically depressed area lies between, and upon contiguous corners of, Africa, Europe, and Asia, and includes the oil fields bordering the Red, Mediterranean, Caspian, and Black Seas, and the Persian Gulf where the Middle East oil deposits are being exploited so actively today. The third of Pratt's intercontinental troughs includes "the environs of the shallow island-studded seas which lie between the continents of Asia and Australia in the Far East." Production has been obtained for a good many years from parts of this province including Borneo, Sumatra, Java, and New Guinea.

The unproved intercontinental depression described by Pratt is the border zone of the landlocked Arctic Sea, surrounding the North Pole and lying upon the northern extremities of the continents of North America, Europe, and Asia. This region is destined for much exploration in the future. Pratt's intercontinental depressed areas can be noted on the hemisphere maps, prepared by Weeks, [14] showing sedimentary basins (Fig. 15-1). Various smaller basins lying upon continental platforms, which are also important sources of oil and gas, can be noted as well. Many of the interior basins are, like the Arctic basin, still unexplored but are potential sources for hydrocarbons in the future.

A breakdown of the sedimentary basins within the United States is given by 6 geographic divisions, and of the oil and gas fields and districts by states in *Petroleum Geology of the United States,* Wiley Interscience (New York), 1970.

[13] Wallace E. Pratt, "Distribution of Petroleum in the Earth's Crust," *Bull. Am. Assoc. Petrol. Geol.,* Vol. 28 (October, 1944), pp. 1506–1509.

[14] L. G. Weeks, "Highlights on 1947 Developments in Foreign Petroleum Fields," *Bull. Am. Assoc. Petrol. Geol.,* Vol. 32 (June, 1948).

THE LITERATURE OF PETROLEUM GEOLOGY

Bibliographies

U.S. Geological Survey, *Bibliography of North American Geology.* Published annually, with periodic cumulative compilations, Washington, D.C. Series terminated, 1970.

Geological Society of America, *Bibliography and Index of Geology,* issued monthly, with cumulative bibliography at the end of year, Boulder, Colorado.

American Association of Petroleum Geologists, Tulsa, OK. December issue of *Bulletin.* Index covers both the bulletin and other publications of the Association published during the year. Also *Comprehensive Indexes* for 1917-1945; 1946-1955, and 1956-1965.

Glossary

American Geological Institute, *Glossary of Geology and Related Sciences,* 1972, Washington, D.C.

Scientific and Technical Journals

Am. Assoc. Petroleum Geologists *Bulletin,* monthly, Tulsa OK

Canadian Soc. Of Petroleum Geologists, *Bulletin* of Canadian Petroleum Geology, quarterly, Calgary, Alberta

Society of Economic Paleontologists and Mineralogists, Journal of Sedimentary Petrology, *quarterly,* Tulsa, OK

Society of Petroleum Engineers, *Journal of Petroleum Technology,* Monthly, Dallas, Texas

Institute of Petroleum, *Monthly,* London

Society of Exploration Geophysicists, *Geophysics,* Monthly,

Geologiya Nefti i Gaza (*Petroleum Geology*), monthly, translated into English and published by Petroleum Geology, P.O. Box 171, McLean, Virginia

Various other foreign language publications, not translated into English.

421

Symposiums

World Petroleum Congress *Proceedings,* with one volume of papers devoted to Geology and Geophysics
Third, The Hague, 1951
Fourth, Rome, 1955
Fifth, New York, 1959
Sixth, Frankfurt am Main, 1963
Seventh, Mexico City, 1967
Eighth, Moscow, 1971 (proceedings published in English in London by Applied Science Publishers, Ltd.)
Ninth, Tokyo, 1975

Wallace E. Pratt and Dorothy Good, Editors and 20 authors, *World Geography of Petroleum,* American Geographical Society, Spec. Pub. 31, 1950
Max W. Ball, Douglas Ball, and Daniel S. Turner, *This Fascinating Oil Business* Bobbs-Merrill (Indianapolis), 1965
Kenneth K. Landes, *Petroleum Geology of the United States,* Wiley Interscience (New York), 1970
E. H. Tiratsoo, *Oil Fields of the World,* Scientific Press, Ltd. (Beaconsfield, England), 1973
Peter Hepple, editor, *The Exploration for Petroleum in Europe and North Africa,* Inst. of Petroleum (London), 1969

For recent exploration and possible discoveries the reader is referred to the following:

Daily newspapers
   Wall Street Journal
   New York Times
Weekly trade journal
   Oil and Gas Journal (Tulsa, OK)
   (Periodic production statistics)
Monthly journals
   World Oil (Houston, TX)
   (Periodic production statistics)
   Offshore (Tulsa, OK)
   (Offshore production statistics)
   Petroleum Review (Inst. of Petroleum, London)
   Petroleum International (London)
   Petroleum Engineer (Dallas, TX)
   Bulletin of the American Association of Petroleum Geologists publishes two issues in the fall of each year (such as August and October) covering developments the preceding calendar year in North America and for foreign countries.
   Geophysics. Journal of the Society of Exploration Geophysicists (Tulsa, OK).

Production Statistics

U.S. Bureau of Mines, *Minerals Yearbook.* Published annually, but with a two or three year lag. Both domestic and foreign.

Special Publications of the American Association of Petroleum Geologists. A wealth of geological information, including especially, but by no means exclusively, petroleum geology is in the publications listed below:

Books
Structure of Typical American Oil Fields, Vols. 1 and 2 (1929), Vol. 3 (1948).
Geology of California (1933).
Problems of Petroleum Geology (1934).
Geology of Tampico Region, Mexico (1936)
Stratigraphic Type Oil Fields (1941).
Geology of Salt Dome Oil Fields (1926).
Gulf Coast Oil Fields (1936).
Geology of Natural Gas (1935).
Source Beds of Petroleum (1942).
Possible Future Petroleum Provinces of North America (1951).
Western Canada Sedimentary Volume (1954).
Petroleum Geology of Southern Oklahoma (1956).
Habitat of Oil (1958).
Jurassic and Carboniferous of Western Canada (1958).
Recent Sediments, Northwest Gulf of Mexico (1960).
Geometry of Sandstone Bodies (1961).
Pennsylvanian System of the United States (1962).
Classification of the Carbonate Rocks (1962).
Backbone of the Americas—Tectonic History from Pole to Pole (1963).
Marine Geology of the Gulf of California (1964).
Fluids in Subsurface Environments (1965).
International Tectonic Dictionary—English Terminology (1967).
Diapirism and Diapirs (1968).
Natural Gases of North America. 2 Vols. (1968).
Subsurface Disposal in Geologic Basins—A Study of Reservoir Strata (1968).
Sourcebook for Petroleum Geology (1969).
Carbonate Sediments and Reefs, Yucatan Shelf, Mexico; Tectonic Relations of Northern
    Central America and the Western Caribbean; and other papers on Florida and British
    Honduras (1969).
North Atlantic—Geology and Continental Drift (1969).
Carbonate Sedimentation and Environments, Shark Bay, Western Australia (1970).
Geology of Giant Petroleum Fields (1970).
Future Petroleum Provinces of the United States—Their Geology and Potential. 2 Vols.
    (1971)
Stratigraphic Oil and Gas Fields—Classification, Exploration Methods, and Case Histories
    (1972).
Western North Atlantic Ocean: Topography, Rocks, Structure, Water, Life, and Sedi-
    ments (1972).
Underground Waste Management and Environmental Implications (1972).
Arctic Geology (1973).
The Black Sea—Geology, Chemistry and Biology—A Bibliography (1974)
Stratigraphic Traps in Sandstones—Exploration Techniques (1974).
Trek of the Oil Finders: A History of Petroleum (1975).

Reprint Series
1. Origin of Petroleum (1971). 15 papers reprinted from 1950–1969 Bulletins.
2. Origin of Evaporites (1971). 8 papers reprinted from 1937–1970 Bulletins.
3. Continental Shelves—Origin and Significance (1972). 14 papers reprinted from 1946–
    1970 Bulletins.

4. Carbonate Rocks I: Classifications–Dolomite–Dolomitization (1972). 13 papers reprinted from 1946–1970 Bulletins.
5. Carbonate Rocks II: Porosity and Classification of Reservoir Rocks (1970). 12 papers reprinted from 1946–1970 Bulletins.
6. Paleoecology (1973). 9 papers reprinted from 1940–1971 Bulletins.
7. Sandstone Reservoirs and Stratigraphic Concepts I (1973). 8 papers reprinted from 1966–1971 AAPG Bulletins and Memoir 18 (1972).
8. Sandstone Reservoirs and Stratigraphic Concepts II (1973). 11 papers reprinted from 1963–1972 AAPG Bulletins and Geometry of Sandstone Bodies (1961).
9. Origin of Petroleum II (1974). 17 papers reprinted from 1950–1969 AAPG Bulletins.
10. Facies and the Reconstruction of Environments (1974). 8 papers reprinted from 1942–1968 AAPG Bulletins.
11. Abnormal Subsurface Pressure (1974). 14 papers from 1953–1973 AAPG Bulletins.
12. Jurassic of the Gulf Coast I: Regional Stratigraphy (1974). 3 papers from 1943–1949 AAPG Bulletins.
13. Jurassic of the Gulf Coast II: Local Stratigraphy and Deposition (1974). 9 papers from 1942–1973 AAPG Bulletins.

Geological Highway Maps
(Prepared with the Cooperation of the U.S. Geological Survey)
1. Mid-Continent-Kans., Mo., Okla., Ark.
2. Southern Rockies-Ariz., Colo., New Mex., Utah
3. Pacific Southwest-Calif., Nev.
4. Mid-Atlantic Region-Ky., W. Va., Va., Md., Del., Tenn., N. Car., S. Car.
5. Northern Rockies-Ida., Mont., Wyo.
6. Pacific Northwest-Wash., Ore.
7. Texas
8. Alaska-Hawaii
(Maps are in preparation for other regions)

Other Maps
Tectonic Map of the United States (1962)
Basement Map of North America (1967)
Tectonic Map of Gulf Coast Region, U.S.A. (1972).

INDEX